Gmelin Handbook of Inorganic and Organometallic Chemistry

8th Edition

Gmelin Handbook of Inorganic and Organometallic Chemistry

8th Edition

Gmelin Handbuch der Anorganischen Chemie

Achte, völlig neu bearbeitete Auflage

PREPARED AND ISSUED BY

Gmelin-Institut für Anorganische Chemie
der Max-Planck-Gesellschaft
zur Förderung der Wissenschaften

Director: Ekkehard Fluck

FOUNDED BY Leopold Gmelin

8TH EDITION 8th Edition begun under the auspices of the Deutsche Chemische Gesellschaft by R. J. Meyer

CONTINUED BY E. H. E. Pietsch and A. Kotowski, and by Margot Becke-Goehring

Springer-Verlag
Berlin · Heidelberg · New York · London · Paris · Tokyo · Hong Kong · Barcelona 1991

GMELIN HANDBOOK

Dr. J. von Jouanne

Dr. L. Berg, Dr. H. Bergmann, Dr. J. Faust, J. Füssel, Dr. H. Katscher, Dr. R. Keim, Dr. E. Koch, Dipl.-Phys. D. Koschel, Dr. A. Kubny, Dr. P. Merlet, Dr. M. Mirbach, Prof. Dr. W. Petz, Dr. F.A. Schröder, Dr. A. Slawisch, Dr. W. Töpper

Dr. R. Albrecht, Dr. G. Bär, D. Barthel, Dr. N. Baumann, Dr. K. Behrends, Dr. W. Behrendt, D. Benzaid, Dr. R. Bohrer, K.D. Bonn, Dr. U. Busch, Dipl.-Ing. V.A. Chavizon, E. Cloos, Dipl.-Phys. G. Czack, A. Dittmar, Dipl.-Geol. R. Ditz, R. Dowideit, Dipl.-Chem. M. Drößmar, P. Dürr, Dr. H.-J. Fachmann, B. Fischer, Dipl.-Ing. N. Gagel, Dipl.-Phys. D. Gras, Dr. K. Greiner, Dipl.-Bibl. W. Grieser, Dr. I. Haas, Dr. R. Haubold, Dipl.-Min. H. Hein, Dipl.-Phys. C. Heinrich-Sterzel, H.-P. Hente, H.W. Herold, U. Hettwer, Dr. G. Hönes, Dr. W. Hoffmann, G. Horndasch, Dr. W. Huisl, Dr. M. Irmler, B. Jaeger, Dr. R. Jotter, Dipl.-Chem. P. Kämpf, Dr. B. Kalbskopf, Dipl.-Chem. W. Karl, H.-G. Karrenberg, Dipl.-Phys. H. Keller-Rudek, B. Kirchner, Dipl.-Chem. C. Koeppel, R. Kolb, Dr. M. Kotowski, E. Kranz, Dipl.-Chem. I. Kreuzbichler, Dr. V. Kruppa, Dr. W. Kurtz, M. Langer, Dr. B. Ledüc, Dr. A. Leonard, H. Mathis, E. Meinhard, M. Meßer, C. Metz, K. Meyer, E. Mlitzke, Dipl.-Chem. B. Mohsin, Dr. U. Neu-Becker, K. Nöring, Dipl.-Min. U. Nohl, Dr. U. Ohms-Bredemann, I. Rangnow, Dipl.-Phys. H.-J. Richter-Ditten, E. Rudolph, G. Rudolph, Dipl.-Chem. S. Ruprecht, Dr. B. Sarbas, Dr. H. Schäfer, Dr. R. Schemm, Dr. D. Schiöberg, V. Schlicht, Dipl.-Chem. D. Schneider, A. Schwärzel, Dr. B. Schwager, Dipl.-Ing. H. M. Somer, Dr. C. Strametz, G. Strauss, Dr. G. Swoboda, Dr. D. Tille, A. Tuttas, H.-M. Wagner, Dipl.-Phys. J. Wagner, R. Wagner, Dr. E. Warkentin, Dr. C. Weber, Dr. A. Wietelmann, Dr. M. Winter, Dr. B. Wöbke, K. Wolff

GMELIN ONLINE

Dr. R. Deplanque

Dr. P. Kuhn, Dr. G. Olbrich

Dr. R. Baier, Dr. B. Becker, Dipl.-Chem. E. Best, Dr. H.-U. Böhmer, Dipl.-Phys. R. Bost, Dr. A. Brandl, Dr. R. Braun, Dipl.-Chem. R. Durban, R. Hanz, Dr. A. Kirchhoff, Dipl.-Chem. H. Köttelwesch, Dipl.-Ing. W. Korba, Dr. M. Kunz, Dr. L. Leichner, Dipl.-Chem. R. Maass, Dr. A. Nebel, Dipl.-Chem. R. Nohl, Dr. B. Rempfer, Dr. U. Tölle, Dipl.-Ing. H. Vanecek

Volumes Published on Halogens in the Gmelin Handbook

F Fluorine (Syst. No. 5)
"Fluor" Main Vol. — 1926
"Fluor" Suppl. Vol. 1 — 1959
"Fluorine" Suppl. Vol. 2 (Element) — 1980
"Fluorine" Suppl. Vol. 3 (Compounds with Hydrogen) — 1982
"Fluorine" Suppl. Vol. 4 (Compounds with Oxygen or Nitrogen) — 1986
"Fluorine" Suppl. Vol. 5 (Compounds with Nitrogen and Oxygen) — 1987

Cl Chlorine (Syst. No. 6)
"Chlor" Main Vol. — 1927
"Chlor" Suppl. Vol. A (Element) — 1968
"Chlor" Suppl. Vol. B 1 (Compounds with Hydrogen) — 1968
"Chlor" Suppl. Vol. B 2 (Compounds with Oxygen, Nitrogen, and Fluorine) — 1969

Br Bromine (Syst. No. 7)
"Brom" Main Vol. — 1931
"Bromine" Suppl. Vol. A (Element) — 1985
"Bromine" Suppl. Vol. B 1 (Compounds with Rare Gases and Hydrogen) — 1990
"Bromine" Suppl. Vol. B 2 (Compounds with Oxygen and Nitrogen) — in preparation
"Bromine" Suppl. Vol. B 3 (Compounds with Fluorine and Chlorine) — 1991 **present volume**

I Iodine (Syst. No. 8)
"Jod" Main Vol. 1 (Element) — 1931
"Jod" Main Vol. 2 (Compounds) — 1933

At Astatine (Syst. No. 8a)
"Astatine" Main Vol. (Element and Compounds) — 1985

Gmelin Handbook of Inorganic and Organometallic Chemistry

8th Edition

Br
Bromine

Supplement Volume B 3

Compounds with Fluorine and Chlorine

With 9 illustrations

AUTHORS Reinhard Haubold, Jörn v. Jouanne, Hannelore Keller-Rudek, Peter Merlet, Ulrike Ohms-Bredemann, Carol Strametz, Joachim Wagner, Astrid Wietelmann

EDITORS Reinhard Haubold, Jörn v. Jouanne, Hannelore Keller-Rudek, Peter Merlet, Ulrike Ohms-Bredemann, Joachim Wagner, Astrid Wietelmann

CHIEF EDITOR Peter Merlet

System Number 7

Springer-Verlag
Berlin · Heidelberg · New York · London · Paris · Tokyo · Hong Kong · Barcelona 1991

LITERATURE CLOSING DATE: END OF 1990
IN MANY CASES MORE RECENT DATA HAVE BEEN CONSIDERED

Library of Congress Catalog Card Number: Agr 25-1383

ISBN 3-540-93642-4 Springer-Verlag, Berlin · Heidelberg · New York · London · Paris · Tokyo
ISBN 0-387-93642-4 Springer-Verlag, New York · Heidelberg · Berlin · London · Paris · Tokyo

Typesetting, printing, and bookbinding: Universitätsdruckerei H. Stürtz AG, Würzburg

Preface

"Bromine" Suppl. Vol. B3 describes compounds of bromine with fluorine and/or chlorine which may additionally contain the remaining elements with a lower Gmelin System Number, namely noble gases, hydrogen, oxygen, and nitrogen.

The largest chapter of this volume covers binary compounds of bromine with fluorine, including such technically important substances as BrF_3 and BrF_5. Of spectroscopic interest is BrF. Other Br-F compounds dealt with are neutral species and ions which only exist under special conditions or as salts. Ternary compounds of bromine, fluorine, and oxygen follow. The systematic study of the bromine fluoride oxides began only in the seventies. There are only a few known ternary compounds with hydrogen or nitrogen. Examples of compounds composed of bromine, fluorine, nitrogen, and oxygen are fluorobromates of nitrosyl and nitryl cations.

Among the binary compounds of bromine with chlorine, BrCl is well-characterized by its molecular and spectroscopic properties, less by its chemical properties. Some information on $BrCl_2^-$ is also available. Among the few ternary compounds of bromine with chlorine and hydrogen, oxygen or fluorine, the hydrogen-bonded complex ion $HBrCl^-$ and the bromine perchlorates are of scientific interest. Bromine, chlorine, nitrogen, and hydrogen form the bromochloroamines which play a role in chlorinated seawater used in cooling systems of power plants.

The present volume, "Bromine" Suppl. Vol. B3, completes the series B of the supplement volumes on bromine. Supplement Volumes B1 and B2 cover compounds of bromine with rare gases and/or hydrogen and with oxygen and/or nitrogen, respectively.

Frankfurt am Main
Oktober 1991

Peter Merlet

Table of Contents

Compounds of Bromine

(continued)

10 Compounds of Bromine with Fluorine

Compounds composed of bromine and fluorine described in this chapter are arranged with decreasing bromine content: Br_2F, Br_2F^-, BrF, BrF^+, BrF^-, Br_2F_2, BrF_2, BrF_2^+, BrF_2^-, BrF_3, BrF_3^+, $Br_2F_7^-$, $Br_3F_{10}^-$, BrF_4, BrF_4^+, BrF_4^-, BrF_5, BrF_5^+, BrF_6, BrF_6^+, BrF_6^-, BrF_7, and BrF_7^+.

The best characterized compounds are BrF_3 and BrF_5 which are important as fluorinating agents. A large amount of spectroscopic data on BrF are available, but there is little information about its chemistry. Other neutral species, such as Br_2F, Br_2F_2, BrF_2, and BrF_6, only exist under matrix-isolated conditions at low temperatures. The existence of BrF_4 and BrF_7 is even doubted. Some of the ions described including BrF_2^+, BrF_2^-, $Br_2F_7^-$, $Br_3F_{10}^-$, BrF_4^+, BrF_4^-, BrF_6^+, and BrF_6^- can be stabilized as salts.

General References:

Shamir, J.; Polyhalogen Cations, Struct. Bonding [Berlin] **37** [1979] 141/210, 163/78.

Shamir, J.; Fluorohalogenates, Israel J. Chem. **17** [1978] 37/47.

Martin, D.; Rousson, R.; Weulersse, J.-M.; The Interhalogens, Chem. Non-Aqueous Solvents B **5** [1978] 157/95.

Christe, K. O.; Halogen Fluorides, Plenary Main Sect. Lect. 24th Intern. Congr. Pure Appl. Chem., Hamburg 1973 [1974] pp. 115/41.

Downs, A. J.; Adams, C. J.; Chlorine, Bromine, Iodine and Astatine, in: Bailar, J. C.; Emeléus, H. J.; Nyholm, R.; Trotman-Dickenson, A. F.; Comprehensive Inorganic Chemistry, Vol. 2, Pergamon, Oxford 1973, pp. 1107/594, 1476/563.

Nikolaev, N. S.; Sukhoverkhov, V. F.; Shishkov, Yu. D.; Alenchikova, I. F.; Khimiya Galoidnykh Soedinenii Ftora, Nauka, Moscow 1968, pp. 99/245.

Meinert, H.; Interhalogenverbindungen, Z. Chem. [Leipzig] **7** [1967] 41/57.

Stein, L.; Physical and Chemical Properties of Halogen Fluorides, in: Gutmann, V.; Halogen Chemistry, Vol. 1, Academic, London-New York 1967, pp. 133/224, 151/74.

Opalovskii, A. A.; Alkali Metal Fluorohalogenates, Usp. Khim. **36** [1967] 1673/700, Russ. Chem. Rev. **36** [1967] 711/25.

Schmeisser, M.; Schuster, E.; Compounds of Bromine with Non-Metals, in: Jolles, Z. E.; Bromine and its Compounds, Benn, London 1966, pp. 179/252, 179/85.

Wiebenga, E. H.; Havinga, E. E.; Boswijk, K. H.; Structures of Interhalogen Compounds and Polyhalides, Advan. Inorg. Chem. Radiochem. **3** [1961] 133/69.

10.1 Dibromine Fluoride, Br_2F

CAS Registry Number: *[64973-52-0]*

The asymmetric Br_2F radical was observed in a mixture of other bromine fluorides after simultaneous deposition of a Br_2-Ar sample together with low concentrations of F_2, followed by UV photolysis of moderate duration. The radical concentration increased slightly after a thermal cycle at 26 K. Codeposition of a Br_2-F_2 mixture in Ar after applying an MW discharge also yielded Br_2F [1]. A threshold of 47.3 kJ/mol for Br_2F formation was deduced from a crossed molecular beam study of the reaction of Br_2 and F_2. The radical is assumed to be the precursor of excited BrF in a chemiluminescent reaction [2]. Formation of Br_2F is considered to be the initial reaction in the gas-phase fluorination of Br_2 [3]. A band at 507 cm^{-1} was tentatively assigned to the Br-F stretching vibration of matrix-isolated Br_2F. The radical decomposes during a thermal cycle at 30 K or during prolonged photolysis [1].

References:

[1] Prochaska, E. S.; Andrews, L.; Smyrl, N. R.; Mamantov, G. (Inorg. Chem. **17** [1978] 970/7).

[2] Kahler, C. C.; Lee, Y. T. (J. Chem. Phys. **73** [1980] 5122/30).

[3] Arutyunov, V. S.; Buben, S. N.; Trofimova, E. M.; Chaikin, A. M. (Kinet. Katal. **21** [1980] 337/42; Kinet. Catal. [USSR] **21** [1980] 258/62).

10.2 Bromofluorobromate(1 −) and Dibromofluorate(1 −), $BrBrF^-$ and $BrFBr^-$

CAS Registry Number: Br_2F^- *[25773-79-9]*

Br_2F^- did not form after adsorbing gaseous Br_2 on dried NaF [1]. Attempts to identify Br_2F^- failed in aqueous or CH_3CN solutions containing Br_2 and F^- at temperatures between 370 to 100 K [2]. However, the HMO method predicts that $BrBrF^-$ is stable with respect to the reference system $Br_2 + F^-$, which in turn is slightly more stable than the $BrFBr^-$ isomer [3].

References:

[1] Kortüm, G.; Vögele, H. (Ber. Bunsenges. Physik. Chem. **72** [1968] 401/7).

[2] Delhaye, M.; Dhanelincourt, P.; Merlin, J.-C.; Wallart, F. (Compt. Rend. B **272** [1971] 1003/6).

[3] Wiebenga, E. H. (Stereochim. Inorg. Acad. Nazl. Lincei 9th Corso Estivo Chim., Rome 1965 [1967], pp. 319/31; C.A. **71** [1969] No. 128926), Wiebenga, E. H.; Kracht, D. (Inorg. Chem. **8** [1969] 738/46).

10.3 Bromine Monofluoride, BrF

CAS Registry Numbers: BrF *[13863-59-7]*, ^{79}BrF *[35489-01-1]*, ^{81}BrF *[35489-02-2]*, ^{80}BrF *[70786-52-6]*

Bromine monofluoride, first detected in 1931, is formed in the reaction of F_2, BrF_3, or BrF_5 with Br_2. The product, described as a reddish brown gas at room temperature or an orange solid below 240 K, cannot be isolated as a pure compound since it disproportionates to Br_2 and BrF_3. Because of this instability, mainly spectroscopic and only few chemical properties are known. BrF reacts similarly as other bromine fluorides, but it is more reactive.

10.3.1 Formation

In gaseous mixtures of bromine and bromine trifluoride, the equilibrium $Br_2 + BrF_3 \rightleftharpoons 3\,BrF$ is rapidly established at temperatures of and above 25 °C; bromine monofluoride cannot be isolated as a pure substance since it disproportionates to the initial components. The liquid-vapor equilibrium of the Br_2-BrF_3 system also suggests the presence of BrF, whereas liquid-liquid and solid-liquid equilibria showed only the presence of Br_2 and BrF_3, see pp. 103/4, "Bromine" Suppl. Vol. A, 1985, p. 415, and [1 to 4]. A gas chromatographic analysis of a gaseous Br_2-BrF_3 equilibrium mixture also revealed the formation of BrF [5].

Traces of BrF and BrF_5 were observed after storing liquid BrF_3 in Ni vessels, which was explained by a spontaneous disproportionation of BrF_3. At room temperature, the disproportionation proceeds very slowly with increasing temperature (150 °C or higher), however, the equilibrium $2\,BrF_3 \rightleftharpoons BrF + BrF_5$ shifts to the right [6]. In the presence of caesium fluoride, the disproportionation of BrF_3 (suspension in $CFCl_3$) is accelerated even at low temperatures (−40 °C) and, upon dropwise addition of pyridine, BrF can be isolated as its pyridine complex [7].

The existence of BrF was first suspected by Ruff and Menzel [8] when they noted during distillation of a Br_2-BrF_5 mixture that the reaction product had a higher vapor pressure and a stronger etching action on vitreous quartz than BrF_3, BrF_5, or Br_2. Ruff and Braida [9, 10] then prepared the compound by the reaction of Br_2 with F_2 and carried out an analysis as well as vapor-pressure and density measurements on the resulting mixture of BrF_3, BrF_5, Br_2, and BrF. Their stability and thermal property data on BrF, however, have been doubted and partially refuted; see [1, 11]. For data and references concerning the fluorination of bromine, including the $Br_2 + F_2 \rightarrow 2\,BrF$ reaction, see "Fluorine" Suppl. Vol. 2, 1980, pp. 143/4, and "Bromine" Suppl. Vol. A, 1985, p. 373.

The reactions $Br + F_2 \rightarrow BrF + F$ and $Br_2 + F \rightarrow BrF + Br$ were treated in "Fluorine" Suppl. Vol. 2, 1980, p. 144, and "Bromine" Suppl. Vol. A, 1985, pp. 373/4, the reactions of O_2F with both molecular and atomic bromine resulting in electronically excited BrF are described in "Bromine" Suppl. Vol. A, 1985, p. 412; see also pp. 29/30. The possible formation of BrF molecular beams from the ion-molecule reaction $Br^+ + F_2 \rightarrow BrF + F^+$ was theoretically (scattering theory) shown [12].

The exchange reactions $Br + FC \rightarrow BrF + C$, $Br + FH \rightarrow BrF + H$, and $F + BrH \rightarrow BrF + H$ were theoretically studied (canonical variational transition state theory) [13].

For spectroscopic studies of the BrF molecule, described in some of the following chapters, the compound was generally obtained from the $Br_2 + F_2$ or $Br_2 + BrF_3$ reactions and from the $Br_2 + F$ reaction with various methods of F atom generation. Details and other methods to form excited BrF molecules are given in the respective chapters.

References:

[1] Stein, L. (in: Gutmann, V.; Halogen Chemistry, Vol. 1, Academic, London-New York 1967, pp. 133/224, 151/5).
[2] Steunenberg, R. K.; Vogel, R. C.; Fischer, J. (J. Am. Chem. Soc. **79** [1957] 1320/3).
[3] Fischer, J.; Bingle, J.; Vogel, R. C. (J. Am. Chem. Soc. **78** [1956] 902/4).
[4] Fischer, J.; Steunenberg, R. K.; Vogel, R. C. (J. Am. Chem. Soc. **76** [1954] 1497/8).
[5] Sukhoverkov, V. F.; Podzolko, L. G. (Zh. Analit. Khim. **38** [1983] 715/21; C.A. **99** [1983] No. 32425).
[6] Stein, L. (J. Am. Chem. Soc. **81** [1959] 1273/6).
[7] Naumann, D.; Lehmann, E. (J. Fluorine Chem. **5** [1975] 307/21).
[8] Ruff, O.; Menzel, W. (Z. Anorg. Allgem. Chem. **202** [1931] 49/61).
[9] Ruff, O.; Braida, A. (Z. Anorg. Allgem. Chem. **214** [1933] 81/90).
[10] Braida, A. (Diss. T.H. Breslau 1933, pp. 1/42).

[11] Booth, H. S.; Pinkston, J. T., Jr. (Chem. Rev. **41** [1947] 421/39, 429).
[12] Baz', A. I.; Gol'danskii, V. I. (Dokl. Akad. Nauk SSSR **186** [1969] 1110/3; Dokl. Phys. Chem. Proc. Acad. Sci. USSR **184/189** [1969] 375/8).
[13] Garrett, B. C.; Truhlar, D. G. (J. Am. Chem. Soc. **101** [1979] 5207/17).

10.3.2 The BrF Molecule

10.3.2.1 Electron Configuration. Electronic States

10.3.2.1.1 Ground State

The ground state X $^1\Sigma^+$ of the 44-electron molecule correlating with the ground-state atoms, $Br(^2P_{3/2})+F(^2P_{3/2})$, is represented by the electron configuration $\{KLM\}$ $(8\sigma)^2$ $(9\sigma)^2$ $(10\sigma)^2$ $(4\pi)^4$ $(5\pi)^4$, where $\{KLM\}$ stands for the core orbitals 1σ to 7σ, 1π to 3π, and 1δ arising from the K, L, and M shells of the Br atom and the K shell of the F atom. The valence orbitals are essentially the Br4s and F2s AO's, the Br4pσ-F2pσ bonding orbital, and the bonding and antibonding Br4pπ-F2pπ combinations. The energetical ordering of the highest three MO's is predicted either to be $10\sigma<4\pi<5\pi$ or $4\pi<10\sigma<5\pi$; see e.g. [1 to 3] (photoelectron spectroscopy of BrF could not yet clearify this question; see pp. 7/8).

The ground state X $^1\Sigma^+$ has been identified with the lower state of the absorption and emission systems B $^3\Pi(0^+)\leftrightarrow$X $^1\Sigma^+$ in the visible and near IR (see pp. 27/31); their analyses gave a number of ground-state properties.

A few quantum chemical ab initio calculations within the Hartree-Fock approximation [1, 4 to 8] or including electron correlation [9 to 11], a relativistic Hartree-Fock-Slater [2], and some 30 semiempirical [2, 12 to 42] calculations deal with the electronic ground state giving total molecular and orbital energies, charge distribution and/or dipole moments, bond properties, spectroscopic and other molecular constants. Bond properties of the Br−F bond, for example the values of ionic character or s character, were also empirically estimated [43 to 51].

Values for the total energy E_T and the corresponding equilibrium internuclear distance r_e obtained by the ab initio calculations are as follows:

$-E_T$ in au	r_e in Å [a)]	basis set [b)]	method [c)]	Ref.
2671.800702	1.799	ext. GTO	Full POL-CI	[9]
2671.77361	1.7192	DZ STO+pol.	SCF MO (near Hartree-Fock limit)	[1]
2664.2980	(1.759)	STO-nG	SCF MO	[4]
2660.4891	(1.76)	ext. GTO	SCF MO	[5]
2642.63613	1.771	STO-3G	SCF MO	[6]
2570.72783	1.760	SV+pol.	SCF MO	[7]

a) Optimized values; in parentheses experimental r_e. — b) GTO = Gaussian-type orbital, STO = Slater-type orbital, DZ = double zeta, ext. = extended, pol. = polarization functions, SV = split valence GTO. — c) SCF MO = self-consistent field molecular orbital, POL-CI = polarization configuration interaction.

References:

[1] Straub, P. A.; McLean, A. D. (Theor. Chim. Acta **32** [1974] 227/42).
[2] Dyke, J. M.; Josland, G. D.; Snijders, J. G.; Boerrigter, P. M. (Chem. Phys. **91** [1984] 419/24).

[3] Grodzicki, M.; Männing, V.; Trautwein, A. X.; Friedt, J. M. (J. Phys. B **20** [1987] 5595/625).
[4] Findlay, R. H. (J. Chem. Soc. Faraday Trans. II **72** [1976] 388/97).
[5] Kirillov, Yu. B.; Klimenko, N. M. (Zh. Strukt. Khim. **21** [1980] 9/14; J. Struct. Chem. [USSR] **21** [1980] 262/6).
[6] Pietro, W. J.; Levi, B. A.; Hehre, W. J.; Stewart, R. F. (Inorg. Chem. **19** [1980] 2225/9).
[7] Andzelm, J.; Klobukowski, M.; Radzio-Andzelm, E. (J. Computat. Chem. **5** [1984] 146/61).
[8] Dobbs, K. D.; Hehre, W. J. (J. Computat. Chem. **7** [1986] 359/78).
[9] Eades, R. A. (Diss. Univ. Minnesota 1983, pp. 1/214, 73/5, 89, 102, 128; Diss. Abstr. Intern. B **44** [1984] 3418).
[10] Kucharski, S. A.; Noga, J.; Bartlett, R. J. (J. Chem. Phys. **88** [1988] 1035/40).

[11] Urban, J.; Klimo, V.; Tiňo, J. (Chem. Phys. Letters **128** [1986] 203/7).
[12] Colbourn, E. A.; Dyke, J. M.; Fayad, N. K.; Morris, A. (J. Electron Spectrosc. Relat. Phenom. **14** [1978] 443/52).
[13] Bowmaker, G. A.; Boyd, P. D. W. (J. Mol. Struct. **150** [1987] 327/44).
[14] Grodzicki, M.; Lauer, S.; Trautwein, A. X.; Vera, A. (Advan. Chem. Ser. No. 194 [1981] 3/37).
[15] Gázquez, J. L.; Ortiz, E. (J. Chem. Phys. **81** [1984] 2741/8).
[16] Cheesman, G. H.; Finney, A. J. T.; Snook, I. K. (Theor. Chim. Acta **16** [1970] 33/42).
[17] Ewig, C. S.; Van Wazer, J. R. (J. Chem. Phys. **63** [1975] 4035/41).
[18] Pohl, H. A.; Raff, L. M. (Intern. J. Quantum Chem. **1** [1967] 577/89).
[19] Hyde, R. G.; Peel, J. B. (J. Chem. Soc. Faraday Trans. II **72** [1976] 571/8).
[20] Dewar, M. J. S.; Zoebisch, E. G. (J. Mol. Struct. **180** [1988] 1/21).

[21] Dewar, M. J. S.; Healy, E. (J. Computat. Chem. **4** [1983] 542/51).
[22] Deb, B. M.; Coulson, C. A. (J. Chem. Soc. A **1971** 958/70).
[23] Rhee, C. H.; Metzger, R. M.; Wiygul, F. M. (J. Chem. Phys. **77** [1982] 899/915).
[24] Hase, H. L.; Schweig, A. (Theor. Chim. Acta **31** [1973] 215/20).
[25] Scharfenberg, P. (Z. Chem. [Leipzig] **17** [1977] 388/9).
[26] Scharfenberg, P. (Theor. Chim. Acta **49** [1978] 115/22).
[27] Scharfenberg, P. (Theor. Chim. Acta **67** [1985] 235/43).
[28] Spurling, T. H.; Winkler, D. A. (Australian J. Chem. **39** [1986] 233/7).
[29] Bhattacharyya, S. P.; Chowdhury, M. (J. Phys. Chem. **81** [1977] 1602/4).
[30] Bhattacharyya, S. P. (Indian J. Chem. A **16** [1978] 4/6).

[31] Hyde, R. G.; Peel, J. B. (J. Chem. Soc. Faraday Trans. II **72** [1976] 571/8).
[32] Sichel, J. M.; Whitehead, M. A. (Theor. Chim. Acta **11** [1968] 220/38).
[33] Sichel, J. M.; Whitehead, M. A. (Theor. Chim. Acta **11** [1968] 239/53).
[34] Sichel, J. M.; Whitehead, M. A. (Theor. Chim. Acta **11** [1968] 254/62).
[35] Sichel, J. M.; Whitehead, M. A. (Theor. Chim. Acta **11** [1968] 263/70).
[36] Wiebenga, E. H.; Kracht, D. (Inorg. Chem. **8** [1969] 738/46).
[37] Andreev, S. (God. Vissh. Khim. Tekhnol. Inst. Sofia **20** No. 2 [1972/74] 43/56; C.A. **83** [1975] No. 198079).
[38] Lippincott, E. R. (J. Chem. Phys. **26** [1957] 1678/85).
[39] Kang, Y. K. (Bull. Korean Chem. Soc. **6** [1985] 107/11).
[40] Durakov, V. I.; Batsanov, S. S. (Zh. Strukt. Khim. **2** [1961] 456/61; J. Struct. Chem. [USSR] **2** [1961] 424/9).

[41] Iczkowski, R. (J. Am. Chem. Soc. **86** [1964] 2329/32).
[42] Boyd, R. J. (Diss. McGill Univ., Montreal, Canada, 1970 from Cornford, A. B.; Diss. Univ. Brit. Columbia, Canada, 1972, pp. 1/169, 72; Diss. Abstr. Intern. B **33** [1972] 2541).
[43] Whitehead, M. A.; Jaffé, H. H. (Trans. Faraday Soc. **57** [1961] 1854/62).
[44] Whitehead, M. A.; Jaffé, H. H. (Theor. Chim. Acta **1** [1962/63] 209/21).

[45] Wilmshurst, J. K. (J. Chem. Phys. **30** [1959] 561/50).
[46] Wilmshurst, J. K. (J. Chem. Phys. **33** [1960] 813/20).
[47] Dailey, B. P. (J. Phys. Chem. **57** [1953] 490/6).
[48] Dailey, B. P.; Townes, C. H. (J. Chem. Phys. **23** [1955] 118/23).
[49] Gordy, W. (J. Chem. Phys. **19** [1951] 792/3).
[50] Gordy, W. (J. Chem. Phys. **22** [1954] 1470/1).

[51] Gordy, W. (Discussions Faraday Trans. No. 19 [1955] 14/29).

10.3.2.1.2 Excited States

Excitations from the highest occupied into the lowest unoccupied MO give rise to the states ...$(4\pi)^4$ $(10\sigma)^2$ $(5\pi)^3$ $(11\sigma)^1$ $^3\Pi_i$ (2, 1, 0^+) and $^1\Pi(1)$ (notation for (Λ, S) and, in parentheses, (Ω,ω) coupling) which correlate with the neutral atoms, Br($^2P_{3/2,1/2}$) + F($^2P_{3/2,1/2}$). The triplet states have been labeled A′ $^3\Pi(2)$, A $^3\Pi(1)$, and B $^3\Pi(0^+)$.

The bound B $^3\Pi(0^+)$ state has been experimentally observed and well characterized by detailed studies of the B $^3\Pi(0^+)\leftrightarrow$X $^1\Sigma^+$ absorption and emission spectra (see pp. 27/9 and 29/31). Clyne et al. [1] obtained the term value $T_e = 18272.0 \pm 0.5$ cm^{-1} from the B→X emission spectrum and confirmed and slightly improved a value of 18281.2 cm^{-1} [2] derived from emission [3] and absorption [2] data; thus an earlier, higher T_e value [4, pp. 512/3], [5, 6], based on an erroneous vibrational assignment in the B←X system has been revised. The B state correlates either with a ground-state Br($^2P_{3/2}$) and a spin-orbit excited F($^2P_{1/2}$) atom or vice versa with F($^2P_{3/2}$) + Br($^2P_{1/2}$) (see p. 19); predissociation into the ground-state atoms Br($^2P_{3/2}$) + F($^2P_{3/2}$) has been inferred from the B↔X spectra (see pp. 27/31) and from lifetime measurements for B-state rovibrational levels (see pp. 21/2).

The A $^3\Pi(1)$ state has not been observed yet. The assignment of a visible absorption system to the A←X transition [2] was later shown [7] to be erroneous. Thus the term value derived by Brodersen and Sicre [2] and quoted by Huber and Herzberg [8] is insignificant.

The A′ $^3\Pi(2)$ state was identified with the lower state of the ultraviolet emission system D′ $^3\Pi(2)$→A′ $^3\Pi(2)$ (see p. 32) [9 to 13], from which the term value $T_e = 15750 \pm 150$ cm^{-1} for ^{79}BrF was derived [13].

For the $^3\Pi_i$ state without spin-orbit splitting, a configuration interaction (Full POL-CI) calculation has been carried out giving its potential energy function and spectroscopic constants [14].

The $^1\Pi(1)$ state was predicted to be repulsive by the Full POL-CI calculation [14].

Full POL-CI calculations for a number of other excited valence states, 2 $^1\Pi$, 2 $^3\Pi$, and 2 $^1\Sigma^+$, 1 $^3\Sigma^+$, 2 $^3\Sigma^+$, 1 $^1\Sigma^-$, 1 $^3\Sigma^-$, 1 $^1\Delta$, 1 $^3\Delta$ (presumably arising from $11\sigma\leftarrow4\pi$ and $11\sigma\leftarrow10\sigma$ and/or $6\pi\leftarrow5\pi$ and/or $(11\sigma)^2\leftarrow(5\pi)^2$ excitations, see [4, pp. 335/7]) showed these all to be repulsive and to dissociate into Br(^{2}P) + F(^{2}P) (spin-orbit coupling neglected) [14].

An excited ionic state D′ $^3\Pi(2)$, correlating with Br$^+$($^3P_2$) + F$^-$($^1S_0$), was identified with the upper state of the ultraviolet emission system D′ $^3\Pi(2)$→A′ $^3\Pi(2)$ (see p. 32) [9 to 13] from which the term value $T_e = 47450 \pm 150$ cm^{-1} for ^{79}BrF was derived [13].

References:

[1] Clyne, M. A. A.; Coxon, J. A.; Townsend, L. W. (J. Chem. Soc. Faraday Trans. II **68** [1972] 2134/43).
[2] Brodersen, P. H.; Sicre, J. E. (Z. Physik **141** [1955] 515/24).
[3] Durie, R. A. (Proc. Roy. Soc. [London] A **207** [1951] 388/95).

[4] Herzberg, G. (Molecular Spectra and Molecular Structure, Vol. 1, Spectra of Diatomic Molecules, Van Nostrand, Princeton, N. J., 1961).
[5] Brodersen, P. H.; Schumacher, H. J. (Z. Naturforsch. **2a** [1947] 358/9).
[6] Brodersen, P. H.; Schumacher, H. J. (Anales Asoc. Quim. Arg. **38** [1950] 52/60).
[7] Coxon, J. A.; Curran, A. H. (J. Mol. Spectrosc. **75** [1979] 270/87).
[8] Huber, K. P.; Herzberg, G. (Molecular Spectra and Molecular Structure, Vol. 4, Constants of Diatomic Molecules, Van Nostrand Reinhold, New York 1979, pp. 110/1).
[9] Diegelmann, M.; Grieneisen, H. P. (Ger. Offen. 3031954 [1980/82] 1/14).
[10] Diegelmann, M.; Grieneisen, H. P.; Hohla, K.; Hu, X.-J.; Krasinski, J.; Kompa, K. L. (Appl. Phys. **23** [1980] 283/7).
[11] Diegelmann, M.; Hohla, K.; Rebentrost, F.; Kompa, K. L. (J. Chem. Phys. **76** [1982] 1233/47).
[12] Henderson, S. D.; Tellinghuisen, J. (Chem. Phys. Letters **112** [1984] 543/6).
[13] Narayani, R. I.; Tellinghuisen, J. (J. Mol. Spectrosc. **141** [1990] 79/90).
[14] Eades, R. A. (Diss. Univ. Minnesota 1983, pp.1/214, 79/81, 89, 102/3, 129/31; Diss. Abstr. Intern. B **44** [1984] 3418).

10.3.2.2 Ionization Potentials

Electron-impact values for the upper limit of the first ionization potential $E_i \leq 11.9 \pm 0.3$ or 11.8 ± 0.2 eV were obtained from the appearance potentials of BrF^+ in the mass spectra of BrF_5 and BrF_3, respectively [1].

The He I photoelectron spectrum (PES) of BrF, produced in the PE spectrometer by the $Br_2 + F \rightarrow BrF + Br$ reaction [2] or obtained from Br_2-BrF_3 mixtures [3, 4], showed two vibrational series around 12 eV [2 to 4] and a broad band centered at 15.92 eV [2]. Using their own SCF MS Xα calculations [2] and ab initio SCF MO (near Hartree-Fock limit) results [5] and on the basis of Koopmans' theorem, $E_i \approx -\varepsilon_i$, Colbourn et al. [2] assigned the two vibrational series to the removal of the 5π valence electron leading to the spin-orbit split ionic states X $^2\Pi_{3/2}$ and X $^2\Pi_{1/2}$ and the broad band to the 4π and 10σ ionizations; the energetical ordering of the latter ionizations is predicted to be either $E_i(4\pi) < E_i(10\sigma)$ or $E_i(4\pi) > E_i(10\sigma)$ by various quantum chemical calculations. No further BrF features were found in the PES up to 21 eV; the next higher (9σ) ionization is predicted to occur above 24 eV [2]. In Table 1, experimental ionization potentials and calculated E_i $(= -\varepsilon_i)$ values are compared.

Further ab initio SCF MO and semiempirical calculations of the valence electron E_i's are available; see references [4, 5] and [3, 17, 19, 21, 22, 24, 29, 30, 33, 42] of Section 10.3.2.1.1.

References:

[1] Irsa, A. P.; Friedman, L. (J. Inorg. Nucl. Chem. **6** [1958] 77/90).
[2] Colbourn, E. A.; Dyke, J. M.; Fayad, N. K.; Morris, A. (J. Electron Spectrosc. Relat. Phenom. **14** [1978] 443/52).
[3] DeKock, R. L.; Higginson, B. R.; Lloyd, D. R.; Breeze, A.; Cruickshank, D. W. J.; Armstrong, D. R. (Mol. Phys. **24** [1972] 1059/72).
[4] Cornford, A. B. (Diss. Univ. Brit. Columbia, Canada, 1972, pp. 1/169, 66, 68/9, 72, 78; Diss. Abstr. Intern. B **33** [1972] 2541).
[5] Straub, P. A.; McLean, A. D. (Theor. Chim. Acta **32** [1974] 227/42).
[6] Dyke, J. M.; Josland, G. D.; Snijders, J. G.; Boerrigter, P. M. (Chem. Phys. **91** [1984] 419/24).

Table 1
BrF. Experimental and Theoretical Ionization Potentials.
Vertical (vert), adiabatic (ad), and calculated (calc) E_i values in eV.

MO	ionic state	E_i(vert)	E_i(vert)	E_i(ad)	E_i(vert)	E_i(ad)	E_i(calc)	E_i(calc)
5π	$^2\Pi_{3/2}$	11.86 ±0.01	11.87 ±0.01	11.78 ±0.01	11.81 ±0.02	11.72 ±0.02	11.12	11.97
	$^2\Pi_{1/2}$	12.19 ±0.01	12.19 ±0.01	12.09 ±0.01	12.13 ±0.02	12.05 ±0.02	11.42	
4π	$^2\Pi_{3/2}$	15.92	–	–	(16.75±0.4)		16.24	18.85
	$^2\Pi_{1/2}$		–	–			16.28	
10σ	$^2\Sigma^+$		–	–	(18.15±0.4)		17.51	18.03
9σ	$^2\Sigma^+$	–	–	–	–		25.22	27.79
8σ	$^2\Sigma^+$	–	–	–	–		–	43.64
remark		a)	b)		c)		d)	e)
Ref.		[2]	[3]		[4]		[6]	[5]

a) He I PES following the reaction $Br_2+F \rightarrow BrF+Br$. The spin-orbit splitting of the first PE band and thus of the ionic ground state is $\Delta E=2590\pm40\ cm^{-1}$. Each component shows vibrational structure, $\omega_e=750\pm30\ cm^{-1}$, $\omega_e x_e=10\pm5\ cm^{-1}$, corresponding to ionic vibrations which is consistent with the removal of the antibonding 5π electron [2].

b) Partial He I PES of a Br_2-BrF_3 mixture in the 11 to 15 eV region; $\Delta E=2600\pm60\ cm^{-1}$, $\omega=750\pm40\ cm^{-1}$ [3].

c) Partial He I PES of a Br_2-BrF_3 mixture in the 10 to 12.5 eV range; $\Delta E=2620\pm40\ cm^{-1}$, $\omega=750\pm40\ cm^{-1}$. The 4π and 10σ ionization potentials were estimated by comparison with other diatomic halogens and interhalogens and using atomic E_i values [4].

d) Relativistic Hartree-Fock-Slater calculation [6].

e) Ab initio SCF MO calculation near the Hartree-Fock limit which also gives E_i values for all core electrons [5].

The references in this table are given on the preceding page.

10.3.2.3 Dipole Moment. Quadrupole Moment

Stark effect measurements on the hyperfine components of the $J=1\leftarrow0$, $v=0$ transition in the microwave spectrum of ^{79}BrF and ^{81}BrF yielded a **dipole moment** $|\mu|=1.422(16)$ D (two standard deviations in parentheses) [1, 2]. The older value given without error limit, $|\mu|=1.29$ D [3 to 5], was derived in a similar way but without including hyperfine interactions in the analysis of the Stark effect data. According to quantum chemical calculations (see below) and electronegativity arguments, $\mu > 0$, i.e., polarization $Br^+ F^-$, is expected. For ClF, however, a negative dipole moment ($Cl^- F^+$) was obtained from molecular Zeeman effect measurements contrary to theoretical predictions; the Zeeman effect data for BrF are not precise enough to definitely determine its dipole moment sign [6].

A number of quantum chemical studies on BrF include the calculation of μ. The result of a calculation using many-body perturbation theory of fourth order (MBPT(4)), $\mu = 1.397$ D [7], and the result of the most rigorous ab initio SCF MO (near Hartree-Fock limit) calculation, $\mu = 1.440$ D [8], are in reasonable agreement with the experimental value. Values between 0.275 and 4.1 D were obtained from additional ab initio SCF MO and semiempirical calculations; see references [4, 5, 7] and [3, 13, 14, 16, 19, 21 to 28, 34, 39] of Section 10.3.2.1.1.

Zeeman effect measurements on the hyperfine components of the $J = 1 \leftarrow 0$, $v = 0$ transition in the microwave spectrum of ^{79}BrF and ^{81}BrF yielded the parallel components of the **molecular quadrupole moments** $\Theta_{\parallel} = (0.91 \pm 1.0)$ and $(1.23 \pm 1.0) \times 10^{-26}$ esu·cm^2, respectively [6]. An ab initio SCF MO (near Hartree-Fock) calculation for BrF gave $\Theta = 0.677 \times 10^{-26}$ esu·cm^2 [8]. Three other theoretical studies resulted in different values; see references [4, 13, 19] of Section 10.3.2.1.1.

References:

[1] Nair, K. P. R. (Kem. Kozlem. **52** [1979] 431/50).
[2] Nair, K. P. R.; Hoeft, J.; Tiemann, E. (J. Mol. Spectrosc. **78** [1979] 506/13).
[3] Smith, D. F.; Tidwell, M.; Williams, D. V. P. (Phys. Rev. [2] **77** [1950] 420/1).
[4] Lovas, F. J.; Tiemann, E. (J. Phys. Chem. Ref. Data **3** [1974] 609/769).
[5] Huber, K. P.; Herzberg, G. (Molecular Spectra and Molecular Structure, Vol. 4, Constants of Diatomic Molecules, Van Nostrand Reinhold, New York 1979).
[6] Ewing, J. J.; Tigelaar, H. L.; Flygare, W. H. (J. Chem. Phys. **56** [1972] 1957/66).
[7] Kucharski, S. A.; Noga, J.; Bartlett, R. J. (J. Chem. Phys. **88** [1988] 1035/40).
[8] Straub, P. A.; McLean, A. D. (Theor. Chim. Acta **32** [1974] 227/42).

10.3.2.4 Polarizability

Only theoretical results are available for the BrF molecule. The average molecular polarizability $\bar{\alpha} = (\alpha_{\parallel} + 2\alpha_{\perp})/3 = 3.644$ Å^3, the parallel and perpendicular components $\alpha_{\parallel} = 4.759$ Å^3, $\alpha_{\perp} = 3.086$ Å^3, and the anisotropy $k = 0.153$ with $k^2 = [(\alpha_{\parallel} - \bar{\alpha})^2 + 2(\alpha_{\perp} - \bar{\alpha})^2]/6\bar{\alpha}^2$ result from a perturbation calculation within the CNDO approximation.

Reference:

Rhee, C. H.; Metzger, R. M.; Wiygul, F. M. (J. Chem. Phys. **77** [1982] 899/915).

10.3.2.5 Nuclear Quadrupole Coupling Constant eqQ(Br)

The hyperfine splitting of the $J = 1 \leftarrow 0$, $v = 0$, 1 and $J = 2 \leftarrow 1$, $v = 0$ microwave transitions of ^{79}BrF and ^{81}BrF gave the values (two standard deviations in parentheses) [1]

eqQ(^{79}Br) = 1086.80(30) MHz and eqQ(^{81}Br) = 908.09(20) MHz for the $v = 0$ state,

eqQ(^{79}Br) = 1085.66(60) MHz and eqQ(^{81}Br) = 907.41(60) MHz for the $v = 1$ state,

which improve the earlier results eqQ(^{79}Br) = 1089.0 MHz and eqQ(^{81}Br) = 909.2 MHz for the $v = 0$ state [2, 3].

A few ab initio SCF MO and semiempirical studies include the calculation of the electric field gradient at the ^{79}Br nucleus and/or the quadrupole coupling constant; theoretical eqQ(^{79}Br) values generally are lower than the experimental values; see references [1, 4] and [13, 14, 22, 29, 30, 35] of Section 10.3.2.1.1.

References:

[1] Nair, K. P. R.; Hoeft, J.; Tiemann, E. (J. Mol. Spectrosc. **78** [1979] 506/13).
[2] Smith, D. F.; Tidwell, M.; Williams, D. V. P. (Phys. Rev. [2] **77** [1950] 420/1).
[3] Lovas, F. J.; Tiemann, E. (J. Phys. Chem. Ref. Data **3** [1974] 609/769).

10.3.2.6 Spin-Rotation Interaction Constants c(Br) and c(F)

The hyperfine splitting of the $J=1\leftarrow0$, $v=0$ and $J=2\leftarrow1$, $v=0$ microwave transitions of ^{79}BrF and ^{81}BrF gave (in addition to the quadrupole coupling constants, see above) the spin-rotation interaction constants $c(^{79}Br)=75(50)$ kHz and $c(^{81}Br)=83(50)$ kHz for the $v=0$ state (two standard deviations in parentheses). Spin-rotation interaction due to the F nucleus could not be resolved in this study [1].

$c(^{19}F)=23.2\pm0.6$ kHz was calculated from the absolute shielding constant for ^{19}F in BrF which in turn was derived from ^{19}F NMR experiments (cf. p. 11) [2].

References:

[1] Nair, K. P. R.; Hoeft, J.; Tiemann, E. (J. Mol. Spectrosc. **78** [1979] 506/13).
[2] Scheffer, T. J. (Diss. Univ. Wisconsin 1969, pp. 1/197, 146/7; Diss. Abstr. Intern. B **31** [1970] 156/7).

10.3.2.7 Molecular and Nuclear g Factors. Magnetic Susceptibility

The molecular Zeeman effect of the $J=1\leftarrow0$ microwave transitions of ^{79}BrF and ^{81}BrF was measured in magnetic fields near 20 kG and analyzed on the basis of a theory [1] which takes the strong quadrupolar coupling of the nuclear spin of Br to molecular rotation occurring in BrF into account. The Zeeman parameters, molecular (rotational) g factor $g_\perp$, shielded nuclear g factor $g_I\cdot(1-\sigma)$, and magnetic susceptibility anisotropy $\Delta\chi=\chi_\perp-\chi_\parallel$, were derived. These in turn were used to derive the molecular quadrupole moment (see p. 9), anisotropies in the second moment of the electronic charge distribution, and the paramagnetic part of the magnetic susceptibility $\chi_\perp^p$ ($\chi_\parallel^p\approx0$). The diamagnetic tensor elements of the magnetic susceptibility $\chi_\perp^d$ and $\chi_\parallel^d$ were calculated using $\Delta\chi$, $\chi_\perp^p$, and $\chi_\parallel$, the latter having been obtained from an estimated value for the bulk susceptibility, $\chi_{bulk}=(2\chi_\perp+\chi_\parallel)/3$. The results are as follows [2]:

	^{79}BrF	^{81}BrF
$g_\perp$ in μ_N	-0.1008 ± 0.0002	-0.1004 ± 0.0002
$g_I\cdot(1-\sigma)$ in μ_N	1.4045 ± 0.003	1.5134 ± 0.003
$\Delta\chi=\chi_\perp-\chi_\parallel$ in 10^{-6} erg·G^{-2}·mol^{-1}	20.9 ± 0.5	21.2 ± 0.5
$\chi_\perp^p$ in 10^{-6} erg·G^{-2}·mol^{-1}	114.4	
$\chi_\perp^d$ in 10^{-6} erg·G^{-2}·mol^{-1}	-133.3 ± 2.3	
$\chi_\perp$ in 10^{-6} erg·G^{-2}·mol^{-1}	-18.9 ± 2.2	
$\chi_\parallel^d=\chi_\parallel$ in 10^{-6} erg·G^{-2}·mol^{-1}	-39.0 ± 1.7	

An ab initio SCF MO calculation (STO-nG basis) gave $\chi_{\perp}^{d} = -132.01$ and $\chi_{\parallel}^{d} = -37.34$ (both in 10^{-6} erg·G^{-2}·mol^{-1}) [3].

References:

[1] Hüttner, W.; Flygare, W. H. (J. Chem. Phys. **47** [1968] 4137/45).
[2] Ewing, J. J.; Tigelaar, H. L.; Flygare, W. H. (J. Chem. Phys. **56** [1972] 1957/66).
[3] Findlay, R. H. (J. Chem. Soc. Faraday Trans. II **72** [1976] 388/97).

10.3.2.8 Magnetic Shielding. Nuclear Magnetic Relaxation (^{19}F NMR)

The chemical shift $\delta = -359.88 \pm 0.10$ ppm (upfield shift) was measured by ^{19}F NMR in gaseous Br_2-BrF_3 mixtures at 31.5 °C with SiF_4 as an internal standard; therefrom the absolute shielding constant $\sigma_F(BrF) = 723.0 \pm 6.0$ ppm was derived [1] using $\sigma_F = 363.2 \pm 6.0$ ppm [2] for the SiF_4 reference.

The ^{19}F spin-spin relaxation time T_2 was measured in the gaseous Br_2-BrF_3-SiF_4 mixtures (1) as a function of the SiF_4 density with the BrF pressure at its maximum value at 25 °C ($T_2 = 0.622$ to 3.49 ms for $(1.48$ to $34.5) \times 10^{-5}$ mol/cm^3) or (2) as a function of various equilibrium partial pressures of BrF, Br_2, and BrF_3 with the SiF_4 pressure held at a constant value (T_2 between 2.11 and 2.43 ms). The T_2 data indicate the occurrence of more than one ^{19}F relaxation process and can be fitted to a relation of the form $1/T_2 = A/\tau_c + B\tau_c$, where A and B are constants and τ_c is a correlation time for molecular reorientation. Consequently, a two-path relaxation model was found to apply for the ^{19}F nucleus in BrF consisting of spin-rotation interaction and scalar spin-spin coupling to the Br nucleus [1].

References:

[1] Scheffer, T. J. (Diss. Univ. Wisconsin 1969, pp. 1/197, 98/119, 146/7; Diss. Abstr. Intern. B **31** [1970] 156/7).
[2] Hindermann, D. K.; Cornwell, C. D. (J. Chem. Phys. **48** [1968] 4148/54).

10.3.2.9 Rotational and Vibrational Constants. Internuclear Distances. Mean Amplitudes of Vibration

Electronic Ground State. Analysis of the microwave absorption spectra in the region of the $J = 1 \leftarrow 0$ to $13 \leftarrow 12$ transitions, of the high-resolution IR spectra in the region of the fundamental band and the first overtone, and of the B $^3\Pi(0^+) \leftrightarrow X\ ^1\Sigma^+$ absorption and emission spectra in the visible and near-IR region gave for each isotopic species, ^{79}BrF and ^{81}BrF, the rotational constant B_e, the centrifugal stretching constants D_e, H_e, L_e, the corresponding rotation-vibration interaction constants α_e, γ_e, β_e, δ_e, and the vibrational constants ω_e, $\omega_e x_e$, $\omega_e y_e$, $\omega_e z_e$. The equilibrium internuclear distance r_e was obtained by converting the rotational constant. The results derived from the microwave spectra are compiled in Table 2, p. 12, and those from the IR and visible spectra in Table 3, p. 13. Details concerning the various analyses as well as additional references can be found in the remarks below both tables.

Using the atomic weights of Br and F and the fundamental frequency $\omega_e = 671$ cm^{-1}, the mean amplitudes of vibration u have been calculated for T = 0 to 1000 K at 100 K intervals: u = 0.0405, 0.0421, and 0.0604 Å at T = 0, 300, and 600 K, respectively [19].

Table 2
BrF, Ground State X $^1\Sigma^+$. Rotational and Vibrational Constants and Internuclear Distance Derived from Microwave Spectra.

		^{79}BrF	^{81}BrF	^{79}BrF	^{81}BrF	^{79}BrF	^{81}BrF
B_e	in MHz	10668.0970(69)	10616.9882(69)	10667.610(60)	10616.522(70)	10667.9	10616.8
α_e	in MHz	77.8629(79)	77.3040(78)	78.282(60)	77.763(70)	78.3	77.9
γ_e	in kHz	−201.6(45)	−199.7(45)	−	−	−	−
D_e	in kHz	12.049(10)	11.934(10)	−	−	−	−
β_e	in Hz	−66(11)	−66(11)	−	−	−	−
$10^3 \cdot H_e$	in Hz	−3.937(7)	−3.881(7)	−	−	−	−
ω_e	in cm^{-1}	669.679(28)	668.671(28)	−	−	−	−
$\omega_e x_e$	in cm^{-1}	3.863(76)	3.844(78)	−	−	−	−
r_e	in Å	1.758987(4)	1.758987(4)	1.758981(50)	1.758981(50)	1.759	1.759
remark		a)		b)		c)	
Ref.		[1, 2]		[3]		[4, 5]	

a) Analysis of the $J=4\leftarrow3$ to $13\leftarrow12$, $v=0$ to 3 transitions using Dunham's [6] energy level equation $E(v, J)=\Sigma_{ik}(v+1/2)^i \cdot J^k(J+1)^k$; the values given for α_e, γ_e, D_e, β_e, and H_e are the Dunham coefficients $-Y_{11}$, Y_{21}, $-Y_{02}$, Y_{12}, and Y_{03}, respectively; further constants: $Y_{31}=-6.2(8)$ and $-6.1(8)$ kHz, Y_{13} ($\approx-\delta_e$) $=-3.8(4)\times10^{-4}$ and $-3.7(4)\times10^{-4}$ Hz, Y_{04} ($\approx L_e$) $=1.5(2)\times10^{-8}$ and $1.5(2)\times10^{-8}$ Hz, and $Y_{00}=24.5(6)$ and 24.4(6) GHz for ^{79}BrF and ^{81}BrF, respectively; one standard deviation is given in parentheses, Willis and Clark [1, 2].

b) Analysis of the $J=1\leftarrow0$, $2\leftarrow1$, $v=0$, 1 transitions with the centrifugal stretching constants D_0 and D_1 being constrained to the optical values [12]; $B_e\approx3/2\,B_0-1/2\,B_1$ and $\alpha_e\approx B_0-B_1$ using the measured values for B_0 and B_1 and neglecting γ_e (estimate ~ −120 kHz); two standard deviations in parentheses, Nair et al. [3].

c) Analysis of the $J=1\leftarrow0$, $v=0$ transition by Smith et al. [4] and recalculation of the constants by Calder and Ruedenberg [5] (recalculation of α_e also in [7]). Huber and Herzberg [8] quote the converted values, $B_e=0.35584_3$ cm^{-1} and $\alpha_e=0.00261_2$ cm^{-1} for ^{79}BrF, add as an uncertain value $D_e=4.01\times10^{-7}$ cm^{-1}, and derive $r_e=1.75894$ Å.

Table 3

BrF, Ground State X $^1\Sigma^+$. Rotational and Vibrational Constants and Internuclear Distance from Spectra in the IR and Visible Region.

		^{79}BrF	^{81}BrF	^{79}BrF	^{81}BrF	^{79}BrF	^{81}BrF
B_e	in cm^{-1}	0.3558215(30)	0.3541197(23)	0.355832(8)	0.354121(10)	0.3558193(12)	0.3541193(12)
$10^3 \cdot \alpha_e$	in cm^{-1}	2.5953(10)	2.5743(7)	2.5920(13)	2.5740(14)	2.5826(12)	2.5703(13)
$10^6 \cdot \gamma_e$	in cm^{-1}	−8.1(5)	−8.8(3)	−8.8(5)	−8.8(5)	−10.60(11) −	10.25(11)
$10^7 \cdot D_e$	in cm^{-1}	4.002(7)	3.977(5)	3.95(6)	3.90(10)	−	−
$10^9 \cdot \beta_e$	in cm^{-1}	−1.94(9)	−2.12(6)	−2.0(6)	−1.8(11)	−	−
ω_e	in cm^{-1}	669.9011(23)	668.2956(8)	669.9002(5)	668.2937(6)	669.823(4)	668.227(4)
$\omega_e x_e$	in cm^{-1}	3.7983(6)	3.7811(8)	3.7988(2)	3.7804(2)	3.753(2)	3.739(2)
$\omega_e y_e$	in cm^{-1}	−	−	−	−	$-8.7(3) \times 10^{-3}$	$-7.8(3) \times 10^{-3}$
$\omega_e z_e$	in cm^{-1}	−	−	−	−	$-1.6(1) \times 10^{-4}$	$-2.0(1) \times 10^{-4}$
r_e	in Å	−	−	−	−	1.758989(5)	
remark		a)		b)		c)	
Ref.		[9]		[10]		[11]	

a) Fourier transform IR spectrum in the region of the fundamental band (v = 1←0 and 2←1) and the first overtone (v = 2←0) analyzed by Bürger et al. [9]; one standard deviations in parentheses.

b) Diode laser IR spectrum in the region of the v = 1←0 and 2←1 bands analyzed by Nakagawa et al. [10]; fit of the wave numbers to the seven Dunham coefficients, Y_{01}, $-Y_{11}$, Y_{21}, $-Y_{02}$, Y_{12}, Y_{10}, and $-Y_{20}$, which are set equal to the constants given above; one standard deviations in parentheses.

c) Analysis of the high-resolution B $^3\Pi(0^+) \rightarrow$ X $^1\Sigma^+$ emission spectrum in the region of the v = 0→6 to 0→10, 1→4 to 1→7, 1→11, 2→10, 3→5, 8, 9, 12, and 4→12 bands by Coxon and Wickramaaratchi [11]; derivation of the constants B'_v, D'_v, B''_v and merging with B←X absorption [12] and microwave [3] data; one standard deviation in parentheses. From their high-resolution B←X absorption spectrum in the region of the v = 3←0 to 8←0, 1←1 to 8←1, and 2←2 to 4←2 bands and by use of the earlier microwave data for B''_0 [4, 5], Coxon and Curran [12] derived similar values for α_e, ω_e, and $\omega_e x_e$ of somewhat lower accuracy. Earlier results of Clyne et al. [13] for ω_e and $\omega_e x_e$ of ^{79,81}BrF obtained from the vibrational analysis of the B→X emission spectrum of lower resolution are improved, as well as those from the earliest B↔X absorption [14 to 16] and emission [17] studies. Rotational constants, B_v and D_v, for the v = 3 state were derived from a laser excitation spectrum in the B←X, v = 3←3 band [18].

A few quantum chemical calculations deal with the derivation of ground-state spectroscopic constants. The results from a polarization configuration interaction (Full POL-CI) calculation for ^{79}BrF, $B_e = 0.3402\ cm^{-1}$, $\alpha_e = 0.00250\ cm^{-1}$, $\omega_e = 672.8\ cm^{-1}$, $\omega_e x_e = 4.43\ cm^{-1}$, $r_e = 1.799$ Å [20] are in reasonable agreement with the experimental results (except for r_e). Ab initio and pseudopotential SCF MO calculations of ω_e and/or r_e and semiempirical calculations of r_e have been carried out; see references [1, 6 to 8, 17] and [18, 21, 22, 24 to 27] of Section 10.3.2.1.1.

Excited State B $^3\Pi(0^+)$. Constants for the B state result from rotational and vibrational analyses of the B $^3\Pi(0^+) \leftrightarrow$ X $^1\Sigma^+$ absorption and emission spectra. High-resolution rotational analyses of the 3←0 to 8←0, 1←1 to 8←1, and 2←2 to 4←2 bands in the B←X absorption spectrum [12, 21], of the 3←0 to 8←0 [22], 0←1 [23], and 3←3 [18] bands in the LIF (laser induced fluorescence) spectrum, and high-resolution analyses of the 0→6 to 0→10, 1→4 to 1→7, 1→11, 2→10, 3→5, 8, 9, 12, and 4→10 bands in the B→X emission spectrum [11] yielded the rotational and centrifugal stretching constants B'_v and D'_v for v′=0 to 8 and H'_v for v′=7 and 8. The derivation of equilibrium values, however, is complicated by the predissociation of the B state, which causes the usual representation of B'_v by a polynomial in (v+1/2) to become unsatisfactory. Using an alternative approach, a perturbational, nonlinear least-squares fitting procedure, which allows for the intersecting repulsive 0^+ state [24], the values

$$B_e = 0.25724\ cm^{-1} \text{ and } r_e = 2.0688\ \text{Å for } ^{79}\text{BrF},$$

$$B_e = 0.25594\ cm^{-1} \text{ and } r_e = 2.0690\ \text{Å for } ^{81}\text{BrF}$$

were derived by Coxon and Wickramaaratchi [11].

By vibrational analyses (v′=0 to 10, v″=0 to 14) of their own B→X emission spectrum [13] and earlier absorption data [16], Clyne et al. [13] derived

$$\omega_e = 372.2\ cm^{-1},\ \omega_e x_e = 3.49\ cm^{-1},\ \omega_e y_e = -0.22\ cm^{-1}$$

and thus improved the earlier results of [14 to 17].

A polarization configuration interaction calculation (Full POL-CI), however without including spin-orbit interaction, gave for the lowest bound $^3\Pi(0^+, 1, 2)$ state of ^{79}BrF the constants B_e, α_e, ω_e, $\omega_e x_e$, and r_e [20].

Other Excited States. Since it was proven (see [12]) that Brodersen and Sicre [16] erroneously assigned a number of band heads in the visible region to the A $^3\Pi(1) \leftarrow$ X $^1\Sigma^+$ system, the constants ω_e and $\omega_e x_e$ derived by them and adopted by Huber and Herzberg [8] become useless.

Vibrational [25] and rotational [26] analysis of one ultraviolet emission system assigned to a transition between an ionic ($Br^+ + F^-$) state and the lowest excited valence state, D′ $^3\Pi(2) \rightarrow$ A′ $^3\Pi(2)$, resulted in the following spectroscopic constants for ^{79}BrF (standard deviation in parentheses) [26]:

constant		A′ $^3\Pi(2)$	D′ $^3\Pi(2)$
B_e	in cm^{-1}	0.2573[a)]	0.15524(1)
$10^3 \cdot \alpha_e$	in cm^{-1}	3.038(69)	0.598(3)
$10^4 \cdot \gamma_e$	in cm^{-1}	1.314(184)[b)]	–
$10^7 \cdot D_e$	in cm^{-1}	4.41(3)	1.578(1)
$10^8 \cdot \beta_e$	in cm^{-1}	3.427(179)[b)]	0.303(19)

constant		$A'\ ^3\Pi(2)$	$D'\ ^3\Pi(2)$
$10^{12}\cdot H_e$	in cm^{-1}	$(-0.92)^{c)}$	0.116(1)
ω_e	in cm^{-1}	393.011(1.510)	307.975(92)
$\omega_e x_e$	in cm^{-1}	7.153(198)	1.53(5)
$\omega_e y_e$	in cm^{-1}	0.0753(115)	–
$\omega_e z_e$	in cm^{-1}	−0.0044(2)	–
r_e	in Å	$2.06852^{a)}$	2.66304(13)

a) Values fixed using those of the B $^3\Pi(0^+)$ state. – b) The constants of next two higher orders were also derived. – c) H_0 from RKR calculation; $L_0 = -1.6\times10^{-16}$ cm^{-1}.

Preliminary results from vibrational ($v' \leq 4$, $v'' \leq 17$) and partial rotational analyses (0→9 and 1→9 bands) are reported by [25]. Approximation of the D′ state potential function by a Rittner potential (cf. p. 18) resulted in $\omega_e = 332$ cm^{-1} and $r_e = 2.575$ Å [27].

Upper state vibrational constants, ω_e and $\omega_e x_e$, for two unidentified emission systems in the Schumann region were derived [28].

References:

[1] Willis, R. E., Jr. (Diss. Duke Univ. 1979, pp. 1/158, 97, 105/6; Diss. Abstr. Intern. B **40** [1980] 4880).
[2] Willis, R. E., Jr.; Clark, W. W., III (J. Chem. Phys. **72** [1980] 4946/50).
[3] Nair, K. P. R.; Hoeft, J.; Tiemann, E. (J. Mol. Spectrosc. **78** [1979] 506/13).
[4] Smith, D. F.; Tidwell, M.; Williams, D. V. P. (Phys. Rev. [2] **77** [1950] 420/1).
[5] Calder, G. V.; Ruedenberg, K. (J. Chem. Phys. **49** [1968] 5399/415).
[6] Dunham, J. L. (Phys. Rev. [2] **41** [1932] 721/31).
[7] Lovas, F. J.; Tiemann, E. (J. Phys. Chem. Ref. Data **3** [1974] 609/769).
[8] Huber, K. P.; Herzberg, G. (Molecular Spectra and Molecular Structure, Vol. 4, Constants of Diatomic Molecules, Van Nostrand Reinhold, New York 1979, pp. 100/1).
[9] Bürger, H.; Schulz, P.; Jacob, E.; Fähnle, M. (Z. Naturforsch. **41a** [1986] 1015/20).
[10] Nakagawa, K.; Horiai, K.; Konno, T.; Uehara, H. (J. Mol. Spectrosc. **131** [1988] 233/40).

[11] Coxon, J. A.; Wickramaaratchi, M. A. (J. Mol. Spectrosc. **87** [1981] 85/100).
[12] Coxon, J. A.; Curran, A. H. (J. Mol. Spectrosc. **75** [1979] 270/87).
[13] Clyne, M. A. A.; Coxon, J. A.; Townsend, L. W. (J. Chem. Soc. Faraday Trans. II **68** [1972] 2134/43).
[14] Brodersen, P. H.; Schumacher, H. J. (Z. Naturforsch. **2a** [1947] 358/9).
[15] Brodersen, P. H.; Schumacher, H. J. (Anales Asoc. Quim. Arg. **38** [1950] 52/60).
[16] Brodersen, P. H.; Sicre, J. E. (Z. Physik **141** [1955] 515/24).
[17] Durie, R. A. (Proc. Roy. Soc. [London] A **207** [1951] 388/95).
[18] Takehisa, Y.; Ohashi, N. (J. Mol. Spectrosc. **128** [1988] 304/5).
[19] Baran, E. J. (Z. Physik. Chem. [Leipzig] **255** [1974] 1022/6).
[20] Eades, R. A. (Diss. Univ. Minnesota 1983, pp.1/214; Diss. Abstr. Intern. B **44** [1984] 3418).

[21] Coxon, J. A.; Curran, A. H. (J. Photochem. **9** [1978] 183/5).
[22] Clyne, M. A. A.; Curran, A. H.; Coxon, J. A. (J. Mol. Spectrosc. **63** [1976] 43/59).
[23] Clyne, M. A. A.; McDermid, I. S. (J. Chem. Soc. Faraday Trans. II **74** [1978] 664/80).
[24] Coxon, J. A. (J. Mol. Spectrosc. **50** [1974] 142/65).

[25] Colbourn, E. A.; Dyke, J. M.; Fayad, N. K.; Morris, A. (J. Electron Spectrosc. Relat. Phenom. **14** [1978] 443/52).
[26] Narayani, R. I.; Tellinghuisen, J. (J. Mol. Spectrosc. **141** [1990] 79/90).
[27] Diegelmann, M.; Hohla, K.; Rebentrost, F.; Kompa, K. L. (J. Chem. Phys. **76** [1982] 1233/47).
[28] Brodersen, P. H.; Mayo, S. (Z. Physik **143** [1955] 477/8).

10.3.2.10 Potential Energy Functions

Ground State X $^1\Sigma^+$. High-resolution absorption and emission data for the B $^3\Pi(0^+)\leftrightarrow$ X $^1\Sigma^+$ system were used to evaluate the Rydberg-Klein-Rees (RKR) potential curves. The so far most reliable results derived by Coxon and Wickramaaratchi [1] are based on their emission data, on refitted absorption data of [2], and on the microwave data of [3]; listed are the energy values $E(v) = G(v) + Y_{00}$ ($Y_{00} = 0.018$ cm^{-1} is the Dunham correction to the vibrational energy levels) up to $\sim$9400 cm^{-1} ($v'' = 0$ to 15) and the classical turning points r_{min} and r_{max} for ^{79}BrF and ^{81}BrF; the curves for both isotopic species are essentially identical as expected. These data slightly improve the earlier results for ^{79}BrF ($v'' = 0$ to 15) of Clyne et al. [4] based on emission [5] and earlier microwave absorption [6, 7] data. They presumably improve also the results of Clyne and McDermid [8] for ^{79,81}BrF ($v'' = 0$ to 13) based on absorption [2, 9] and laser-induced fluorescence (LIF) [4] data which are given in a graphical representation only (no numerical values). **Fig. 1** is reproduced from [8].

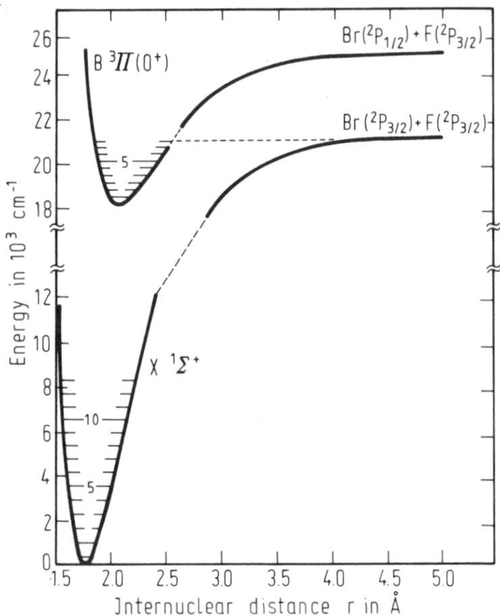

Fig. 1. RKR potential energy curves for the X $^1\Sigma^+$ and B $^3\Pi(0^+)$ states of BrF [8]. Vertical bars and integral numbers indicate vibrational levels; the dotted horizontal line indicates the first dissociation limit.

The coefficients a_i of Dunham's power series expansion of the potential energy around the equilibrium internuclear distance r_e, $V(x) = a_0 x^2(1 + a_1 x + a_2 x^2 + a_3 x^3 + ...)$ with $x = (r - r_e)/r_e$, were derived from the Dunham coefficients Y_{ik} (cf. p. 12) which resulted from the analysis of the microwave spectrum: $a_0=3.1503(26)\times10^5$ cm^{-1}, $a_1=-3.2893(10)$, $a_2=6.29(15)$, and $a_3=-9.5(11)$ [10, 11].

A few empirical potential functions were compared and/or tested for their abilities to reproduce a certain molecular parameter for the equilibrium position or a certain part of the potential curve [12 to 15].

Quantum chemical ab initio calculations for the ground state and several excited states (see below) using GVB(pp) (perfect pairing generalized valence bond), POL-CI (polarization configuration interaction), and Full POL-CI wave functions resulted in numerical energy values for r=2.2000 to 25.0000 au (1.16 to 13.2 Å); a graphical representation of the Full POL-CI potential curve at r=1.25 to 3.25 Å was given [16]. Semiempirical calculations also gave ground-state potential energy functions; see references [16, 18, 38] of Section 10.3.2.1.1.

Excited State B $^3\Pi(0^+)$. Predissociation. The evaluation of B-state RKR potential functions from spectroscopic data is complicated by the occurrence of predissociation, which on the basis of earlier absorption [17] and emission [18] data was explained to arise from a strongly avoided crossing (above v'=10) between the potential curve of the bound B state correlating with a spin-orbit-excited and a ground-state atom, $Br(^2P_{1/2})+F(^2P_{3/2})$, and a repulsive 0^+ state correlating with two ground-state atoms, $Br(^2P_{3/2})+F(^2P_{3/2})$; predissociation then was thought to occur by rotation over a potential maximum [19]. More recent lifetime measurements for individual rovibrational levels of the B state [8, 20, 21] ruled out this possibility, however; they were interpreted as Herzberg's case I(b) predissociation [22] arising from a weak interaction at a crossing between the B state and a shallow-bound $C(0^+)$ state correlating with $Br(^2P_{3/2})+F(^2P_{1/2})$, where the latter is predissociated by a third state correlating with $Br(^2P_{3/2})+F(^2P_{3/2})$. High-resolution absorption measurements [2] are also consistent with a predissociation due to curve crossing. As a consequence, the vibrational terms of the B state are perturbed even below the crossing point; thus the usual polynomial repesentation in (v+1/2) had to be extended by perturbational terms which describe the interaction between the crossing potential curves; see [1, 4].

As in the case of the X $^1\Sigma^+$ ground state, high-resolution B↔X absorption and emission data and ground-state microwave data were used to derive RKR potential functions (see above) for the B state. Coxon and Wickramaaratchi [1] give the most reliable results, i.e., numerical $E(v)=G(v)+Y_{00}$ values up to ~2770 cm^{-1} (with $Y_{00}=-0.232$ cm^{-1} for ^{79}BrF and $Y_{00}=-0.279$ cm^{-1} for ^{81}BrF) and r_{min} and r_{max} values for v'=0 to 8 of both ^{79}BrF and ^{81}BrF, which improve the results of [4] (numerical values for v'=0 to 8 of ^{79}BrF) and [8] (graphical representation for v'=0 to 8 and extrapolation up to the dissociation limit for $^{79,81}BrF$). The latter is reproduced in **Fig. 1**.

Ab initio calculations using the Full POL-CI wave function (cf. above) resulted in numerical energy values for r=2.8000 to 25.0000 au (1.48 to 13.2 Å) and in a graphical representation at r=1.5 to 3.5 Å for the bound 1 $^3\Pi(0^+, 1, 2)$ state (the calculations do not include spin-orbit effects) [16].

Other Excited States. Using emission data of [18] and their own absorption data for the A $^3\Pi(1)$↔X $^1\Sigma^+$ system, Brodersen and Sicre [17] evaluated a Morse potential curve for the A $^3\Pi(1)$ state. Since it was shown that the A↔X assignment was erroneous (cf. p. 27), this result is obsolete.

Thus the potential curve approximated for the A' $^3\Pi(2)$ state by Diegelmann et al. [23, 24] using the A $^3\Pi(1)$ data [17] is questionable. More recently, however, Narayani and Tellinghuisen [25] calculated an RKR potential function for the A' $^3\Pi(2)$ state of ^{79}BrF, i.e., G(v) values between 194.726 and 4677.975 cm^{-1} for v=0 to 17 and the corresponding r_{min} and r_{max} values, using their high-resolution D′ $^3\Pi(2)\rightarrow$A′ $^3\Pi(2)$ emission data. In a study on the reliability of RKR potentials and calculation of statistical uncertainties in the turning points, Tellinghuisen [26] used these A′ $^3\Pi(2)$ data.

The potential curve for the ion-pair excited state D′ $^3\Pi(2)$ ($Br^+(^3P_2)+F^-(^1S_0)$) was approximated by a Rittner potential [27] and using literature values for the relevant ionic parameters [23, 24].

For several low-lying (i.e., arising from ground state $Br(^2P)$ and $F(^2P)$ atoms) repulsive singlet (1 $^1\Pi$, 2 $^1\Pi$, 1 $^1\Delta$, 2 $^1\Sigma^+$, 1 $^1\Sigma^-$) and triplet (2 $^3\Pi$, 1 $^3\Sigma^+$, 1 $^3\Sigma^-$, 1 $^3\Delta$, 2 $^3\Sigma^+$) valence states, numerical energy values for r=2.8000 to 25.0000 au (1.48 to 13.2 Å) and potential curves at r=1.25 to 3.25 Å (singlets) and 1.5 to 3.5 Å (triplets) were derived from ab initio Full POL-CI (cf. above) calculations [16].

References:

[1] Coxon, J. A.; Wickramaaratchi, M. A. (J. Mol. Spectrosc. **87** [1981] 85/100).
[2] Coxon, J. A.; Curran, A. H. (J. Mol. Spectrosc. **75** [1979] 270/87).
[3] Nair, K. P. R.; Hoeft, J.; Tiemann, E. (J. Mol. Spectrosc. **78** [1979] 506/13).
[4] Clyne, M. A. A.; Curran, A. H.; Coxon, J. A. (J. Mol. Spectrosc. **63** [1976] 43/59).
[5] Clyne, M. A. A.; Coxon, J. A.; Townsend, L. W. (J. Chem. Soc. Faraday Trans. II **68** [1972] 2134/43).
[6] Smith, D. F.; Tidwell, M.; Williams, D. V. P. (Phys. Rev. [2] **77** [1950] 420/1).
[7] Calder, G. V.; Ruedenberg, K. (J. Chem. Phys. **49** [1968] 5399/415).
[8] Clyne, M. A. A.; McDermid, I. S. (J. Chem. Soc. Faraday Trans. II **74** [1978] 664/80).
[9] Coxon, J. A.; Curran, A. H. (J. Photochem. **9** [1978] 183/5).
[10] Willis, R. E., Jr. (Diss. Duke Univ. 1979, pp. 1/158, 101; Diss. Abstr. Intern. B **40** [1980] 4880).

[11] Willis, R. E., Jr.; Clark, W. W., III (J. Chem. Phys. **72** [1980] 4946/50).
[12] Lippincott, E. R.; Schroeder, R. (J. Chem. Phys. **23** [1955] 1131/41).
[13] Lippincott, E. R.; Steele, D.; Caldwell, P. (J. Chem. Phys. **35** [1961] 123/41).
[14] Borkman, R. F.; Simons, G.; Parr, R. G. (J. Chem. Phys. **50** [1969] 58/65).
[15] Varshni, Y. P. (Can. J. Chem. **66** [1988] 763/6).
[16] Eades, R. A. (Diss. Univ. Minnesota 1983, pp. 1/214, 102/3, 128/31; Diss. Abstr. Intern. B **44** [1984] 3418).
[17] Brodersen, P. H.; Sicre, J. E. (Z. Physik **141** [1955] 515/24).
[18] Durie, R. A. (Proc. Roy. Soc. [London] A **207** [1951] 388/95).
[19] Child, M. S.; Bernstein, R. B. (J. Chem. Phys. **59** [1973] 5916/25).
[20] Clyne, M. A. A.; McDermid, I. S. (J. Chem. Soc. Faraday Trans. II **74** [1978] 644/63).

[21] Clyne, M. A. A.; McDermid, I. S. (J. Chem. Soc. Faraday Trans. II **74** [1978] 1376/92).
[22] Herzberg, G. (Molecular Spectra and Molecular Structure, Vol. 1, Spectra of Diatomic Molecules, Van Nostrand, Princeton, N. J., 1961, pp. 413/4, 420/4).
[23] Diegelmann, M.; Hohla, K.; Rebentrost, F.; Kompa, K. L. (J. Chem. Phys. **76** [1982] 1233/47).
[24] Diegelmann, M. (Max-Planck-Ges. Foerd. Wiss. Projektgruppe Laserforsch. Ber. PLF-3 [1980] 1/124).
[25] Narayani, R. I.; Tellinghuisen, J. (J. Mol. Spectrosc. **141** [1990] 79/90).
[26] Tellinghuisen, J. (J. Mol. Spectrosc. **141** [1990] 258/64).
[27] Rittner, E. S. (J. Chem. Phys. **19** [1951] 1030/5).

10.3.2.11 Dissociation Energy

The heat of formation, based on thermochemical data for the $Br_2 + BrF_3 \rightleftharpoons 3\,BrF$ equilibrium [1] and given in the JANAF Tables [2], corresponds to the ground-state dissociation energy

$$D_0 = 20540 \pm 140\ cm^{-1}\ (2.547 \pm 0.017\ eV\ or\ 245.7 \pm 1.7\ kJ/mol)$$

Coxon et al. [3, 4] suggested this value to be a lower limit. The thermochemical value $D_0^\circ = 2.54_8$ eV was adopted by Huber and Herzberg [5].

Spectroscopic values are based on absorption and emission studies of the B $^3\Pi(0^+) \leftrightarrow$ X $^1\Sigma^+$ system. The earlier results of Brodersen et al. [6 to 8] and Durie [9, 10] suffer from deficiencies of the Birge-Sponer extrapolations of B-state vibrational levels and from the inexact localization of the onset of predissociation in the B state; furthermore, the convergence limit of the A $^3\Pi(1) \leftarrow$ X $^1\Sigma^+$ system (the assignment of which was later proven to be erroneous) was included in the analysis [8], and the question whether the B state dissociates into a ground-state Br atom and a spin-orbit excited F atom or vice versa ($Br(^2P_{3/2}) + F(^2P_{1/2})$ or $Br(^2P_{1/2}) + F(^2P_{3/2})$) lead to a higher and a lower value, respectively, for the ground-state dissociation energy. B-state dissociation into Br + F* was assumed by [8]; their result $D_0 = 2.384$ eV (compared to the Br* + F value, 2.79 eV) was recommended by Gaydon [11]. Durie [9, 10] also favored his lower value ($D_0 = 2.17$ and 2.58 eV), whereas Evans et al. [12], assuming B-state dissociation into Br* + F adopted the higher one for calculating the thermodynamic properties of BrF.

An accurate value for the ground-state dissociation energy could not be derived until Clyne and co-workers had thoroughly studied the B-state predissociation by high-resolution absorption and laser-induced fluorescence measurements of the B↔X system and by lifetime measurements of B-state rovibrational levels. In case I(b) of predissociation which applies to the B state of BrF (cf. p. 17), the onset of predissociation nearly coincides with the dissociation limit, thus D_0 can be established within narrow bounds. Analysis of the energy data of the highest stable levels (v′, J′ = 7, 29 and 6, 48) and lowest unstable levels (v′, J′ = 7, 30 and 6, 49) [13, 14] results in

$$D_0(^{79}BrF) = 20622 \pm 20\ cm^{-1} = 2.557 \pm 0.002\ eV = 246.69 \pm 0.24\ kJ/mol\ and$$

$$D_0(^{81}BrF) = 20616 \pm 20\ cm^{-1} = 2.556 \pm 0.002\ eV = 246.62 \pm 0.24\ kJ/mol,$$

which by including zero-point energies results in (for either isotopic species)

$$D_e(BrF) = 20953 \pm 20\ cm^{-1} = 2.598 \pm 0.002\ eV = 250.65 \pm 0.24\ kJ/mol\ [14]$$

These confirm the less precise estimate, $D_0(BrF) = 20600 \pm 50\ cm^{-1}$ [15], which was obtained by combining the spectroscopic ($D_0 \leq 20622\ cm^{-1}$; onset of predissociation at v′ = 7, J′ = 28 ± 1) [15] and thermochemical [2] values. Preliminary estimates $D_0 \approx 20880\ cm^{-1}$ [16] (based on the energy of the v′ = 8, J′ = 28 level for which the onset of predissociation was observed in these lifetime measurements) and $D_{298} < 20500\ cm^{-1}$ [17] (based on the energy of the highest vibrational level, v′ = 8, observed in the emission spectrum) are thus improved.

A few theoretical values of D_e are available. A Møller-Plesset perturbation calculation of second order (UMP2) gives $D_e = 2.555$ eV [18] and polarization configuration interaction calculations (POL-CI or Full POL-CI) give $D_e = 2.55$ or 2.56 eV [19] which are in fair agreement with the experimental result. SCF MO calculations with basis sets of various sizes give too low D_e values or even are not able to predict a stable molecular ground state ($D_e < 0$); semiempirical methods and simple models of bonding have also been applied; see references [7, 18, 32, 39] of Section 10.3.2.1.1.

For the dissociation of BrF into $Br^+ + F^-$ or $Br^- + F^+$, D = 1188 and 883 kJ/mol, respectively, were estimated from dipole moment and polarizability data [20].

References:

[1] Steunenberg, R. K.; Vogel, R. C.; Fischer, J. (J. Am. Chem. Soc. **79** [1957] 1320/3).
[2] Stull, D. R.; Prophet, H. (JANAF Thermochemical Tables, 2nd Ed., NSRDS-NBS-37 [1971]), Chase, M. W., Jr.; Davies, C. A.; Downey, J. R., Jr.; Frurip, D. J.; McDonald, R. A.; Syverud, A. N. (JANAF Thermochemical Tables, 3rd Ed., J. Phys. Chem. Ref. Data **14** Suppl. No. 1 [1985] 425).
[3] Coxon, J. A. (Chem. Phys. Letters **33** [1975] 136/40).
[4] Clyne, M. A. A.; Curran, A. H.; Coxon, J. A. (J. Mol. Spectrosc. **63** [1976] 43/59).
[5] Huber, K. P.; Herzberg, G. (Molecular Spectra and Molecular Structure, Vol. 4, Constants of Diatomic Molecules, Van Nostrand Reinhold, New York 1979, pp. 110/1).
[6] Brodersen, P. H.; Schumacher, H. J. (Z. Naturforsch. **2a** [1947] 358/9).
[7] Brodersen, P. H.; Schumacher, H. J. (Anales Asoc. Quim. Arg. **38** [1950] 52/60).
[8] Brodersen, P. H.; Sicre, J. E. (Z. Physik **141** [1955] 515/24).
[9] Durie, R. A. (Proc. Roy. Soc. [London] A **207** [1951] 388/95).
[10] Durie, R. A.; Gaydon, A. G. (J. Phys. Chem. **56** [1952] 316/9).
[11] Gaydon, A. G. (Dissociation Energies and Spectra of Diatomic Molecules, 3rd Ed., Chapman & Hall, London 1968, p. 265).
[12] Evans, W. H.; Munson, T. R.; Wagman, D. D. (J. Res. Natl. Bur. Std. **55** [1955] 147/64).
[13] Coxon, J. A.; Curran, A. H. (J. Mol. Spectrosc. **75** [1979] 270/87).
[14] Clyne, M. A. A.; McDermid, I. S. (J. Chem. Soc. Faraday Trans. II **74** [1978] 1376/92).
[15] Clyne, M. A. A.; McDermid, I. S. (J. Chem. Soc. Faraday Trans. II **74** [1978] 644/63).
[16] Clyne, M. A. A.; Curran, A. H.; Coxon, J. A. (J. Mol. Spectrosc. **63** [1976] 43/59).
[17] Clyne, M. A. A.; Coxon, J. A.; Townsend, L. W. (J. Chem. Soc. Faraday Trans. II **68** [1972] 2134/43).
[18] Urban, J.; Klimo, V.; Tiňo, J. (Chem. Phys. Letters **128** [1986] 203/7).
[19] Eades, R. A. (Diss. Univ. Minnesota 1983, pp. 1/214, 74, 89; Diss. Abstr. Intern. B **44** [1984] 3418).
[20] Cantacuzène, J. (J. Chim. Phys. Physicochim. Biol. **65** [1968] 502/15).

10.3.2.12 Relaxation Processes. Deactivation of B $^3\Pi(0^+)$ (v′, J′) Levels

General. Systematic studies on the dynamic properties of the B $^3\Pi(0^+)$ state of BrF have been carried out by Clyne and co-workers [1 to 8] by using laser excitation techniques. The BrF molecules were generated in a discharge flow system (He carrier gas) by the rapid $F + Br_2 \rightarrow BrF + Br$ reaction and the time-dependent decay of the fluorescence (LIF), which is due to the B $^3\Pi(0^+) \rightarrow X\ ^1\Sigma^+$ system, was monitored after selective excitation of individual B state rovibrational (v′, J′) levels by tunable dye lasers. The various contributions to the fluorescence decay, that are noncollisional processes such as spontaneous emission and predissociation, and collisional quenching such as electronic quenching, rotational and vibrational energy transfer, had to be deconvoluted; this was achieved by choosing the appropriate experimental conditions (variation of pressure, modifications of the optical detection system) and by computer modeling of the fluorescence decay curves. Each of the publications [3 to 7] gives very detailed descriptions of the experimental equipment, the procedure of lifetime measurements, and analysis of data as well as detailed discussions on the various mechanisms which contribute to the B-state deactivation.

Summaries of these experiments and results are given by Clyne and McDermid [8] and by Heaven [9]. A compilation of lifetimes, quenching rate constants, and rotational and vibrational energy-transfer rate constants is presented by Steinfeld [10].

Collision-Free Lifetimes τ_0. Radiative Lifetimes τ_{rad}. Predissociation Rate Constants. Fluorescence decay studies were first carried out at reasonably high pressures, such as p(BrF+He)≈50 to 900 mTorr, where molecular collisions have to be taken into consideration. The collision-free lifetimes τ_0 were derived from the measured lifetimes τ by Stern-Volmer plots following the relation $1/\tau=1/\tau_0+\Sigma k_Q[Q]$, where k_Q is the rate constant for collision-induced deactivation and [Q] is the concentration of the quenching molecules (here Q=He, BrF, Br_2; see below) [1 to 5]. For the rovibrational levels (v', J') with v'=1 to 6 and J' up to 41 and for the v'=7, J'≤21 levels, the extrapolated τ_0 values show no significant variation with J' indicating these levels to be stable with respect to predissociation. The following mean values of τ_0 for the (v'=1, J') to (v'=7, J') manifolds of ^{79}BrF were derived (which, however, have been later shown to be underestimated, see below):

v'	J'	τ_0 in μs	Ref.	v'	J'	τ_0 in μs	Ref.
1	0, 11, 16, 21, 26	28.5±8.7	[5]	5	0, 6, 11, 21, 31, 41	26.8±2.1	[5]
2	0, 11, 16, 21	32.8±5.0	[5]	6	0, 6, 11, 21, 31, 41	29.2±1.2	[5]
3	0, 11, 16, 19, 26	30.8±9.4	[5]	6	20	24.7±4.0*)	[3]
4	0, 5 to 31, 35, 36	40.6±6.9	[5]	7	4 to 21	22.4±3.3	[4]

*) ^{81}BrF

Predissociation occurring in the higher rotational levels (J'≥28) of the v'=7 state and in all accessible rotational levels (J'≤31) of the v'=8 state drastically shortens the lifetimes: τ_0(v'=7)=1.81 to 0.37 μs at J'=28 to 34, τ_0(v'=8)=1.74 to 0.11 μs at J'=1 to 31 for ^{79}BrF (slightly different values for the few v'=8, J'≤31 levels studied in ^{81}BrF). For the intermediate rotational levels (J'=16 to 27) of the v'=7 state, rotational energy transfer by collisions with the He atoms (R-T upward transfer into predissociated rotational levels) was found to be responsible for the successive shortening of τ_0 to values below 21 μs [4] (preliminary results [1]).

Improved τ_0 values were derived from measurements at collision-free conditions, i.e., at low pressures of 0.5 to 1.0 mTorr. For all rotational levels of the v'=0 to 5 states and for the stable levels of the v'=6 (J'≤48) and v'=7 (J'≤27) states, τ_0 values are in the 42 to 56 μs range [6]. Some of them were improved once more to give values in the 55 to 64 μs range by using an experimental equipment that carefully minimized the effects of diffusion which cause an increase of the decay rate with time [7]. The following table compares mean values of τ_0 for the (v'=0, J') to (v'=7, J') manifolds of ^{79}BrF and ^{81}BrF from [6] and [7]:

v′	J′	τ_0 in μs [6][a)]	J′	τ_0 in μs [7]
0	16 to 26	43.0 ± 0.7 44.2 ± 2.5	–	–
1	5 to 31	44.0 ± 2.1 43.4 ± 1.8	–	–
2	7 to 39	46.0 ± 2.3 46.0 ± 2.4	–	–
3	9 to 42	43.9 ± 2.0 43.9 ± 2.0	21	55.5[d)]
4	3 to 24,[b)] 27 to 45	44.7 ± 3.2 45.6 ± 3.1	21	59.0[d)]
5	3 to 38[c)]	44.2 ± 0.5 44.2 ± 0.4	21	58.9[d)]
6	3 to 44	46.3 ± 0.3 46.4 ± 0.4	10, 21, 45	61.8 ± 1.9[e)]
7	3 to 27	48.1 ± 1.6 48.5 ± 1.6	11, 20	64.2 ± 1.4[e)]

[a)] Upper row: ^{79}BrF; lower row: ^{81}BrF; average value over all v′ = 0 to 7 states of ^{79}BrF and ^{81}BrF: $\tau_0 = 45.2 \pm 2.4$ μs. – [b)] For J′ = 25 to 28, increase of τ_0 to a maximum of ~57 μs at J′ = 25. – [c)] For J′ = 38 to 46, increase of τ_0 to a maximum of ~56 μs at J′ = 40 (^{79}BrF) and J′ = 39 (^{81}BrF). – [d)] Values for ^{79}BrF. – [e)] Average values for ^{79}BrF and ^{81}BrF.

The τ_0 values for the stable (i.e., not predissociated), rovibrational levels should be equal to the radiative lifetimes τ_{rad}. The successive improvement of the τ_0 values and the assumption that there is no isotopic dependence of τ_{rad} lead one to conclude that the most reliable radiative lifetimes of the B $^3\Pi(0^+)$ state are those listed in the last column of the table above [7]. The trend of τ_{rad} to increase with increasing v′ was noted by [6, 7].

The low-pressure studies indicate that predissociation commences abruptly at the (v′ = 6, J′ = 49) and (v′ = 7, J′ = 30) states, where the lifetimes fall below 0.1 μs [6]. Collisional rotational relaxation was found to be responsible for significantly shortening the lifetimes for the two or three rotational levels immediately below the onset of predissociation [6, 7].

For the predissociated (v′ = 8, J′ = 4 to 27) manifold, lifetimes decrease from $\tau_0 = 1.77$ to 0.28 μs for ^{79}BrF and from 1.12 to 0.35 μs for ^{81}BrF. For the J′ = 4 to 21 range, a linear variation of $1/\tau_0$ with rotational energy, i.e., $1/\tau_0 = 1/\tau_{00} + k_{v'} \cdot J'(J'+1)$, was found; plots of $1/\tau_0$ vs. $J'(J'+1)$ yielded the lifetimes $\tau_{00} = \tau_{rad} = 1.75 \pm 0.20$ μs for ^{79}BrF and 1.33 ± 0.15 μs for ^{81}BrF as well as the predissociation rate constants $k_{v'} = (4.02 \pm 0.40) \times 10^4$ s^{-1} for ^{79}BrF and $(3.22 \pm 0.40) \times 10^4$ s^{-1} for ^{81}BrF [6].

Collisional Quenching. Lifetime measurements for the B-state (v′ = 6, J′ = 20) level of ^{81}BrF in flowing BrF-Br_2-He mixtures of various relative concentrations yielded linear Stern-

Volmer plots, the slopes of which gave the quenching rate constants k_Q(in 10^{-10} $cm^3 \cdot molecule^{-1} \cdot s^{-1}$) = 2.6 for BrF (self-quenching), 1.5 for Br_2, and ≤0.02 for He [3]. These have been confirmed by measurements for the (v' = 7, J' = 4 to 15) levels of ^{79}BrF and the (v' = 7, J' = 6, 11, 16) levels of ^{81}BrF [4].

For the higher rotational levels of the v' = 7 state, nonexponential decay curves were obtained with an initial rapid decay rate Γ^i and a delayed slower decay rate Γ^d; these were interpreted by rapid rotational energy transfer (R-T) involving collisions of BrF with He followed by electronic quenching. Computer simulations of the fluorescence decay curves were performed, which indicate that collisional rotational relaxation occurs with a high probability of multiquantum transitions ($\Delta J \leq \pm 10$); assuming the J' = 15 to 27 levels to be stable (i.e., not predissociated), the best fit to the experimental results was obtained with the R-T rate constant $k_R = 2.5 \times 10^{-11}$ $cm^3 \cdot molecule^{-1} \cdot s^{-1}$ which is reliable within a factor of two [4].

Nonexponential fluorescence decay curves observed for v' = 8 rotational levels were explained by vibrational relaxation (V-V) of the predissociated (v' = 8, J') manifold into the stable (v' = 7, J' ≤ 27) manifold. Computer simulations of the (v' = 8, J' = 6) decay yielded a V-V rate constant of $k_V = (2.5 \pm 1.0) \times 10^{-11}$ $cm^3 \cdot molecule^{-1} \cdot s^{-1}$ for BrF diluted in He; assuming that self-relaxation is the major vibrational relaxation process and that 7% of the total gas is BrF, the self-relaxation rate constant for V-V transfer between BrF(B $^3\Pi(0^+)$) and BrF($X^1\Sigma^+$) becomes $k_{V,BrF} \approx 3.6 \times 10^{-10}$ $cm^3 \cdot molecule^{-1} \cdot s^{-1}$ [4].

Collisional quenching of BrF B $^3\Pi(0^+)$ rovibrational levels by a number of foreign collision partners, such as M = O_2, Cl_2, Ar, $CHFCl_2$, and HCl, was the subject of further detailed studies. Nonexponential decay curves were observed, the analysis of which showed vibrational energy transfer to be the dominant deactivation process for all M except HCl. The following rate constants $k_{Q,V}$ for vibrational energy transfer and $k_{Q,M}$ for electronic quenching (both ±30% or upper limit) at 293 K were derived [7]:

collision partner M	v'(BrF)	$k_{Q,V}$ in 10^{-11} $cm^3 \cdot molecule^{-1} \cdot s^{-1}$	$k_{Q,M}$ in 10^{-12} $cm^3 \cdot molecule^{-1} \cdot s^{-1}$
O_2	3	3.6	2.6*) (v' = 3 to 7)
	4	2.2	
	5	2.1	
	6	3.4	
	7	12.5	
Cl_2	3	0.5	<0.1
Ar	3	≤0.25	<0.1
$CHFCl_2$	3	2	6
HCl	3	5	12

*) Vibrational Boltzmann distribution.

References:

[1] McDermid, I. S.; Clyne, M. A. A. (Lasers Chem. Proc. Conf., London 1977, pp. 122/7).
[2] Clyne, M. A. A.; Curran, A. H.; McDermid, I. S. (NBS-SP 526 [1978] 77/8).
[3] Clyne, M. A. A.; McDermid, I. S. (J. Chem. Soc. Faraday Trans. II **73** [1977] 1094/106).
[4] Clyne, M. A. A.; McDermid, I. S. (J. Chem. Soc. Faraday Trans. II **74** [1978] 644/63).
[5] Clyne, M. A. A.; McDermid, I. S. (J. Chem. Soc. Faraday Trans. II **74** [1978] 664/80).

[6] Clyne, M. A. A.; McDermid, I. S. (J. Chem. Soc. Faraday Trans. II **74** [1978] 1376/92).
[7] Clyne, M. A. A.; Liddy, J. P. (J. Chem. Soc. Faraday Trans. II **76** [1980] 1569/85).
[8] Clyne, M. A. A.; McDermid, I. S. (Advan. Chem. Phys. **50** [1982] 1/104, 58/62).
[9] Heaven, M. C. (Chem. Soc. Rev. **15** [1986] 405/48).
[10] Steinfeld, J. I. (J. Phys. Chem. Ref. Data **13** [1984] 445/553).

10.3.3 Spectra

10.3.3.1 Microwave Absorption Spectrum

The MW spectrum of BrF was measured in the 1.43 to 0.1 cm range (21 to 276 GHz) which pertains to the J=1←0 to 13←12 rotational transitions. BrF was obtained by decomposing BrF_3 [1, 2] or BrF_5 [1], by mixing gaseous F_2 and Br_2 [1] or Br_2 and BrF_3 [3, 4] in the absorption cell, and from mixtures of Br_2 and perfluoro-1,3-dimethylcyclohexane ($C_6F_{10}(CF_3)_2$) passing through an MW discharge before entering the absorption cell [5, 6].

For the J=1←0 transition of ^{79}BrF and ^{81}BrF in both their lowest vibrational levels, v=0 and 1, three hyperfine (hf) lines each were observed near 21 GHz, which arise from hf interaction (quadrupole coupling and spin-rotation) with the Br nuclei ($J+I_{Br}=F_1$; $I_{Br}=3/2$).The hf splitting due to the spin-rotation interaction of the ^{19}F nucleus ($F_1+I_F=F$; $I_F=1/2$) could not be resolved. For the J=2←1, v=0 transition near 42 GHz, five of the expected six hf components (due to the Br nuclei) were observed. Measured hf components ν (in MHz) with two standard deviations in parentheses and assignments are as follows [6]:

$J'←J''$	v	$F_1'←F_1''$	ν(^{79}BrF)	ν(^{81}BrF)
1←0	0	1/2←3/2	20986.120(50)	20928.819(50)
		5/2←3/2	21203.090(50)	21110.294(50)
		3/2←3/2	21475.090(50)	21337.375(50)
	1	1/2←3/2	20829.875(100)	20773.420(100)
		5/2←3/2	21046.666(100)	21181.710(100)
		3/2←3/2	21318.286(100)	21181.710(100)
2←1	0	1/2←1/2	42513.310(100)	42309.968(100)
		3/2←1/2	42786.375(100)	42537.977(100)
		7/2←5/2, 5/2←5/2	42490.337(100)	42290.725(100)
		3/2←3/2	42297.465(100)	42129.403(100)
		5/2←5/2	42762.100(100)	42517.907(100)

The frequencies given above slightly refine the earlier results for ^{79}BrF and ^{81}BrF, J=1←0, v=0, given with uncertainties of 0.50 MHz [1, 7].

Measurements with Br_2-BrF_3 mixtures at 25 to 450 °C and 86 to 276 GHz enabled the higher rotational transitions up to J=13←12 of ^{79}BrF and ^{81}BrF in their vibrational levels v=0 to 3 to be observed. The hf components with $\Delta F_1=+1$ dominate in strength and appear as doubly degenerate doublets corresponding to the transitions $F_1''+1←F_1''$ with $F_1''=J-I_{Br}$ and $J-I_{Br}+1$ and $F_1''=J+I_{Br}-1$ and $J+I_{Br}$. The frequencies of the unsplit rotational lines were determined from the measured hf splitting and by using the more accurate eqQ values from [1], and the average of both results was chosen to yield the final values for the reduced rotational frequencies [3, 4]. The measured and reduced rotational frequencies are published in [3, pp. 120/4]; the latter are compiled in the following table (ν in MHz):

$J'\leftarrow J''$	v	$\nu(^{79}BrF)$	$\nu(^{81}BrF)$
$4\leftarrow3$	0	85024.67	84618.10
$6\leftarrow5$	0	127531.19	–
$7\leftarrow6$	0	148781.94	148070.59
	1	147685.90	146982.40
	2	146583.30	145887.83
$8\leftarrow7$	0	170030.63	169217.85
	1	168778.04	167974.07
	2	167518.01	166723.16
$9\leftarrow8$	0	–	190362.72
	1	–	188963.4
	3	187023.68	186139.89
$10\leftarrow9$	0	212521.04	211505.02
	1	210955.06	209950.30
	2	209379.84	–
	3	207794.80	–
$12\leftarrow11$	0	254999.69	253780.78
$13\leftarrow12$	0	276234.01	–

Rotational and vibrational constants of BrF derived from the MW spectra are given on p. 12, the hf constants are given on pp. 9/10.

Stark effect measurements for the $J=1\leftarrow0$, $v=0$ hf transitions in electric fields up to 500 V/cm gave Stark coefficients of 8.09(50) to 9.88(50) $Hz\cdot cm^2\cdot V^{-2}$ and the dipole moment of BrF (see p. 8) [5, 6] (earlier value [1]).

Zeeman effect measurements for the $J=1\leftarrow0$, $v=0$ hf transitions in magnetic fields near 20 kG gave the molecular Zeeman parameters (g factor, susceptibility anisotropy) which in turn were used to evaluate additional molecular properties (see p. 10) [2].

References:

[1] Smith, D. F.; Tidwell, M.; Williams, D. V. P. (Phys. Rev. [2] **77** [1950] 420/1).

[2] Ewing, J. J.; Tigelaar, H. L.; Flygare, W. H. (J. Chem. Phys. **56** [1972] 1957/66).

[3] Willis, R. E., Jr. (Diss. Duke Univ. 1979, pp. 1/158, 120/4; Diss. Abstr. Intern. B **40** [1980] 4880).

[4] Willis, R. E., Jr.; Clark, W. W., III (J. Chem. Phys. **72** [1980] 4946/50).

[5] Nair, K. P. R. (Kem. Kozlem. **52** [1979] 431/50).

[6] Nair, K. P. R.; Hoeft, J.; Tiemann, E. (J. Mol. Spectrosc. **78** [1979] 506/13).

[7] Lovas, F. J.; Tiemann, E. (J. Phys. Chem. Ref. Data **3** [1974] 609/769).

10.3.3.2 IR Absorption Spectrum

The Fourier transform (FT) IR spectrum of **gaseous** BrF was measured at 298 K and ~5 mbar in the range of the fundamental band (15 μm) and at 13 mbar in the range of the first overtone (7.6 μm); the resolution was 0.04 cm^{-1}. BrF was synthesized in the absorption cell by slowly adding F_2 to a tenfold excess of Br_2. (After 20 min BrF was detected to be the only absorbing species except some CF_4 impurity.) The fundamentals ($v=1 \leftarrow 0$) and overtones ($v=2 \leftarrow 0$) of both ^{79}BrF and ^{81}BrF appear each with intensity ratios of 1:1 and exhibit well-resolved P and R branches (J'' up to 63) and a gap at the band origin. The fundamental band is accompanied by a hot band ($v=2 \leftarrow 1$), which, based on the Boltzmann factor for the v=1 level at 298 K, was expected to have an intensity of 7.2% of the main band. The rotational lines were fitted to a quintic polynomial of m ($m=-J''$ for the P branch, $m=J''+1$ for the R branch) which resulted in the following band origins ν_0 (in cm^{-1}) [1]:

	^{79}BrF	^{81}BrF
ν_0 (1←0)	662.3046(20)	660.7334(20)
ν_0 (2←1)	654.71	653.17
ν_0 (2←0)	1317.0095(40)	1313.9092(40)

A complete list of the observed wave numbers is available from a data bank; see [1, reference 22].

The diode laser spectrum of BrF (generated by MW discharges through Br_2-SF_6 mixtures at p=0.5 Torr) was observed and analyzed in the fundamental region with a more than tenfold higher resolution. Out of the 90 lines recorded between 631 and 679 cm^{-1}, 32 lines could be assigned to the 1←0 (P(3) to P(38), R(0) to R(25)) and 2←1 (P(7) to P(24), R(5) to R(40)) bands of ^{79}BrF, 39 lines to the 1←0 (P(1) to P(31), R(2) to R(28)) and 2←1 (P(5) to P(28), R(3) to R(43)) bands of ^{81}BrF; the estimated uncertainties of the wave numbers were 0.001 to 0.003 cm^{-1}. Least-squares fits to a Dunham expansion with seven parameters resulted in the Dunham coefficients Y_{ik} (i, k=0, 1, and 2) for each of the isotopic species (cf. p. 13) [2].

The earliest IR study on BrF at 298 K, which was carried out upon mixing of gaseous Br_2 and BrF_3 in the IR cell, had assigned the fundamental vibration to a band at 669 cm^{-1} and the first overtone to a band at 1326 cm^{-1} [3].

The IR spectra of **matrix-isolated** BrF at low temperatures were recorded in the region of the fundamental vibration: $\nu=650.5$ cm^{-1} was measured upon vacuum-UV photolysis of SF_5Br in an Ar matrix (1:100, 1:400) at 8 K ($SF_5Br+h\nu \rightarrow SF_4+BrF$) [4, 5]. Weak (presumably isotopic) doublets at 660.0/661.5 cm^{-1} and 660/662 cm^{-1} were observed upon codeposition of F_2-Ar (1:400) mixtures with CsBr and NaBr vapor, respectively, at 15 K and photolysis with an Hg arc [6]. The 600.0/661.5 cm^{-1} doublet appeared also upon condensation of a BrF_3-Ar (1:2000) mixture (before and after UV photolysis), upon cocondensation of a Br_2-BrF-BrF_3 equilibrium mixture in Ar (1:50) with F_2-Ar (1:170), and upon condensation of premixed (with or without MW discharge) Br_2-F_2 in Ar (1:2:200) at (presumably) 15 K [7].

References:

[1] Bürger, H.; Schulz, P.; Jacob, E.; Fähnle, M. (Z. Naturforsch. **41a** [1986] 1015/20).
[2] Nakagawa, K.; Horiai, K.; Konno, T.; Uehara, H. (J. Mol. Spectrosc. **131** [1988] 233/40).
[3] Stein, L. (J. Am. Chem. Soc. **81** [1959] 1273/6).
[4] Smardzewski, R. R.; Fox, W. B. (J. Chem. Phys. **67** [1977] 2309/16).
[5] Smardzewski, R. R.; Fox, W. B. (J. Fluorine Chem. **7** [1976] 453/5).
[6] Miller, J. H.; Andrews, L. (Inorg. Chem. **18** [1979] 988/92).
[7] Prochaska, E. S.; Andrews, L.; Smyrl, N. R.; Mamantov, G. (Inorg. Chem. **17** [1978] 970/7).

10.3.3.3 Near-IR, Visible, and UV Spectra

10.3.3.3.1 Near-IR and Visible Region. The B $^3\Pi(0^+)\leftrightarrow X\ ^1\Sigma^+$ System

Absorption Spectrum. Two different techniques were employed for the study of the absorption spectrum of BrF: Conventional absorption measurements in Br_2-F_2 gaseous mixtures at temperatures above 100 °C [1 to 6] and laser-induced fluorescence (LIF), i.e., measurement of the total fluorescence intensity as a function of laser wavelength (excitation spectrum), in combination with a discharge-flow system in which known concentrations of BrF were produced by the rapid $F + Br_2$ reaction [7 to 15].

A discrete absorption spectrum was observed in the visible region between 450 and 610 nm; vibrational and rotational analyses resulted in the unambiguous assignment to the B $^3\Pi(0^+)\leftarrow X\ ^1\Sigma^+$ system. Early attempts to fit 13 band heads between 525 and 603 nm (and three others at 627.5, 641, and 668 nm observed in the emission spectrum [16]) to a second, weaker transition, A $^3\Pi(1)\leftarrow X\ ^1\Sigma^+$ ($\nu_0 = 17572\ cm^{-1}$) [2, 3] were already questioned by [17] and later disproved by high-resolution absorption and excitation spectra: Rotational structure with P and R branches only and not with PQR structure as expected for $\Omega = 1\leftarrow 0$ transitions in Hund's case (c) was observed in the entire region.

The first vibrational analyses of the B←X system recorded at low resolution were carried out by Brodersen and co-workers [1 to 3]; 39 band heads between 450 and 605 nm were assigned to the vibrational progressions with v″=0 (v′=2 to 10, 14), v″=1 (v′=1 to 9, 15 to 20), v″=2 (v′=0 to 6), and v″=3 (v′=2 to 5) [3] (vibrational assignment of [1, 2] revised). Rotational analysis of the v′←v″=4←0 to 9←0 and 8←1 bands followed [4].

A high-resolution absorption spectrum of the B←X system was recorded by Coxon and Curran [5, 6] at 160 °C in the 477 to 570 nm range (inverse dispersion 1.0 to 0.9 Å/mm); it comprised 18 bands of the v″=0 (v′=3 to 9), v″=1 (v′=1 to 8), and v″=2 (v′=2 to 4) progressions. Each band exhibits simple PR structure (as expected for an $\Omega = 0\leftarrow 0$ transition) with J up to 57 and isotopic splitting due to the almost equally abundant ^{79}BrF and ^{81}BrF species. Line broadening occurs in the 9←0 band for $J' \geq 23$ in the P branches and for $J' \geq 21$ in the R branches of both isotopic species, which is explained by predissociation of the B $^3\Pi(0^+)$, v′=9 level. Rotational analyses of all bands (except 9←0) by using the ground-state rotational constant B_0'', derived from the microwave spectrum [18, 19], lead to the ground- and excited-state rotational constants B_v'' (v=1, 2) and B′, D_v' (v=1 to 8), H_v' (v=7, 8), and vibrational term values G_v'' (v=1, 2) and T_v' (v=1 to 8). The extensive list of vacuum wave numbers of all assigned lines used for the evaluation of the molecular constants is available on request from the authors [5, 6]; the band origins ν_0 are as follows (standard deviation in parentheses) [5]:

band	$\nu_0(^{79}BrF)$ in cm^{-1}	$\nu_0(^{81}BrF)$ in cm^{-1}
3←0	19187.251(5)	19185.191(6)
4←0	19521.347(3)	19518.609(3)
5←0	19842.262(6)	19838.915(5)
6←0	20148.408(3)	20144.556(3)
7←0	20437.394(4)	20433.162(4)
8←0	20704.821(5)	20700.435(4)
1←1	17821.37(2)	17822.40(2)
2←1	18178.807(7)	18179.038(5)
3←1	18524.978(5)	18524.480(4)

band	$\nu_0(^{79}BrF)$ in cm^{-1}	$\nu_0(^{81}BrF)$ in cm^{-1}
4←1	18859.058(4)	18857.889(5)
5←1	19179.929(6)	19178.150(5)
6←1	19486.096(3)	19483.812(5)
7←1	19775.108(5)	19772.440(4)
8←1	20042.557(5)	20039.734(5)
2←2	17524.05(2)	17525.88(4)
3←2	17870.329(5)	17871.343(5)
4←2	18204.381(4)	18204.734(5)

These stem from least-squares fits of ν_0 and the rotational constants (B'_v, D'_v, H'_v) to the respective individual band; slightly different ν_0 values (mostly ± 0.001 to ± 0.01 cm^{-1}) were obtained by fitting the constants of all bands simultaneously [5]. A refit (by allowing B''_0 to vary instead of being fixed on the MW values), presented later [20], again slightly changed the ν_0's in the last one or two digits.

For the origins of the 9←0 band the wave numbers $\nu_0(^{79}BrF) = 20939.94(4)$ or 20940.02(1) cm^{-1} and $\nu_0(^{81}BrF) = 20935.98(4)$ or 20936.02(1) cm^{-1} were derived by fitting the $J' \leq 28$ or $J' \leq 8$ rotational levels, respectively [5].

A high-resolution excitation spectrum of the B←X system has been observed in the 485 to 520 nm region which comprised the $v''=0$ ($v'=3$ to 8) progression. LIF was excited in a discharge-flow system, containing streams of F atoms (in He) and Br_2 molecules, by using a tunable dye laser with a 0.1 Å band width. Analogous to the absorption spectrum, P and R branches (J generally up to ~40 or ~50) for both ^{79}BrF and ^{81}BrF were identified; in contrast to the absorption spectrum, predissociation occurs already in the 8←0 band with the onset near $J'=28$ and no detectable transitions above $J'=35$; no fluorescence could be observed near the 9←0 band head in the 475 to 480 nm region [7] (short note [8]). For further and more detailed information on the B $^3\Pi(0^+)$ state predissociation, see Section 10.3.2.10, p. 17 and Section 10.3.2.12, pp. 20/2.

According to Coxon and Curran [5] the frequency measurements of Clyne et al. [7] suffered from a systematic error due to technical deficiencies of the spectrometer; the band origins listed in [7] have slightly higher wave numbers (at an average by ~1.7 cm^{-1} for ^{79}BrF and ~1.6 cm^{-1} for ^{81}BrF) than those given by [5].

The LIF studies on BrF have been repeated and extended by Clyne and co-workers [9 to 13] at a tenfold higher resolution using a narrow-band tunable dye laser with a 0.01 Å excitation band width. These studies, which mainly concentrated on the fluorescence decay dynamics of selected rovibrational (v', J') levels of the B $^3\Pi(0^+)$ state (radiative lifetimes, predissociation and collisional relaxation processes, see pp. 20/4), dealt with the $v''=0$ progression, 8←0, 7←0 [9, 11], 6←0 [10], 5←0 to 3←0 [12], and with the $v''=1$ (2←1 to 0←1) and $v''=2$ (4←2 to 2←2) progressions [12]. A rotational analysis of the 0←1 band is presented giving band origins of $\nu_0(^{79}BrF) = 17453.42$ cm^{-1} and $\nu_0(^{81}BrF) = 17455.29$ cm^{-1} [12].

Employing an improved fluorescence detection system, the complete $v''=0$ progression with $v'=0$ to 8 and the $v''=1$ progression up to $v'=3$ could be observed with the weak 0←0 (origin expected at 551.7 nm = 18126 cm^{-1} [12]), 1←0, and 2←0 bands being overlapped by the stronger $v''=1$ and 2 bands [13].

More recently the high-resolution excitation spectrum (580 to 585 nm) of the 3←3 band and a rotational analysis (J up to 27) have been reported yielding $\nu_0(^{79}BrF) = 17223.25876(62)$ cm^{-1} and $\nu_0(^{81}BrF) = 17225.79604(73)$ cm^{-1} [14].

The electronic transition dipole moment $|R_e|^2$ for the B↔X transition was calculated from radiative lifetimes measured for the vibrational manifolds v′=0 to 7 (cf. pp. 21/2): Successive improvement of the lifetime measurements lead to successive downward revisions of the magnitude from $|R_e|^2 = (5.5 \pm 0.8) \times 10^{-2}$ D^2 [12] to $(3.55 \pm 0.15) \times 10^{-2}$ D^2 [13] and finally to $|R_e|^2 = (2.8 \pm 0.1) \times 10^{-2}$ D^2 [21, 22].

The derivation of the absorption coefficient for rovibrational lines of the B←X system by use of the transition dipole moment was illustrated for the relatively strong R(11) line of the 7←0 band [12, 13].

Emission Spectrum. The emission spectrum of BrF observed between 450 and 975 nm was found to belong to the B $^3\Pi(0^+)$→X $^1\Sigma^+$ system only; transitions originating from the vibrational levels v′=0 to 9 and terminating on ground-state vibrational levels up to v″=14 were identified which give rise to a system of more than 60 red-degraded bands. No evidence for a second emission system, A $^3\Pi(1)$→X $^1\Sigma^+$, was found.

The B→X emission was first detected by Durie [16] who impinged a jet of F_2 gas on the surface of liquid bromine and observed a pale reddish yellow flame; spectral analysis in the 473 to 670 nm region revealed some 30 bands which could be assigned to the vibrational progressions with v′=0 to 4, v″ up to 6; v′=5 to 8, v″=0 to 2; and v′→v″=9→0 (diffuse); the partially resolved rotational structure (single P and R branches) and a comparison with the absorption spectrum [1, 2] supported the B→X assignment [16].

Very intense and extended B→X emission was observed from the recombination of the ground state $F(^2P_{3/2})$ and $Br(^2P_{3/2})$ atoms [20, 23] and from the reaction of Br atoms with O_2F radicals [24], both, however, only in the presence of metastable singlet oxygen, O_2^* (a $^1\Delta_g$, b $^1\Sigma_g^+$). The excitation processes were envisaged to occur via intermediately formed BrF molecules in a lower electronically excited state, A $^3\Pi(1, 2, 0^-)$, [23] or in ground-state vibrational levels, X $^1\Sigma^+$, v″>0, [24] followed by collisional energy transfer from the metastable oxygen to generate the B state.

Using a discharge-flow system and CF_4-Ar and Br_2-O_2 gaseous mixtures, Clyne et al. [23] recorded the B→X spectrum between 483 and 975 nm and measured the wave numbers of 59 band heads between 520 and 975 nm (v′=0, v″=3 to 12; v′=1, v″=2 to 8, 10 to 13; v′=2, v″=2 to 6, 8 to 11, 13, 14; v′=3, v″=3 to 5, 7 to 9, 12 to 14; v′=4, v″=1 to 4, 6 to 8, 10 to 12, 14; v′=5, v″=1, 3, 6, 7; v′=6, v″=2, 3, 12); isotopic splitting due to ^{79}BrF and ^{81}BrF was resolved for five bands only; for the remaining band heads, the assignment to ^{79}BrF was suggested. A Deslandres table for the B→X system, v′=0 to 6, v″=1 to 14, was compiled by Clyne et al. [23]. In addition to their own results, they included for the $\lambda < 520$ nm region the more accurate band-head wave numbers from the absorption spectrum [3] and a few previous emission data for the missing bands [16].

The chemiluminescence spectrum resulting from the $Br + O_2F + O_2^*$ reaction was recorded by Coombe and Horne [24] between 450 and 900 nm at somewhat higher resolution; it consisted of 63 bands with well-resolved rotational structure for many of them and with resolved isotopic ^{79}BrF-^{81}BrF splitting for 22 bands in the long-wavelength region between 640 and 890 nm (see table below). In general, the results of [23] and [24] agree well except those for the long-wavelength region where a number of band heads of [23] have to be reassigned to the ^{81}BrF species.

The first and so far only rotational analysis of the B→X emission system was carried out by Coxon and Wickramaaratchi [20] for sixteen bands of ^{79}BrF and ^{81}BrF (Br+F+O_2^* reaction) at 625 to 870 nm. The extensive list of wave numbers is available on request from the authors. The following band origins ν_0 (in cm^{-1}) were obtained by fitting ν_0 and the rotational constants B'_v, D'_v, B''_v (D''_v held fixed) to the wave numbers of each individual band (for comparison and completion the band heads ν_H from [24] are included in the table):

band	ν_0(^{79}BrF)*)	ν_0(^{81}BrF)*)	ν_H(^{79}BrF)	ν_H(^{81}BrF)
0→4	–	–	15517±2	15522±2
0→5	–	–	14883±2	14893±2
0→6	14256.94(2)	14266.20(1)	14259±2	14269±2
0→7	13641.16(1)	13651.71(1)	13642±2	13654±2
0→8	13033.28(1)	13045.15(1)	13032±2	13044±2
0→9	12433.63(1)	11856.47(2)	12435±2	12448±2
0→10	11842.13(1)	11856.47(2)	11844±2	11858±2
0→11	–	–	11260±2	11275±2
1→4	15880.07(2)	15885.69(2)	–	–
1→5	15248.55(2)	15255.55(2)	15252±2	15259±2
1→6	14624.95(3)	14633.38(1)	14629±2	14636±2
1→7	14009.37(4)	14019.02(3)	14013±2	14021±2
1→8	–	–	13408±4	13417±4
1→10	–	–	12212±3	12227±3
1→11	11626.91(2)	11641.71(2)	11629±3	11644±3
2→9	–	–	13157±2	13170±2
2→10	12567.66(2)	12580.33(2)	12568±2	12581±2
2→11	–	–	11986±2	12000±2
3→5	15951.92(5)	15957.55(4)	–	–
3→8	14105.12(2)	14114.49(2)	–	–
3→9	13505.33(2)	13516.09(3)	13510±4	13521±4
3→12	11755.68(2)	11769.95(3)	11760±3	11774±3
4→7	–	–	15052±2	15059±2
4→11	12664.73(2)	12677.17(2)	12666±2	12680±2
4→12	–	–	12094±3	12106±3

*) Standard deviation in parentheses.

A simultaneous fit for all observed bands resulted in slightly different ν_0's (changes in the last one or two digits) [20].

Further observations of the B→X system: A faint luminescence shown to be B→X was observed during the reaction of molecular bromine with O_2F; it presumably results from the Br_2+F_2 reaction, where F_2 stems from the decay of O_2F [24]. The "weak" band system B→X, v′=1 to 9, v″=2 to 7 resulted from reactions of atomic fluorine with Br_2 or CF_3Br [25]. Admission of a BrN_3-N_2 mixture to a stream of F atoms in a discharge-flow system produced the orange-yellow BrF chemiluminescence, the intensity of which increased linearly with the BrN_3 flow rate; eighteen bands at 500 to 800 nm were observed, the transitions originating from the v′=0, 1, and 2 levels [26].

Laser action of BrF has not been observed in the visible region until now. By considering the radiative lifetimes for several rovibrational levels [13, 21] (cf. pp. 21/2), Davis [27] demonstrated the B→X, v=0→8 transition to be an attractive candidate for visible lasing and predicted the optical gain that would be available from such a BrF laser. Upon examination of possible pumping/lasing pairs of molecules for demonstrating chemically pumped electronic transition lasers, the b $^1\Sigma^+$ state of NF (18887 cm^{-1}) was predicted to be an appropriate pump for BrF B $^3\Pi(0^+)$ levels [28].

Franck-Condon Factors, r-Centroids. Using reliable RKR potentials for the X and B states of ^{79}BrF and ^{81}BrF (cf. pp. 16/7) based on emission data [20] and on refitted absorption data [5], Franck-Condon factors and r-centroids were calculated for B↔X transitions between vibrational levels with v″=0 to 12 and v′=0 to 8 in ^{79}BrF and ^{81}BrF; the tables are available on request from the authors [20]. These data improve the earlier results which were based on absorption data only and which are presented as mean values for ^{79}BrF and ^{81}BrF by [7].

References:

[1] Brodersen, P. H.; Schumacher, H. J. (Z. Naturforsch. **2a** [1947] 358/9).
[2] Brodersen, P. H.; Schumacher, H. J. (Anales Asoc. Quim. Arg. **38** [1950] 52/60).
[3] Brodersen, P. H.; Sicre, J. E. (Z. Physik **141** [1955] 515/24).
[4] Brodersen, P. H.; Cudmani, L. y. C. (Deut. Versuchsanst. Luft-Raumfahrt Ber. No. 250 [1963] 1/18; C.A. **61** [1964] 12805).
[5] Coxon, J. A.; Curran, A. H. (J. Mol. Spectrosc. **75** [1979] 270/87).
[6] Coxon, J. A.; Curran, A. H. (J. Photochem. **9** [1978] 183/5).
[7] Clyne, M. A. A.; Curran, A. H.; Coxon, J. A. (J. Mol. Spectrosc. **63** [1976] 43/59).
[8] Clyne, M. A. A.; Curran, A. H.; McDermid, I. S. (NBS-SP 526 [1978] 77/8).
[9] McDermid, I. S.; Clyne, M. A. A. (Lasers Chem. Proc. Conf., London 1977, pp.122/7).
[10] Clyne, M. A. A.; McDermid, I. S. (J. Chem. Soc. Faraday Trans. II **73** [1977] 1094/106).

[11] Clyne, M. A. A.; McDermid, I. S. (J. Chem. Soc. Faraday Trans. II **74** [1978] 644/63).
[12] Clyne, M. A. A.; McDermid, I. S. (J. Chem. Soc. Faraday Trans. II **74** [1978] 664/80).
[13] Clyne, M. A. A.; McDermid, I. S. (J. Chem. Soc. Faraday Trans. II **74** [1978] 1376/92).
[14] Takehisa, Y.; Ohashi, N. (J. Mol. Spectrosc. **128** [1988] 304/5).
[15] Farthing, J. W.; Fletcher, I. W.; Whitehead, J. C. (J. Phys. Chem. **87** [1983] 1663/5).
[16] Durie, R. A. (Proc. Roy. Soc. [London] A **207** [1951] 388/95).
[17] Coxon, J. A. (Chem. Phys. Letters **33** [1975] 136/40).
[18] Smith, D. F.; Tidwell, M.; Williams, D. V. P. (Phys. Rev. [2] **77** [1950] 420/1).
[19] Calder, G. V.; Ruedenberg, K. (J. Chem. Phys. **49** [1968] 5399/415).
[20] Coxon, J. A.; Wickramaaratchi, M. A. (J. Mol. Spectrosc. **87** [1981] 85/100).

[21] Clyne, M. A. A.; Liddy, J. P. (J. Chem. Soc. Faraday Trans. II **76** [1980] 1569/85).
[22] Heaven, M. C. (Chem. Soc. Rev. **15** [1986] 405/48).
[23] Clyne, M. A. A.; Coxon, J. A.; Townsend, L. W. (J. Chem. Soc. Faraday Trans. II **68** [1972] 2134/43).
[24] Coombe, R. D.; Horne, R. K. (J. Phys. Chem. **83** [1979] 2435/40).
[25] Schatz, G.; Kaufman, M. (J. Phys. Chem. **76** [1972] 3586/90).
[26] Coombe, R. D.; Lam, C. H.-T. (J. Chem. Phys. **78** [1983] 3746/8).
[27] Davis, S. J. (Laser Interact. Relat. Plasma Phenom. **6** [1984] 33/45).
[28] Avizonis, P. V. (Gas Flow Chem. Lasers Proc. 4th Intern. Symp., Stresa, Italy, 1982 [1984], pp. 1/17; C.A. **103** [1985] No. 131316).

10.3.3.3.2 UV Region

Absorption Spectrum. In the study of the B←X absorption system, strong continuum absorption was noticed in the UV (presumably near UV) which, however, was not analyzed in detail [1].

The cross section σ of a continuous absorption spectrum in the 250 to 200 nm region was reported to increase from $\sim 0.3 \times 10^{-19}$ to $\sim 2.5 \times 10^{-19}$ cm^2 [2].

Emission Spectrum. BrF Laser. Diegelmann et al. [3, 4] recorded between 345 and 360 nm a strong emission (peak intensity at 354.5 nm; resolution 0.17 nm) when mixtures of CF_3Br (as a bromine donor) and NF_3 (as a fluorine donor) in helium gas were excited by short pulses of high-energy electrons; by analogy to Br_2 and I_2, they assigned the spectrum to a charge-transfer transition in BrF between the ion-pair excited bound state D′ $^3\Pi(2)$ ($Br^+(^3P_2)+F^-(^1S_0)$) and the lowest excited valence state A′ $^3\Pi(2)$. Using F_2-Br_2 mixtures in argon (isotopically pure $^{79}Br_2$ and "natural" Br_2) and Tesla discharges for excitation, Henderson and Tellinghuisen [5] presented a vibrational analysis of the D′→A′ emission system of BrF which they photographed in the 340 to 360 nm region at a reciprocal dispersion of 5 Å/mm: sixteen bands for ^{79}BrF and eight bands for ^{81}BrF were assigned to vibrational transitions with v′=0, v″=8 to 13; v′=1, v″=8 to 10, 12, 13; v′=2, v″=8, 15; v′=3, v″=15, 17; v′=4, v″=16; ν_{00}=31605 cm^{-1}; the rotational structure was analyzed in the 0→9 and 1→9 bands. Employing a Morse potential function for the D′ state and a Morse-RKR function for the A′ state, Franck-Condon factors were calculated for all vibrational transitions in ^{79}BrF with v′=0 to 4 and v″=6 to 19 [5]. More recently, a rotational analysis of 904 lines of ^{79}BrF and ^{81}BrF belonging to the 0→8 to 12 and 1→9, 12, 13 bands (J″ range from 7 to 75) was presented by Narayani and Tellinghuisen [6] upon generating the high-resolution D′→A′ spectrum (reciprocal dispersion ~1.35 Å/mm) by Tesla discharges in Br_2-SF_6-Ar mixtures; the extensive list of assigned lines is available on request from J. Tellinghuisen or from the Editorial Office of the Journal of Molecular Spectroscopy.

Evidence for at least two other electronic transitions is given by the emission spectrum in the shorter wavelength region down to ~300 nm [5].

As pointed out already by Diegelmann et al. [3] and confirmed by the Franck-Condon factors [5], the strong variation of band intensity (and hence FC factors) with v″ indicates the BrF D′→A′ transition, mainly the 0→11 band, to be an appropriate candidate for lasing in the UV. Such a BrF laser has indeed been realized by use of TEA discharges (transverse electric discharges at near atmospheric pressures) in gaseous mixtures of BBr_3, CF_3Br, or HBr as bromine donors, NF_3 as a fluorine donor, and He or Ne as buffer gas; laser oscillation occurs on the 354.5 nm peak (0→11 band) with an output energy of 0.5 mJ at pulse lengths of 10 ns [4, 7, 8].

In addition to a laser transition at 350 nm (upper ionic state here correlated with $Br^+(^3P_{1,0})+F^-(^1S_0)$), two others originating from ionic states of BrF and terminating in the lowest excited valence states $^3\Pi(2)$ or $^3\Pi(1)$ have been predicted by Parks [9]: one transition at 400 nm, where the upper state correlates with $Br^+(^3P_2)+F^-(^1S_0)$, and one at 150 nm, where the upper state correlates with $F^+(^3P_{1,0})+Br^-(^1S_0)$; these, however, have yet not been experimentally realized.

In the Schumann region at 175 to 156 nm, two groups of red-degraded emission bands were observed from hollow-cathode discharges in BrF_3 vapor; ordering of the band heads into vibrational systems (v′=0 to 7, v″=0 to 3) resulted in the lower state vibrational frequency ω_e''=673 cm^{-1} which proved the BrF species to be the emitter [10].

References:

[1] Coxon, J. A.; Curran, A. H. (J. Mol. Spectrosc. **75** [1979] 270/87).
[2] Buben, S. N.; Chaikin, A. M. (Kinetika Kataliz **21** [1980] 1591/2; C.A. **94** [1981] No. 73863).
[3] Diegelmann, M.; Hohla, K.; Rebentrost, F.; Kompa, K. L. (J. Chem. Phys. **76** [1982] 1233/47).
[4] Diegelmann, M. (Max-Planck-Ges. Foerd. Wiss. Projektgruppe Laserforsch. Ber. PLF-33 [1980] 1/124).
[5] Henderson, S. D.; Tellinghuisen, J. (Chem. Phys. Letters **112** [1984] 543/6).
[6] Narayani, R. I.; Tellinghuisen, J. (J. Mol. Spectrosc. **141** [1990] 9/90).
[7] Diegelmann, M.; Grieneisen, H. P.; Hohla, K.; Hu, X.-J.; Krasinski, J.; Kompa, K. L. (Appl. Phys. **23** [1980] 283/7).
[8] Diegelmann, M.; Grieneisen, H. P. (Ger. Offen. 3031954 [1980/82] 1/14).
[9] Parks, J. H. (AD-A085520 [1979] 1/142, 119/31; C.A. **93** [1980] No. 212997).
[10] Brodersen, P. H.; Mayo, S. (Z. Physik **143** [1955] 477/8).

10.3.4 Heat of Formation. Free Energy of Formation. Thermodynamic Functions. Transition Points

In a recent review on chemical thermodynamic properties, the following values for the standard heat of formation at temperatures of 298.15 and 0 K and a pressure of 0.1 MPa are recommended [1]:

$\Delta_fH^\circ_{298.15} = -93.85$ kJ/mol and $\Delta_fH^\circ_0 = -86.2$ kJ/mol

The corresponding free energy of formation is $\Delta_fG^\circ_{298.15} = -109.18$ kJ/mol [1]. The quoted Δ_fH and Δ_fG values were taken from "analyses of literature data"; no references are given in [1] which would explain the rather large discrepancies from all previously published values.

The JANAF Tables [2] give

$\Delta_fH^\circ_{298.15} = -58.463 \pm 1.7$ kJ/mol and $\Delta_fH^\circ_0 = -50.815 \pm 1.7$ kJ/mol,

$\Delta_fG^\circ_{298.15} = -73.809$ kJ/mol

for the formation of BrF(g) via 1/2 Br_2(l) + 1/2 F_2(g) → BrF(g) from the standard states at 298.15 and 0 K and 0.1 MPa; Δ_fH° and Δ_fG° values are tabulated for T = 0 to 6000 K at 100 K intervals (these are the old JANAF data [3] converted from a standard state pressure of 1 atm to that of 0.1 MPa); the $\Delta_fH^\circ_{298.15}$ value is based on measurements of the equilibrium reaction $Br_2(g) + BrF_3(g) \rightleftharpoons 3\,BrF(g)$ at 328 to 380 K of Steunenberg et al. [4] ($\Delta_rH_{298.15} = 49.29 \pm 2.1$ kJ/mol) and on the heat of formation of BrF_3 from gaseous Br_2 and F_2, measured by Stein [5] and adjusted to the liquid Br_2 reference state. Stein [5], [6, pp. 151/5] had derived $\Delta_fH^\circ_{298.15} = [\Delta_rH_{298.15} + \Delta_fH^\circ_{298.15}(BrF_3)]/3 = -74.1$ kJ/mol, $\Delta_fG^\circ_{298.15} = -75.3$ kJ/mol and also $\Delta_fH^\circ_{298.15} = -58.6$ kJ/mol for the liquid Br_2 reference state.

Using the dissociation energies of BrF, Br_2, and F_2, the heat of formation of BrF from the gaseous elements is $\Delta_fH^\circ_0 = 1/2\,D^\circ_0(Br_2) + 1/2\,D^\circ_0(F_2) - D^\circ_0(BrF)$. The early $\Delta_fH^\circ_0$ values, -76.6 ± 2.1 kJ/mol [7], -81.96 or -39.50 kJ/mol [5], and -58.2 kJ/mol [5], based on the spectroscopic D°_0(BrF) values of [8], [9], and [10], respectively, suffer from the ambiguity of the latter (cf. p. 19). Taking the so far most reliable spectroscopic value based on lifetime measurements of B-state rovibrational levels, $D^\circ_0(BrF) = 246.66$ kJ/mol (cf. p. 19), as well as $D^\circ_0(Br_2) = 190.16$ kJ/mol (cf. "Bromine" Suppl. Vol. A, 1985, p. 191) and $D^\circ_0(F_2) = 154.0$ kJ/mol (cf. "Fluorine" Suppl. Vol. 2, 1980, p. 66), then $\Delta_fH^\circ_0 = -74.58$ kJ/mol.

Two semiempirical SCF MO calculations gave much too low values of $-\Delta_f H_0$; see references [20, 21] of Section 10.3.2.1.1.

The heat capacity C_p^o, thermodynamic functions S^o, $-(G^o-H_{298}^o)/T$, $H^o-H_{298}^o$, and the logarithm of the equilibrium constant K_f for the formation of BrF as an ideal gas from the elements have been calculated for a standard state pressure of 1 atm and tabulated for T=298.15 K and T=0 to 6000 K at 100 K intervals in the old JANAF Tables [3] (see also [6, p. 209]); rotational [11] and vibrational [8] constants, and the ground-state configuration $^1\Sigma^+$ were used to establish the partition function. The old values were converted to a standard state pressure of 0.1 MPa and tabulated in the more recent JANAF Tables [2]; excerpted values are as follows:

T in K	C_p^o	S^o	$-(G^o-H_{298}^o)/T$	$H^o-H_{298}^o$ in kJ/mol	log K_f
	in J·mol⁻¹·K⁻¹				
0	0.00	0.00	∞	−9.019	∞
100	29.162	195.677	256.772	−6.109	30.877
200	30.760	216.261	231.914	−3.131	17.534
298.15	32.956	228.967	228.967	0.000	12.931
400	34.537	238.891	230.283	3.443	9.906
600	36.148	253.245	235.688	10.535	6.686
800	36.900	263.759	241.450	17.847	5.075
1000	37.330	272.043	246.769	25.274	4.108
1500	37.934	287.306	257.902	44.106	2.819
2000	38.323	298.275	266.688	63.174	2.173
3000	38.946	313.934	279.995	101.817	1.523
4000	39.511	325.216	289.955	141.047	1.197
5000	40.060	334.092	297.925	180.833	1.005
6000	40.602	341.444	304.583	221.164	0.885

$C_{p,298.15}^o = 32.97\ J \cdot mol^{-1} \cdot K^{-1}$, $S_{298.15}^o = 228.97\ J \cdot mol^{-1} \cdot K^{-1}$, and $H_{298.15}^o - H_0^o = 9.021$ kJ/mol for a standard pressure of 0.1 MPa are recommended by Wagman et al. [1].

The heat capacity, equilibrium constant, and thermodynamic functions S^o, $-(G^o-H_0^o)/T$, $-(H^o-H_0^o)/T$, and $(H^o-H_0^o)$ based on rotational [11] and vibrational [8] constants are tabulated for T=0, 250, 273.16, 298.16, and 300 to 1500 K at 100 K intervals in [7]. Heat capacity and thermodynamic functions S^o, $-(G^o-H_0^o)/T$, $(H^o-H_0^o)/T$ based on rotational [11] and vibrational [9] constants are tabulated for T=298.16 K, T=300 to 1000 K at 100 K intervals, and T=1200 to 2000 K at 200 K intervals in [12]. Heat capacities C_p/R at a pressure of 1 atm are tabulated for T=100 to 5000 K at 100 K intervals in [13]. The function $L(T) = -(G^o-H_0^o)/RT$ was tabulated for T=400, 1000, 3000, and 6000 K, and a plot of log $K_f = L(T) - D_0/RT$ vs. D_0 was constructed for a number of monofluorides, including BrF [14].

More current spectroscopic data for ground-state and electronically excited BrF [15] were used to give polynomial expansions of the partition function and the equilibrium constant ($Br + F \rightleftharpoons BrF$) for the temperature range T=1000 to 9000 K (astrophysical interest) [16].

The transition points reported for BrF upon studying the $Br_2 + F_2$ reaction, i.e., melting point $T_m \approx 240$ K and boiling point $T_b \approx 293$ K [17, 18], are "of questionable value" according to Stein [6, p. 152]. An estimate of the entropy of vaporization from the boiling point was given by [19].

References:

[1] Wagman, D. D.; Evans, W. H.; Parker, V. B.; Schumm, R. H.; Halow, I.; Bailey, S. M.; Churney, K. L.; Nuttall, R. L. (J. Phys. Chem. Ref. Data Suppl. **11** No. 2 [1982] 2-1/2-392, 2-51).
[2] Chase, M. W., Jr.; Davies, C. A.; Downey, J. R., Jr.; Frurip, D. J.; McDonald, R. A.; Syverud, A. N. (JANAF Thermochemical Tables, 3rd Ed., J. Phys. Chem. Ref. Data Suppl. **14** No. 1 [1985] 426).
[3] Stull, D. R.; Prophet, H. (JANAF Thermochemical Tables, 2nd Ed., NSRDS-NBS-37 [1971]).
[4] Steunenberg, R. K.; Vogel, R. C.; Fischer, J. (J. Am. Chem. Soc. **79** [1957] 1320/3).
[5] Stein, L. (J. Phys. Chem. **66** [1962] 288/91).
[6] Stein, L. (in: Gutmann, V., Halogen Chemistry, Vol. 1, Academic, London-New York, 1967, pp. 133/224).
[7] Evans, W. H.; Munson, T. R.; Wagman, D. D. (J. Res. Natl. Bur. Std. **55** [1955] 147/64).
[8] Durie, R. A. (Proc. Roy. Soc. [London] A **207** [1951] 388/95).
[9] Brodersen, P. H.; Schumacher, H. J. (Z. Naturforsch. **2a** [1947] 358/9).
[10] Brodersen, P. H.; Sicre, J. E. (Z. Physik **141** [1955] 515/24).
[11] Smith, D. F.; Tidwell, M.; Williams, D. V. P. (Phys. Rev. [2] **77** [1950] 420/1).
[12] Cole, L. G.; Elverum, G. W., Jr. (J. Chem. Phys. **20** [1952] 1543/51).
[13] Svehla, R. (NASA-TR-R-132 [1962] 1/139, 53; N.S.A. **16** [1962] No. 13210).
[14] Godnev, I. N.; Aleshonkova, Yu. A. (Teplofiz. Vys. Temp. **5** [1967] 272/7; High. Temp. [USSR] **5** [1967] 239/44).
[15] Huber, K. P.; Herzberg, G. (Molecular Spectra and Molecular Structure, Vol. 4, Constants of Diatomic Molecules, Van Nostrand Reinhold, New York 1979, pp. 110/11).
[16] Sauval, A. J.; Tatum, J. B. (Astrophys. J. Suppl. Ser. **56** [1984] 193/209).
[17] Ruff, O.; Braida, A. (Z. Anorg. Allgem. Chem. **214** [1933] 81/90).
[18] Braida, A. (Diss. T.H. Breslau 1933, pp. 1/42, 21/2, 29).
[19] Stølevik, R. (Acta Chem. Scand. **43** [1989] 860/7).

10.3.5 Transport Properties

Viscosities and thermal conductivities were estimated for BrF gas at 1 atm and between 100 and 5000 K using a Lennard-Jones (12-6) potential and are tabulated at 100 K intervals.

Reference:

Svehla, R. (NASA-TR-R-132 [1962] 1/139, 53; N.S.A. **16** [1962] No. 13210).

10.3.6 Chemical Behavior

Few chemical properties of BrF are known because it is instable: continuously disproportionating ($3\,BrF \rightleftharpoons Br_2 + BrF_3$), it exists only in equilibrium with bromine and BrF_3 so that a demarcation of its chemical properties is difficult [1 to 5]. In the systems Br_2-F_2 (see "Fluorine" Suppl. Vol. 2, 1980, pp. 143/4, and "Bromine" Suppl. Vol. A, 1985, p. 373) and Br_2-BrF_3 (see "Bromine" Suppl. Vol. A, 1985, p. 415), the conversion of BrF to the more stable higher fluorides BrF_3 and BrF_5 has been proposed to proceed via reactions such as $BrF + F_2 \rightleftharpoons BrF_2 + F$ ($k = (2 \pm 0.6) \times 10^{-12} \exp[-(48.1 \pm 2.1)\ kJ \cdot mol^{-1}/RT]\ cm^3 \cdot molecule^{-1} \cdot s^{-1}$ [6]), $BrF + F_2 \rightleftharpoons BrF_3$ [1 to 3, 5], $2\,BrF + BrF_3 \rightleftharpoons Br_2 + BrF_5$, $5\,BrF \rightleftharpoons 2\,Br_2 + BrF_5$ [7], and $8\,BrF \rightleftharpoons 3\,Br_2 + BrF_3 + BrF_5$ ($-\Delta H \leq 247$ kJ/mol) [8].

Self-ionization according to $2\,BrF \rightleftharpoons Br^{+} + BrF_2^{-}$ can be expected [9] on the basis of a partial ionic character of approximately 33% [10].

In general BrF is reported to have chemical properties similar to other bromine fluorides but to be more reactive than these. An early pioneering work reported that BrF attacks Pt as well as Au and quickly destroys vitreous quartz [1]. A BrF intermediate was proposed to explain processes such as the dissolution of Pt in Br_2-BrF_5 solution [11] and the dissolution of Au and Ag, separately, in Br_2-BrF_3 solution [12]. A polar BrF intermediate can be added to TeF_4 to yield $TeBrF_5$ in the TeF_4-Br_2-F_2 system [13] and to SO_2 to yield SO_2BrF in the Br_2-BrF_3-SO_2 system [14, 15]. The liquid-phase direct fluorination of $SOBr_2$ was claimed to proceed in part via the reaction $BrF + SOFBr \rightarrow SOF_2 + Br_2$ [16]. The bromine catalyzed oxidation of arsenic trifluoride by antimony pentafluoride also proceeds via a BrF intermediate: $2\,SbF_5 + Br_2 \rightleftharpoons 2\,BrF + SbF_3 + SbF_5$; $2\,BrF + AsF_3 \rightarrow Br_2 + AsF_5$ [17]. The reaction of BrF with SF_4 was used to prepare SF_5Br [18, 19]. Attempts to obtain $CsBrF_2$ product from the reaction of BrF with CsF in CCl_3F solution were unsuccessful. Bromine monofluoride was found to form an insoluble complex with pyridine in a trichlorofluoromethane solution and could thus be isolated from the Br_2-BrF_3 system [5].

The rate constant $k \approx 1.5 \times 10^{-13}\ cm^3 \cdot molecule^{-1} \cdot s^{-1}$ for the gas-phase reaction $O + BrF \rightarrow OBr + F$ was estimated in a study of the recombination of O and Br atoms in the presence of F_2 [21].

Product energy distributions from the gas-phase reaction $H + BrF \rightarrow HF(v \leq 7) + Br$ were obtained by observing the HF chemiluminescence. BrF was formed via the $F + Br_2$ reaction in the flow system, immediately ahead of the reaction vessel [22].

The activation energies E_a for the atom-transfer and exchange reactions involving BrF were calculated by means of empirical [23, 24], semiempirical [25, 26], and quantum mechanical [27] methods. Summarized below are the calculated values, some of which [25 to 27] include the correction for the zero-point energy. Ab initio calculations (SOGVB + POL-CI) on the triatomic potential energy hypersurface were also used to predict the geometries and normal mode frequencies of the H + FBr and H + BrF transition states and the reaction exoergicities [27].

reaction	E_a in kJ/mol	Ref.	reaction	E_a in kJ/mol	Ref.
$H + BrF \rightarrow HF + Br$ [a]	12	[25]	$C + BrF \rightarrow CF + Br$	23	[25]
	13	[26]	$HBr + BrF \rightarrow HF + Br_2$	242	[24]
	37	[27]		108	[23]
$H + BrF \rightarrow HBr + F$ [b]	11	[25]	$H_2 + BrF \rightarrow HBr + HF$	261	[24]
	6	[26]		147	[23]
	9	[27]			

[a] $\Delta H = -295$ kJ/mol (calculated) [27], -341 kJ/mol (experimental) [27, 28]. –
[b] $\Delta H = -98.3$ kJ/mol (calculated) [27], -128 kJ/mol (experimental) [27, 28].

The frequency factor A in the modified Arrhenius expression $k = AT^{0.67}\exp(-E_a/RT)$ $cm^3 \cdot mol^{-1} \cdot s^{-1}$ was calculated to be 8.1×10^{-11}, 3.3×10^{-11}, and 11.0×10^{-11} for the H + BrF and C + BrF fluorine atom transfer and the H + BrF bromine atom-transfer reactions, respectively [25].

BrF prepared in situ by the reaction of F_2 with Br_2 in $CFCl_3$ at 198 K is readily and efficiently added to the double bonds of organic compounds in the presence of a proton donor. Transaddition of Br and F was observed [29, 30]. Various types of alkynes react to give bromofluoroalkenes by the same procedure [31]. BrF (from Br_2 and BrF_3) also adds readily to the double bond of fluoroalkenes [20]. (In contrast, in other procedures, for example, in a mixture of anhydrous HF and N-bromoacetamide [32, 33] or in Br_2 and AgF [34], the addition of bromine and fluorine across double bonds is considered to proceed via two consecutive reactions each using a separate reagent [30].)

References:

[1] Ruff, O.; Braida, A. (Z. Anorg. Allgem. Chem. **214** [1933] 81/91).
[2] Stein, L. (J. Am. Chem. Soc. **81** [1959] 1269/73).
[3] Stein, L. (J. Am. Chem. Soc. **81** [1959] 1273/6).
[4] Steunenberg, R. K.; Vogel, R. C.; Fischer, J. (J. Am. Chem. Soc. **79** [1957] 1320/3).
[5] Naumann, D.; Lehmann, E. (J. Fluorine Chem. **5** [1975] 307/21).
[6] Arutyunov, V. S.; Buben, S. N.; Trofimova, E. M.; Chaikin, A. M. (Kinet. Katal. **21** [1980] 337/42; Kinet. Catal. [USSR] **21** [1980] 258/62).
[7] Stein, L. (Halogen Chem. **1** [1967] 151/5).
[8] Slutsky, L.; Bauer, S. H. (J. Am. Chem. Soc. **76** [1954] 270/5).
[9] Meinert, H. (Z. Chem. [Leipzig] **7** [1967] 41/57).
[10] Iczkowski, R. P. (J. Am. Chem. Soc. **86** [1964] 2329/32).

[11] Chernyaev, I. I.; Nikolaev, N. S.; Ippolitov, E. G. (Dokl. Akad. Nauk SSSR **130** [1960] 104113; Proc. Acad. Sci. USSR Chem. Sect. **130/135** [1960] 167/9).
[12] Mit'kin, V. N.; Zemskov, S. V. (Zh. Prikl. Khim. **54** [1981] 2180/6; J. Appl. Chem. USSR **54** [1981] 1913/8).
[13] Lawlor, L.; Passmore, J. (Inorg. Chem. **18** [1979] 2921/3).
[14] Kwasnik, W. (BIOS Final Rept. No. 1595, Item 22, p. 98; see [9]).
[15] Reed, P. R., Jr.; Lovejoy, R. W. (Spectrochim. Acta A **24** [1968] 1795/801).
[16] Ruppert, I. (J. Fluorine Chem. **20** [1982] 75/8).
[17] Passmore, J.; Shantha Nandana, W. A.; Richardson, E. K.; Taylor, P. (J. Fluorine Chem. **15** [1980] 435/9).
[18] Wessel, J.; Kleemann, G.; Seppelt, K. (Chem. Ber. **116** [1983] 2399/407).
[19] Christe, K. O.; Curtis, E. C.; Schack, C. J. (Spectrochim. Acta A **33** [1977] 69/73).
[20] Chambers, R. D.; Musgrave, W. K. R.; Savory, J. (J. Chem. Soc. **1961** 3779/86; Proc. Chem. Soc. **1961** 113).

[21] Arutyunov, V. S.; Buben, S. N.; Chaikin, A. M. (Kinet. Katal. **20** [1979] 570/4; Kinet. Catal. [USSR] **20** [1979] 465/9).
[22] Brandt, D.; Polanyi, J. C. (Chem. Phys. **45** [1980] 65/84).
[23] Noyes, R. M. (J. Am. Chem. Soc. **88** [1966] 4318/24).
[24] Benson, S. W.; Haugen, G. R. (J. Am. Chem. Soc. **87** [1965] 4036/44).
[25] Mayer, S. W.; Schieler, L.; Johnston, H. S. (Symp. Combust. **11** [1967] 837/44).
[26] Karachevtsev, G. V.; Savkin, V. V. (Khim. Fiz. **1983** No. 9, pp. 1286/8; C.A. **99** [1983] No. 164682).
[27] Eades, R. A. (Diss. Univ. Minnesota 1983; Diss. Abstr. Intern. B **44** [1984] 3418).
[28] Huber, K. P.; Herzberg, G. (Molecular Spectra and Molecular Constants, Vol. 4, Constants of Diatomic Molecules, Van Nostrand Reinhold, New York 1979).
[29] Brand, M.; Rozen, S. (J. Fluorine Chem. **20** [1982] 419/24).
[30] Rozen, S.; Brand, M. (J. Org. Chem. **50** [1985] 3342/8).

[31] Rozen, S.; Brand, M. (J. Org. Chem. **51** [1986] 222/5).
[32] Bowers, A. (J. Am. Chem. Soc. **81** [1959] 4107/8).
[33] Bowers, A.; Laura Cuéllar Ibáñez; Denot, E.; Becerra, R. (J. Am. Chem. Soc. **82** [1960] 4001/7).
[34] Hall, L. D.; Jones, D. L. (Can. J. Chem. **51** [1973] 2902/13).

10.4 The BrF^+ Ion

CAS Registry Number: *[37354-78-2]*

The BrF^+ ion (m/e=98) is the strongest component in the mass spectrum of BrF_3 with an intensity of 202 relative to 100.0 for BrF_3^+, whereas in the mass spectrum of BrF_5, the intensity of BrF^+ is 18.8 relative to 100.0 for BrF_4^+; the appearance potentials gave upper limits for the ionization energy of the BrF molecule (see p. 7) [1].

The heats of formation from the elements in their standard states and from the Br^+ ion and the F atom were derived to be $\Delta_fH_0^o = 1104$ kJ/mol and $\Delta_fH_0^{at} = 911$ kJ/mol (presumably from dissociation and ionization energies); a semiempirical MO LCAO calculation resulted in $\Delta_fH_0^{at} = 895$ kJ/mol [2].

The He I photoelectron spectrum of BrF covers the region of 5π and 10σ and/or 4π electron removal, which corresponds to the formation of the BrF^+ ion in its spin-orbit split electronic ground state $X\ ^2\Pi_{3/2,1/2}(...(10\sigma)^2\ (4\pi)^4\ (5\pi)^3$; $\Delta E(^2\Pi_{3/2} - {}^2\Pi_{1/2}) = 2590 \pm 40\ cm^{-1}$) and in the exited states $A\ ^2\Sigma^+(...(10\sigma)^1\ (4\pi)^4\ (5\pi)^4)$ and/or $B\ ^2\Pi_{3/2,1/2}$ $(...(10\sigma)^2\ (4\pi)^3\ (5\pi)^4)$. The vibrational structure in the first photoelectron band resulted in the ground-state vibrational constants $\omega_e = 750 \pm 30\ cm^{-1}$ and $\omega_ex_e = 10 \pm 5\ cm^{-1}$ [3]. The earlier results, $\Delta E(^2\Pi_{3/2} - {}^2\Pi_{1/2}) = 2600 \pm 60\ cm^{-1}$ and $\omega_e = 750 \pm 40\ cm^{-1}$ [4], are quoted by Huber and Herzberg [5]. The latter authors derived the dissociation energy $D_0(Br^+ - F) = D_0(BrF) + E_i(Br) - E_i(BrF) = 2.61$ eV (E_i = ionization potential) [5].

Semiempirical SCF MO calculations of the electronic structure of BrF^+ were reported, but no results were published [6].

References:

[1] Irsa, A. P.; Friedman, L. (J. Inorg. Nucl. Chem. **6** [1958] 77/90).
[2] Andreev, S. (God. Vissh. Khim. Tekhnol. Inst. Sofia **20** No. 2 [1972/74] 65/76; C.A. **83** [1975] No. 183865).
[3] Colbourn, E. A.; Dyke, J. M.; Fayad, N. K.; Morris, A. (J. Electron Spectrosc. Relat. Phenom. **14** [1978] 443/52).
[4] DeKock, R. L.; Higginson, B. R.; Lloyd, D. R.; Breeze, A.; Cruickshank, D. W. J.; Armstrong, D. R. (Mol. Phys. **24** [1972] 1059/72).
[5] Huber, K. P.; Herzberg, G. (Molecular Spectra and Molecular Structure, Vol. 4, Constants of Diatomic Molecules, Van Nostrand Reinhold, New York 1979, pp. 110/1).
[6] Charkin, O. P.; Smolyar, A. E.; Klimenko, N. M. (Proc. 16th Intern. Conf. Coord. Chem., Dublin 1974, Abstr. 2.3a, pp. 1/3).

10.5 The BrF^- Ion

CAS Registry Number: *[12272-35-4]*

Semiempirical SCF MO calculations of the electronic structure of the **free** BrF^- ion were reported, but no results were published [1].

A **stabilization** and **orientation** of BrF^- ions as interstitial centers in alkali halide crystals can be achieved at liquid nitrogen temperatures by X or γ irradiation of XCl crystals (X = Na, K, Rb) which have been doped with up to 0.5 mol% of the corresponding fluoride and bromide (and with Pb^{2+} ions acting as electron traps) or by X or γ irradiation of F^- doped KBr. In the former case, the initially formed FCl^- centers have to be destroyed by warming the crystal to above 150 K for several minutes; then high concentrations of BrF^- centers arise upon F atom migration through the crystal. In the latter case, the BrF^- centers are formed immediately. The BrF^- center was identified by electron spin resonance (ESR) and shown to be oriented exactly along a [111] direction and to occupy a single negative-ion vacancy. The ESR spectrum consists of 16 lines; the two hyperfine (hf) lines due to the ^{19}F nucleus (I = 1/2) are split into two times four lines due to the ^{79}Br and ^{81}Br nuclei (I = 3/2) [2 to 6]. The following anisotropic g-tensor elements $g_{\parallel}$ and $g_{\perp}$, hf coupling constants $A_{\parallel}$ and $A_{\perp}$, and the corresponding isotropic values g_{iso} and A_{iso} in different crystals at 77 K were reported by Schoemaker [4] (see also [7]):

	$NaCl:Br^-:F^-$	$KCl:Br^-:F^-$	$RbCl:Br^-:F^-$	$KBr:F^-$
$g_{\parallel}$	1.9897	1.9891	1.9898	1.9903
$g_{\perp}$	2.12	2.125	2.123	2.12
g_{iso}	2.076	2.080	2.079	2.077
$A_{\parallel}(^{81}Br)$ in G	507	601	599.9	591
$A_{\perp}(^{81}Br)$ in G	118 ± 10	123 ± 10	125 ± 10	115 ± 10
$A_{iso}(^{81}Br)$ in G	278	282	283	274
$A_{\parallel}(^{19}F)$ in G	687	735.2	735.8	711
$A_{\perp}(^{19}F)$ in G	−50 ± 40	−50 ± 20	−50 ± 20	−50 ± 30
$A_{iso}(^{19}F)$ in G	196	212	212	204

Similar g and A values (except $A_{\perp}(^{19}F) = +89 \pm 10$ G) for BrF^- in $KCl:Br^-:F^-$ and similar $A_{\parallel}$ values for BrF^- in $NaCl:Br^-:F^-$ and $KBr:F^-$ at 80 K have been derived by Dreybrodt and Silber [5].

Annealing experiments showed the BrF^- center to be stable up to about 230 K in $KCl:Br^-:F^-$ and up to 180 K in $KBr:F^-$ [4]. The temperature dependence (T = 20 to 150 K) of the hyperfine structure has been studied in $NaCl:Br^-:F^-$ and $KCl:Br^-:F^-$ and the interaction of the BrF^- center with the different lattices discussed; this allowed to estimate the vibrational frequencies of BrF^- to be $\omega = 42.5\ cm^{-1}$ in KCl and $\omega = 85\ cm^{-1}$ in NaCl, which may be compared to $\omega = 340\ cm^{-1}$ estimated for the free BrF^- ion and $\omega_e = 363\ cm^{-1}$ [8] for the B $^3\Pi(0^+)$ state of the free BrF molecule [6]. An absorption band in the UV with $\lambda_{max} = 294$ nm has been observed in $KCl:Br^-:F^-$ [2].

Theoretical studies of the ESR parameters and optical transition energies of the BrF^- center in alkali halides by using a simple MO LCAO wave function [9] and a valence-bond wave function [10] are available.

References:

[1] Charkin, O. P.; Smolyar, A. E.; Klimenko, N. M. (Proc. 16th Intern. Conf. Coord. Chem., Dublin 1974, Abstr. 2.3a, pp. 1/3).
[2] Wilkins, J. W.; Gabriel, J. R. (Phys. Rev. [2] **132** [1963] 1950/7).
[3] Schoemaker, D.; Delbecq, C. J.; Yuster, P. H. (Bull. Am. Phys. Soc. **10** [1965] 1201).
[4] Schoemaker, D. (Phys. Rev. [2] **149** [1966] 693/704).
[5] Dreybrodt, W.; Silber, D. (Phys. Status Solidi **16** [1966] 215/23).
[6] Dreybrodt, W. (Phys. Status Solidi **21** [1967] 99/112).
[7] Morton, J. R.; Preston, K. F. (Landolt-Börnstein New Ser. Group II **9** Pt. a [1977] 5/289, 223).

[8] Herzberg, G. (Molecular Spectra and Molecular Structure, Vol. 1, Spectra of Diatomic Molecules, Van Nostrand, Princeton, N.J., 1961, pp. 512/3).
[9] Van Puymbroeck, W.; Lagendijk, A.; Schoemaker, D. (Phys. Status Solidi A **59** [1980] 585/95).
[10] Jette, A. N.; Adrian, F. J. (Phys. Rev. [3] B **14** [1976] 3672/81).

10.6 Dibromine Difluoride, Br_2F_2

CAS Registry Number: *[65414-56-4]*

Cocondensation of F_2-Ar and Br_2-Ar (or Br_2-BrF-BrF_3-Ar) samples and subsequent UV photolysis yields product mixtures containing Br_2F_2. An IR absorption band at 555 cm^{-1} was assigned to the antisymmetric F-Br-F stretching vibration of the symmetric, T-shaped Br-BrF_2 molecule. A Raman band at 603 cm^{-1} could be due to the asymmetric F-BrBrF isomer.

Reference:

Prochaska, E. S.; Andrews, L.; Smyrl, N. R.; Mamantov, G. (Inorg. Chem. **17** [1978] 970/7).

10.7 Bromine Difluoride, BrF_2

CAS Registry Numbers: *[43340-59-6, 64973-51-9]*

Cocondensation of F_2-Ar and precursor molecules with single Br atoms, e.g. BrF or BrF_3, in Ar at about 15 K yields product mixtures containing BrF_2 after UV photolysis. Higher concentrations result after thermally cycling to 25 K. Photolysis of BrF_3 in an Ar matrix followed by a thermal cycle at 25 K also produces some BrF_2 [1]. The formation of BrF_2 during gas-phase fluorination of Br_2 at total pressures below 210 Torr in the temperature range 291 to 310 K by the reaction $BrF + F_2 \rightarrow BrF_2 + F$ was demonstrated in a kinetic investigation. The rate constant of $k = (2 \pm 0.6) \times 10^{-12} \exp[-(48.1 \pm 2.1\ kJ/mol)/RT]\ cm^3/s$ was obtained assuming quasi-steady-state conditions for this reaction [2].

Vibrational frequencies at 568.0 and 570.2 cm^{-1} were assigned to the antisymmetric stretching fundamentals of $^{81}BrF_2$ and $^{79}BrF_2$. The FBrF radical is believed to be nearly linear. A lower limit of the FBrF angle of $152° \pm 8°$ was calculated from the fundamentals [1]. A BrF_2 ionization potential of 11.2 eV was derived from the appearance potential of BrF_2^+ in the mass spectrum of BrF_3 and from the energy of the Br-F bond [3].

Matrix-isolated BrF_2 is destroyed when thermally cycled to 36 K [1]. Proposed gas-phase reactions of BrF_2 are $BrF_2 + F_2 \rightarrow BrF_3 + F$ and $2\ BrF_2 \rightarrow BrF_3 + BrF$ [2].

References:

[1] Prochaska, E. S.; Andrews, L.; Smyrl, N. R.; Mamantov, G. (Inorg. Chem. **17** [1978] 970/7).
[2] Arutyunov, V. S.; Buben, S. N.; Trofimova, E. M.; Chaikin, A. M. (Kinet. Katal. **21** [1980] 337/42; Kinet. Catal. [USSR] **21** [1980] 258/62).
[3] Irsa, A. P.; Friedman, L. (J. Inorg. Nucl. Chem. **6** [1958] 77/90).

10.8 Difluorobromine(1+), BrF_2^+

Other names: Difluorobrominium ion, difluorobromonium ion

CAS Registry Number: *[31159-95-2]*

Formation. The cation BrF_2^+, m/z 117, forms during mass spectrometry of BrF_3 (see also p. 75) with an appearance potential of 13.5±0.3 eV directly from the parent molecule. The appearance potential with BrF_5 as parent molecule (see also p. 128) is 16.1±0.2 eV. The BrF_2^+ ion forms by decomposition of BrF_3^+ [1]. BrF_2^+ is also observed in mass spectra taken after partial hydrolysis of BrF_3 and BrF_5 [2] and in the mass spectrum of BrF_7 [3]; see p. 144.

The formation of BrF_2^+ by self-dissociation of liquid BrF_3 is described on p. 66. The equilibrium 2 $BrF_3 \rightleftharpoons BrF_2^+ + BrF_4^-$ is increasingly shifted towards acidic BrF_2^+ by adding a fluorine compound capable of acting as a fluoride acceptor (Lewis acid) [4]. The formed BrF_2^+ salts were frequently isolated from the solution, usually by evaporating the volatiles. Compounds of BrF_2^+ were obtained with the anions BF_4^- [5 to 8], $B_2F_7^-$ [5 to 7], GeF_6^{2-} [5, 9], SnF_6^{2-} [10 to 12], PF_6^-, $P_2F_{11}^-$ [7], AsF_6^- [7, 9] or $As_2F_{11}^-$ [7], SbF_6^- [12], $Sb_3F_{16}^-$ [13], BiF_6^- [14, 15], SO_3F^- [16], HF_2^- (see "Fluorine" Suppl. Vol. 3, 1982, p. 192), and $Mn_2F_9^-$ [17]. The fluoride acceptors are formed intermediately in BrF_3 solution by adding the appropriate elements or compounds other than fluorides. This method yields the BrF_2^+ salts of SnF_6^{2-} from Sn [12, 18], $SnCl_2$, or $SnCl_4$ [12], SbF_6^- from Sb_2O_3 [12, 19], Sb_2O_5 [20], SbClO [12], or ClO_2SbF_6 [19], NbF_6^- and TaF_6^- from the metals or their pentoxides [14], IrF_6^- from the metal [18], PdF_4^- from $PdCl_2$ or $PdBr_2$ [21], and AuF_4^- from the metal [22]. The formation of BrF_2SO_3F results from reactions of BrF_5 with SO_3, BrF_3 with $Br(SO_3F)_3$, and Br_2 with $FOSO_2F$ [16].

Intermediate formation of BrF_2MnF_5 or $(BrF_2)_2MnF_6$ [23] and of BrF_2RuF_6 [24] in BrF_3 solution was inferred from the formation of ternary fluorides with added fluoride donors. The possible formation of BrF_2^+ salts at least in BrF_3 solution was discussed for the anions TiF_6^{2-} [25], VF_6^- [26], OsF_6^- [27], $CrOF_4^-$ [23], and $ReO_2F_4^-$ [28] and was mentioned for PdF_6^{2-}, VOF_4^-, and $ReOF_4^-$ [29].

BrF_2AsF_6 and BrF_2BF_4 are synthesized by reacting stoichiometric amounts of Br_2, F_2, and AsF_5 or BF_3 at 195 K. Oxidation of Br_2 by O_2AsF_6 yields BrF_2AsF_6 with evolution of O_2 [30]. A mixture of Br_2 and BrF_5 dissolves Pt with formation of $(BrF_2)_2PtF_6$ [31]. $BrOF_2SbF_6$ slowly decomposes to yield BrF_2SbF_6 at ambient temperature [32].

Applications. The "acidic" BrF_2^+ salt BrF_2SbF_6 is a convenient fluorinating agent at temperatures below 770 K, especially for basic oxides such as CaO [33]. This reaction can also be used to determine O_2 in organic substances, including phosphates and sulfones, and in CO_2 [34]. Oxidative fluorination by solid BrF_2SbF_6 at 213 to 373 K [35, 36] and by BrF_2BiF_6 or BrF_2TaF_6 at 298 K [36] removes Rn from dry air; additional details are given in [37]. The hypergolic reaction of solid BrF_2BF_4 [38], BrF_2AsF_6, or BrF_2SbF_6 [39] with amines or boranes is potentially useful in propellant systems.

Structural Investigations. The nearly ionic structure of BrF_2SbF_6 [40] and $(BrF_2)_2GeF_6$ [41] is known from photographic X-ray investigations of single crystals. The Br atom is surrounded by two nearest F neighbors at about 1.70 Å in a distorted square plane and two others at about 2.20 Å. Structural data of the BrF_2^+ ion are given in Table 4. The longer Br-F distances are significantly smaller than the van der Waals distance of 3.30 Å for a Br-F contact [42]. This is indicative of a covalent contribution to the structure [40, 41] which is larger in the case of $(BrF_2)_2GeF_6$ [41].

Table 4
Structural Parameters of BrF_2^+.

compound	distance Br–F in Å	angle FBrF in degree	distance Br···F(bridging) in Å	Ref.
BrF_2SbF_6	1.69(2)	93.5(2.1)	2.29(2)	[40]
$(BrF_2)_2GeF_6$	1.74(2) and 1.69(2)	90.7(9)	2.25(2) and 2.17(2)	[41]

Preliminary X-ray investigations showed that BrF_2NbF_6 [40] and BrF_2IrF_6 [18] are isostructural with BrF_2SbF_6. An intermediate state between ionic and covalent bonding for $(BrF_2)_2SnF_6$ was derived from the Mössbauer spectra [10, 43].

The VSEPR model predicts an angular BrF_2^+ ion in accordance with sp^3 hybridization of Br [44]. The electronic structure of BrF_2^+ was treated by the semiempirical NDDO-2(α, β) method [45].

^{19}F NMR Spectra. The ^{19}F NMR spectra of BrF_2^+ salts (negative values for high field shifts) in SO_2ClF solution exhibit two signals of the cation. Their chemical shifts with respect to CCl_3F are [7]:

compound	BrF_2BF_4	BrF_2PF_6	BrF_2AsF_6
δ in ppm	−60.6, −70.5	−59.9, −69.2	−61.0, −68.3
T in K	153	143	153

Coalescence of the BrF_2^+ signals occurs at 173 K. The spectra are interpreted in terms of nonlinear BrF_2 groups which form T-shaped $F_2Br\cdots F$ units by fluorine bridges to the anions. There is no evidence for the signals to split by spin-spin coupling [7]; its absence is attributed to rapid quadrupole relaxation [46].

Spectra of solid salts at low temperatures contain similar shifts; values in ppm with respect to F_2 are as follows [18]:

compound	BrF_2AuF_4	$(BrF_2)_2SnF_4$	$(BrF_2)_2PtF_6$	BrF_2IrF_6
δ at 150 K	−520	−510	−520	−510
δ at 298 K	−558	−550	−558	–

The chemical shift of BrF_2^+ in liquid BrF_2SO_3F is $\delta = 29 \pm 1$ from the reference CF_3COOH, the corresponding signal in molten BrF_2SbF_6 has a chemical shift of $\delta = 32$ [16]. The different intensity of the two broad lines in the spectrum of solid BrF_2AuF_4 is indicative of its ionic structure [47].

Vibrational and Electronic Spectra. Assignments of vibrational spectra are based on a bent BrF_2^+ ion of C_{2v} symmetry [9, 42, 48]. Two stretching modes and one bending mode are expected being active in the Raman and IR spectra [49]. Bands were assigned in the Raman spectra only, since the resolution of the IR spectra is too low; see for example [42]. Wave numbers are given in Table 5, earlier spectral data are given in [42, 50].

Table 5
Fundamental Vibrations of BrF_2^+ in cm^{-1}.

compound	state	$\nu_1(A_1)$ sym. stretch Raman	IR	$\nu_2(A_1)$ bend Raman	$\nu_3(B_1)$ antisym. stretch Raman	Ref.
$(BrF_2)_2GeF_6$	solid	690 (10)	688 s	344 (0.5)		[5, 9]
BrF_2AsF_6	solid	706 (10)		360 (1.8)	703 (3)	[9]
		709		307, 294		[50]
BrF_2SbF_6	solid	705 (10)	705 s	362 (2.0)	702 sh	[9]
		704 (10)		308 (0.9)		[50]
	in HF	707 p		361	695	[48]
	in BrF_3-HF	700 p		362	689	[48]
	in BrF_3 *)	625				[50]
BrF_2^+	in BrF_3	625				[51]

*) Unassigned IR bands were found at 635 ms and 292 mw.

The ν_1 band was assigned by its polarization [48]. In solids, the BrF_2^+ ion is bonded by weak fluorine bridges to the counterions [9] (see Table 4) which loose their octahedral symmetry by the interaction [5]. The interaction of BrF_2^+ with the solvent HF is small. The interaction of BrF_2^+ and BrF_3 results in lower wave numbers of the stretching vibrations on adding BrF_3 to a HF solution [48, 50] and the loss of polarization of the ν_1 band in a BrF_3 solution [48]. The interaction was also discussed in [52].

The Raman spectrum of $BrF_2Mn_2F_9$ is displayed in [17]. Unassigned IR and Raman bands of BrF_2^+ in BrF_2BiF_6 are given in [15].

A vibrational transition of 730 cm^{-1} was assigned to the excited state of BrF_2^+ based on the fine structure of the band at 380 nm in the UV spectra of $BrF_2B_2F_7$ in SO_2ClF solution. Compounds formed with PF_5 and AsF_5 have similar UV spectra [7].

Mean Amplitudes of Vibration. Vibrational amplitudes of BrF_2^+ from 0 to 1000 K were calculated from force constants based on experimental vibrational and structural data. The method of "characteristic vibrations" yields low mean amplitudes which are similar to those of BrF; results for u_{Br-F} and $u_{F...F}$ at 298.16 K are 0.0409 and 0.072 Å [53] in agreement with a later calculation [54].

Force Constants. Valence force constants of BrF_2^+ were calculated on the basis of experimental vibrational [9] and structural data [40]. The following values in mdyn/Å were obtained:

f_r	f_{rr}	f_α	$f_{r\alpha}$	Ref.
4.60	0.21	0.47	0 (assumed)	[9]
4.4991	0.0945	0.1520	0.05897	[54]

Force constants calculated by the simple bond-charge (SBC) model [55] are in good agreement with the results of the valence force field calculations [54].

Chemical Behavior. Thermal dissociation of BrF_2^+ salts yields BrF_3 and a fluoride derived from the anion. Dissociation temperatures vary widely; BrF_2BF_4 decomposes above

193 K [5], while BrF_2NbF_6 [14] or $(BrF_2)_2PtF_6$ starts to decompose at about 450 K [21]. Decomposition of BrF_2BiF_6 between 370 to 470 K yields BrF_3, BrF_5, BiF_3, and BiF_5 [15].

Rn is oxidized in contact with solid BrF_2SbF_6 [35, 36] and BrF_2MF_6 with M = Bi, Ta [36]. The reaction $2\ BrF_2AsF_6 + 5\ Br_2 + 4\ AsF_5 \rightarrow 6\ Br_2AsF_6$ proceeds with the appropriate amount of Br_2 [30]. BrF_2^+ salts with fluoroantimonate anions hydrolyze readily [35, 56].

Neutralization of the acidic BrF_2^+ cation in BrF_3 solution by compounds containing BrF_4^- anions at 298 K [16, 57] yields BrF_3 and a ternary fluoride. The heat of neutralization of BrF_2SbF_6 is about −19 kJ/mol for reactions with strong bases such as $KBrF_4$ or $AgBrF_4$. Lower values were measured for reactions of BrF_2^+ salts with weaker bases [57]. Neutralization reactions were invoked to prove the ionic structure of BrF_2NbF_6, BrF_2TaF_6, BrF_2BiF_6 [14], $(BrF_2)_2PtF_6$ [21], $(BrF_2)_2TiF_6$ [25], BrF_2AuF_4 [22], BrF_2SbF_6, $(BrF_2)_2SnF_6$ [12], and a BrF_2^+ salt of Mn(IV) [23]. However, it was argued whether these reactions necessarily demonstrate the ionic character of the investigated compounds, because the same products can form from neutral BrF_3 and Lewis acids when a compound of sufficient donor strength is added [52].

Polyfluorinated derivates of benzene, naphthalene, and pyridine react with BrF_2BF_4 and BrF_2SbF_6 in SO_2ClF solution at or below room temperature. Cyclohexadienes are obtained by oxidative fluorination; other products resulted when H in reactants and products is substituted by Br [58, 59]. Reactions of the BrF_2^+ salts with organic liquids usually are less violent than the reactions of BrF_3; some observations for $(BrF_2)_2SnF_6$ were described [12].

References:

[1] Irsa, A. P.; Friedman, L. (J. Inorg. Nucl. Chem. **6** [1958] 77/90).
[2] Sloth, E. N.; Stein, L.; Williams, C. W. (J. Phys. Chem. **73** [1969] 278/80).
[3] Fogle, C. E.; Rewick, R. T.; United Aircraft Corp. (U.S. 3615206 [1971] 4 pp.; C.A. **76** [1972] No. 5447).
[4] Shamir, J. (Israel J. Chem. **17** [1978] 37/47).
[5] Brown, D. H.; Dixon, K. R.; Sharp, D. W. A. (Chem. Commun. **1966** 654/5).
[6] Christe, K. O. (J. Phys. Chem. **73** [1969] 2792/3).
[7] Cyr, T.; Brownstein, S. (J. Inorg. Nucl. Chem. **39** [1977] 2143/5).
[8] Toy, M. S.; Cannon, W. A. (J. Phys. Chem. **70** [1966] 2241/4).
[9] Christe, K. O.; Schack, C. J. (Inorg. Chem. **9** [1970] 2296/9).
[10] Sukhoverkhov, V. F.; Dzevitskii, B. Z. (Dokl. Akad. Nauk SSSR **170** [1966] 1099/102; Dokl. Chem. Proc. Acad. Sci. USSR **166/171** [1966] 983/6).

[11] Sukhoverkhov, V. F.; Dzevitskii, B. Z. (Dokl. Akad. Nauk SSSR **177** [1967] 611/4; Dokl. Chem. Proc. Acad. Sci. USSR **172/177** [1967] 1089/91).
[12] Woolf, A. A.; Emeléus, H. J. (J. Chem. Soc. **1949** 2865/71).
[13] Fischer, J.; Liimatainen, R.; Bingle, J. (J. Am. Chem. Soc. **77** [1955] 5848/9).
[14] Gutmann, V.; Emeléus, H. J. (J. Chem. Soc. **1950** 1046/50).
[15] Popov, A. I.; Scharabin, A. V.; Sukhoverkhov, V. F.; Tchumaevsky, N. A. (Z. Anorg. Allgem. Chem. **576** [1989] 242/54).
[16] Gross, U.; Meinert, H.; Grimmer, A.-R. (Z. Chem. [Leipzig] **10** [1970] 441/3).
[17] Sukhoverkhov, V. F.; Melkumyants, M. V. (Tr. Inst. Mosk. Khim. Tekhnol. Inst. im. D. I. Mendeleeva No. 143 [1986] 62/4; C.A. **108** [1988] No. 197206).
[18] Mit'kin, V. N.; Mironov, Yu. I.; Zemskov, S. V.; Zil'berman, B. D.; Gabuda, S. P. (Koord. Khim. **9** [1983] 20/5; Soviet J. Coord. Chem. **9** [1983] 18/23).
[19] Woolf, A. A. (J. Chem. Soc. **1954** 4113/6).
[20] Woolf, A. A. (Chem. Ind. [London] **1954** 346).

[21] Sharpe, A. G. (J. Chem. Soc. **1950** 3444/50).

[22] Sharpe, A. G. (J. Chem. Soc. **1949** 2901/2).
[23] Sharpe, A. G.; Woolf, A. A. (J. Chem. Soc. **1951** 798/801).
[24] Hepworth, M. A.; Peacock, R. D.; Robinson, P. L. (J. Chem. Soc. **1954** 1197/201).
[25] Sharpe, A. G. (J. Chem. Soc. **1950** 2907/8).
[26] Emeléus, H. J.; Gutmann, V. (J. Chem. Soc. **1949** 2979/82).
[27] Hepworth, M. A.; Robinson, P. L.; Westland, G. J. (J. Chem. Soc. **1954** 4269/75).
[28] Beattie, J. R.; Crocombe, R. A.; Ogden, J. S. (J. Chem. Soc. Dalton Trans. **1977** 1481/91).
[29] Woolf, A. A. (Advan. Inorg. Chem. Radiochem. **9** [1966] 217/314, 273).
[30] Smalc, A. (Inst. Jozef Stefan IJS Rept. R-612 [1972] 1/7; C.A. **79** [1973] No. 13032).

[31] Chernyaev, I. I.; Nikolaev, N. S.; Ippolitov, E. G. (Dokl. Akad. Nauk SSSR **130** [1960] 1041/3; Proc. Acad. Sci. USSR Chem. Sect. **130/135** [1960] 167/9).
[32] Bougon, R.; Bui Huy, T.; Charpin, P.; Gillespie, R. J.; Spekkens, P. H. (J. Chem. Soc. Dalton Trans. **1979** 6/12).
[33] Sheft, I.; Martin, A. F.; Katz, J. J. (J. Am. Chem. Soc. **78** [1956] 1557/9).
[34] Sheft, I.; Katz, J. J. (Anal. Chem. **29** [1957] 1322/5).
[35] Avrorin, V. V.; Nefedov, V. D.; Toropova, M. A. (Radiokhimiya **18** [1976] 518/9; Soviet Radiochem. **18** [1976] 449/50).
[36] Stein, L. (Science **175** [1972] 1463/5, Noble Gases Symp., Las Vegas 1973, pp. 376/85, CONF-730915).
[37] Stein, L. (Can. 952290 [1974] 11 pp.; C.A. **81** [1974] No. 175632, U.S. 3784674 [1974] 3 pp.).
[38] Toy, M. S.; Cannon, W. A. (U.S. 3645702 [1972] 2 pp.; C.A. **76** [1972] No. 143022).
[39] Olah, G. A.; Kuhn, S. J.; Dow Chemical Co. (U.S. 3103782 [1963] 2 pp.; C.A. **60** [1963] 2718).
[40] Edwards, A. J.; Jones, G. R. (J. Chem. Soc. A **1969** 1467/70, Chem. Commun. **1967** 1304/5).

[41] Edwards, A. J.; Christe, K. O. (J. Chem. Soc. Dalton Trans. **1976** 175/7).
[42] Carter, H. A.; Aubke, F. (Can. J. Chem. **48** [1970] 3456/9).
[43] Dzevitskii, B. Z.; Sukhoverkhov, V. F. (Izv. Sibirsk. Otd. Akad. Nauk SSSR Ser. Khim. Nauk **1968** No. 2, pp. 54/7).
[44] Gillespie, R. J.; Nyholm, R. S. (Quart. Rev. Chem. Soc. **11** [1957] 339/80, 373).
[45] Smolyar, A. E.; Charkin, O. P.; Klimenko, N. M. (Zh. Strukt. Khim. **15** [1974] 993/1003; J. Struct. Chem. [USSR] **15** [1974] 885/93).
[46] Shamir, J. (Struct. Bonding [Berlin] **37** [1979] 141/210, 195).
[47] Gabuda, S. P.; Zemskov, S. V.; Mit'kin, V. N.; Obmoin, B. I. (Zh. Strukt. Khim. **18** [1977] 515/24; J. Struct. Chem. [USSR] **18** [1977] 413/20).
[48] Surles, T.; Quarterman, L. A.; Hyman, H. H. (J. Fluorine Chem. **3** [1973] 293/306).
[49] Shamir, J. (from [46, p. 167]).
[50] Surles, T.; Hyman, H. H.; Quarterman, L. A.; Popov, A. I. (Inorg. Chem. **9** [1970] 2726/30).

[51] Surles, T.; Quarterman, L. A.; Hyman, H. H. (J. Fluorine Chem. **3** [1973] 453/6).
[52] Meinert, H.; Gross, U. (J. Fluorine Chem. **2** [1973] 381/6).
[53] Baran, E. J. (Anales Asoc. Quim. Argent. **63** [1975] 239/41).
[54] Rawat, T. S.; Sharma, R. G.; Dixit, L.; Misra, A. K. (Acta Ciencia Indica Phys. **10** [1984] 179/90).
[55] Gazquez, J. L.; Ray, N. K.; Parr, R. G. (Theor. Chim. Acta **49** [1978] 1/11).
[56] Sukhoverkhov, V. F.; Takanova, N. D. (Zh. Anal. Khim. **33** [1978] 1365/9; J. Anal. Chem. [USSR] **33** [1978] 1070/3).
[57] Richards, G. W.; Woolf, A. A. (J. Fluorine Chem. **1** [1971] 129/39).
[58] Bardin, V. V.; Furin, G. G.; Yakobson, G. G. (Zh. Org. Khim. **17** [1981] 999/1004; J. Org. Chem. [USSR] **17** [1981] 879/84).
[59] Bardin, V. V.; Furin, G. G.; Yakobson, G. G. (J. Fluorine Chem. **23** [1983] 67/86).

10.9 Difluorobromate(1−) and Bromofluorofluorate(1−), $FBrF^-$ and $BrFF^-$

CAS Registry Numbers: $FBrF^-$ *[25730-96-5]*, $BrFF^-$ *[96607-01-1]*

Formation. The BrF_2^- ion was observed in the negative ion mass spectrum of BrF_3 at m/z 117 with low relative intensity [1]. The synthesis of CsFBrF starts with CsF and a solution of BrF_3 in Br_2 which is distilled onto the salt held at liquid nitrogen temperature. Upon warming to ambient temperature, a maximum yield of 31% of CsFBrF in CsF is obtained. The synthesis is sensitive to changes of the experimental conditions [2]. An attempt to synthesize CsFBrF in CCl_3F failed as did the reaction of CsF with the BrF-pyridine complex in CH_3CN [3]. Codeposition of alkali bromide vapors and F_2-Ar at 15 K yields $MBrF_2$ (M = Na, K, Rb, Cs) in the product mix. Initially the MBrFF isomer forms and slowly rearranges to the MFBrF isomer. The decomposition products MF·BrF were also identified [4].

Vibrational Spectra. The Raman spectrum of solid CsFBrF in CsF indicates a bent $FBrF^-$ ion instead of the expected linear one; bands (in cm^{-1}) were assigned consistent with C_{2v} symmetry: $\nu_s = 442$ ms, $\delta = 198$ sh, $\nu_{as} = 596$ m, 562 ms. The splitting of ν_{as} is attributed to site effects in the lattice [2].

The observed IR bands of BrF_2^- salts in Ar matrices at 15 K are:

M in MFBrF	Na	K	Rb	Cs	M in MBrFF ...	K	Rb	Cs
ν in cm^{-1}	543	521	524	527	ν in cm^{-1}	364	366	360

The bands at about 530 cm^{-1} were assigned to the antisymmetric stretching mode of linear, symmetric $FBrF^-$ because of the considerable heavy halogen shift upon changing the unique halogen atom. The bands at about 360 cm^{-1} have a small heavy-halogen shift, suggesting that they are predominantly an F-F stretching mode of the $BrFF^-$ ion [4].

Mean Amplitudes of Vibration. For the bent $FBrF^-$ ion, mean amplitudes of vibration of $u_{Br-F} = 0.0505$ and $u_{F\cdots F} = 0.072$ Å and a Bastiansen-Morino shrinking of 0.0070 Å were calculated for 298.16 K [5] on the basis of vibrational bands in [2] and a Br-F distance of 1.93 Å. Results for temperatures between 0 and 1000 K were also listed [5].

Force Constants. The vibrational data from [2] were used to calculate the vibrational force constants $f_r = 2.36$, $f_{rr} = -0.18$, and $f_\alpha = 0.15$ mdyn/Å for a bent $FBrF^-$ ion assuming an angle of 175° [5]. A value of 2.11 mdyn/Å was calculated for a linear anion in CsFBrF from the band at 527 cm^{-1} [4].

Quantum Chemical Calculations. An ab initio calculation for linear $FBrF^-$ using the HF method with a double-zeta basis deals with the nature of the bonding. The three Br lone pairs are in a plane perpendicular to the axis of the ion. A trigonal bipyramidal coordination of Br results. The F lone pairs are approximated by sp^3 hybrid orbitals, and the lone pairs of Br and F form a staggered conformation [6]. Energy levels and the AO population of BrF_2^- were also calculated by the NDDO-2(α, β) method [7]. The HMO method predicts that both isomers of BrF_2^- are stable with respect to the reference system $BrF + F^-$ [8].

References:

[1] Irsa, A. P.; Friedman, L. (J. Inorg. Nucl. Chem. **6** [1958] 77/90).

[2] Surles, T.; Quarterman, L. A.; Hyman, H. H. (J. Inorg. Nucl. Chem. **35** [1973] 668/70).

[3] Naumann, D.; Lehmann, E. (J. Fluorine Chem. **5** [1975] 307/21).

[4] Miller, J. H.; Andrews, L. (Inorg. Chem. **18** [1979] 988/92).

[5] Baran, E. J. (Z. Naturforsch. **28b** [1973] 502/3).

[6] Kirillov, Yu. B.; Klimenko, N. M. (Zh. Strukt. Khim. **21** No. 3 [1980] 9/14; J. Struct. Chem. [USSR] **21** [1980] 262/6).
[7] Smolyar, A. E.; Charkin, O. P.; Klimenko, N. M. (Zh. Strukt. Khim. **15** [1974] 993/1003; J. Struct. Chem. [USSR] **15** [1974] 885/93).
[8] Wiebenga, E. H. (Stereochim. Inorg. Accad. Nazl. Lincei 9th Corso Estivo Chim., Rome 1965 [1967], pp. 319/31; C.A. **71** [1969] No. 128926), Wiebenga, E. H.; Kracht, D. (Inorg. Chem. **8** [1969] 738/46).

10.10 Bromine Trifluoride, BrF_3

CAS Registry Number: *[7787-71-5]*

Bromine trifluoride at ambient temperature is a colorless to straw-colored liquid. It is highly toxic and corrosive and reacts violently with many inorganic and organic substances, thereby setting free toxic fumes.

BrF_3 is prepared from liquid Br_2 and gaseous F_2 at ambient temperature. Autoionization in liquid BrF_3 yields BrF_2^+ and BrF_4^-. The oxidizing and fluorinating properties of BrF_3 make it a convenient reactant for the preparation of inorganic fluorides. Early work on BrF_3 is described in "Brom" 1931, pp. 337/8.

Some characteristic properties are:

molecular weight	136.91
melting point	281.9 K
boiling point	399.0 K
triple point temperature	281.9 K
vapor pressure	17.67 Torr (311.88 K)
standard enthalpy of formation (l)	−314 kJ/mol (298.16 K, 1 atm)
density (l)	2.80 g/cm^3 (298 K)
dipole moment (g)	1.19 D
dielectric constant (l)	106.8 (298 K)
viscosity (l)	2.219 cp (298 K)
electric conductivity (l)	8×10^{-3} S/cm (298 K)

10.10.1 Preparation. Formation

For the preparation of BrF_3, the reaction of elemental bromine and fluorine is usually employed. The batchwise method, bubbling gaseous fluorine through liquid bromine, is used on a laboratory scale, while continuous production on a larger scale is carried out with gaseous reactants.

10.10.1.1 Oxidation of Bromine by Fluorine

The simplest method of BrF_3 preparation is to bubble dry fluorine through dry liquid bromine in a copper flask. The completion of the reaction is indicated by the escape of fluorine from the exit tube [1]. By using a transparent Hostaflon or Pyrex glass reaction vessel, the end of the exothermal reaction can be observed by a brightening of the bromine color to yellow [2 to 4]. Attack of formed BrF_3 on the reaction vessel is reduced by cooling it to or below ambient temperatures, if a Pyrex glass apparatus is used [4 to 6]. The escape of bromine during the reaction is prevented by a condenser kept at 195 K [4]. Very pure BrF_3 is isolated from the reaction of bromine suspended in CCl_3F at 233 K and a cooled, equimolar mixture of fluorine and nitrogen. The color of the formed solid turns to light yellow at the end of the reaction. The solid BrF_3 is filtered and dried at 195 K after driving off excess fluorine with a stream of nitrogen [7]; see also [8]. The low reaction temperature prevents formation of BrF_5 and attack of BrF_3 on the glass apparatus [7].

For the continuous production of BrF_3 from the elements a slant iron condenser kept at 350 K is used. Streams of fluorine and liquid bromine come into contact at the upper end of the condenser. In order to produce nearly colorless liquid BrF_3 bromine is added at a sufficiently slow rate. A reflux condenser kept at 255 K is used to purge BrF_3 and BrF_5 from the escaping F_2 gas. The BrF_3 yield is 90% with respect to fluorine, and the product contains 2% BrF_5 [8, 9]. The short time required for the reaction is attributed to intermediate formation of BrF_5, which reacts with bromine to give BrF_3 [9]. The reaction of gaseous bromine and fluorine in a heated copper column was used in an experimental plant for BrF_3 production. The liquid product is collected in a Teflon vessel, gaseous BrF_3 is scrubbed from the escaping gas in a copper distillation and reflux column packed with copper rings [10]. To prepare BrF_3 at temperatures below the boiling point of bromine (322 K), a nitrogen stream saturated with bromine is mixed with fluorine in a T-shaped copper pipe [11]. BrF_3 was also prepared at 290 to 320 K in a water-cooled, tubular copper reactor which was inclined at a slight angle from the horizontal. Liquid bromine is dropped into the upper end of the reactor and fluorine is introduced near the lower end from which BrF_3 is withdrawn [12].

The reaction of equimolar amounts of fluorine and bromine at 298 K yields BrF besides BrF_3 [13]. The concentration of BrF decreases with increasing temperature [13], while excess fluorine enhances formation of BrF_5 [14]. The reaction of bromine and excess fluorine yields about 30% of BrF_3 in BrF_5 at 315 to 390 K. The BrF_3 concentration decreases linearly with temperature and reaches about 10% at 500 K [15]. At 570 K, formed BrF_5 contains only traces of BrF_3 [16].

The reaction with added bromine to give BrF_3 was proposed to remove excess fluorine from freshly produced UF_6 [17]. The bromine-fluorine reaction to BrF_3 at less than 300 K was proposed as an analytical method to determine fluorine or bromine [18].

Additional information on the Br_2-F_2 reaction system with formation of BrF_n (n=1, 3, 5) can be found in "Fluorine" Suppl. Vol. 2, 1980, pp. 143/4. Subsequent results are given in "Bromine" Suppl. Vol. A, 1985, pp. 373/4.

References:

[1] Simons, J. H. (Inorg. Syn. **3** [1950] 184/6).
[2] Bouy, P. (Ann. Chim. [Paris] [13] **4** [1959] 853/90).
[3] Martin, D. (Rev. Chim. Miner. **4** [1967] 367/97).
[4] Slivnik, J.; Zemljic A. (Vestn. Sloven. Kem. Drustva **9** [1962] 57/9).
[5] DeKock, R. L.; Higginson, B. R.; Lloyd, D. R.; Breeze, A.; Cruickshank, D. W. J.; Armstrong, D. R. (Mol. Phys. **24** [1972] 1059/72).
[6] Hepworth, M. A.; Peacock, R. D.; Robinson, P. L. (J. Chem. Soc. **1954** 1197/1201).
[7] Lehmann, E.; Naumann, D.; Schmeisser, M. (Z. Anorg. Allgem. Chem. **338** [1972] 1/3).
[8] Kwasnik, W. (in: Brauer, G.; Handbuch der Präparativen Anorganischen Chemie, Vol. 1, Enke, Stuttgart 1975, pp. 169/70).
[9] Kwasnik, W. (Ger. Appl. 76585, ref. 3 in: Schwarz, R.; FIAT Rev. Ger. Sci. **23** I [1949] 167/74, 168).
[10] Camozzo, G.; Pizzini, S. (Energia Nucl. [Milan] **7** [1960] 849/61).

[11] Ruff, O.; Braida, A. (Z. Anorg. Allgem. Chem. **206** [1932] 59/64).
[12] Swinehart, C. F. (private communication, ref. 33 in: Booth, H. S.; Pinkston, J. T.; Fluorine Chem. **1** [1950] 189/200, 191).
[13] Sukhoverkhov, V. F.; Podzolko, L. G. (Zh. Analit. Khim. **38** [1983] 715/21; J. Anal. Chem. [USSR] **38** [1983] 551/7).
[14] Meinert, H.; Groß, U. (Z. Chem. [Leipzig] **9** [1969] 455/6).
[15] Iwasaki, M.; Yawata, T.; Suzuki, K.; Tsujimura, S.; Oshima, K. (Nippon Kagaku Zasshi **83** [1962] 36/9, A3).
[16] Banks, A. A.; Maddock, J. J. (J. Chem. Soc. **1955** 2779/81).
[17] Nakumura, J.; Takeuchi, T.; Hayata, T.; Kanto Renko Kogyo Co., Ltd. (Japan. Kokai 74-120 897 [1974] 1/3 from C.A. **82** [1975] No. 173155).
[18] Sheft, J.; Hyman, H. H.; Katz, J. J. (Anal. Chem. **25** [1953] 1877/9).

10.10.1.2 Oxidation of Bromine and Bromide by Fluorine Compounds

Synthesis of BrF_3 from bromine by oxidation with ClF_3 instead of fluorine was used to avoid handling problems with elemental fluorine. Gaseous ClF_3 is passed into bromine in a cooled nickel [1] or quartz vessel [2]. Scrubbing of bromine from the escaping gas by a nickel reflux condenser raises the yield of BrF_3 from about 85 to 100% with respect to Br_2 [3]. The end of the reaction $2\,ClF_3 + Br_2 \rightarrow 2\,BrF_3 + Cl_2$ [4] is indicated by a decoloration of the solution [1, 3]. An excess of ClF_3 is to be avoided because ClF_3 distills with the formed BrF_3 [5] and has to be driven out by a stream of nitrogen before BrF_3 distillation [3]. Alternatively, the reaction can be stopped, when the solution turns transparent, and excess bromine is flushed out with nitrogen at ambient temperature [2]. Products formed besides BrF_3 are chlorine and BrCl which are removed from the bromine-BrF_3 solution by heating until bromine begins to distill [6]. Earlier results on the reaction of bromine and ClF_3 are given in "Chlor" Suppl. Vol. B 2, 1969, p. 565.

Reaction of bromine and ClF at ambient temperature yields a white smoke of BrF_3 [7]. The oxidation of bromine by BrF_5 with formation of BrF_3 and BrF is incomplete at room temperature [8]. The reaction rate is moderate at 420 K and rapid at 570 K [9]. Platinum dissolves in a 1:2 molar mixture of bromine and BrF_5 to form $(BrF_2)_2PtF_6$ in addition to BrF_3 at 313 K [10]. Oxidation of bromine by IF_5 yields BrF_3 and IBr on heating, while IF_7 does not react; see "Jod" 1933, pp. 601, 603. However, the reaction of bromine and IF_5 in the sample manifold of a mass spectrometer yielded only IBr rather than BrF_3 in detectable amounts [11]. A stoichiometric mixture of bromine and OF_2 yields BrF_3 and oxygen upon ignition, see "Fluorine" Suppl. Vol. 4, 1986, p. 50. Reaction of bromine and F_5SOF [12] or F_5SeOF [13] at room temperature yields BrF_3 and SOF_4 or F_5SeOBr, respectively. Fluorination of bromine by NF_3 in an electric discharge gives BrF_3 and BrF_5 as the principal products [14]; see also "Fluorine" Suppl. Vol. 4, 1986, p. 214. The photochemical reaction of XeF_2 in bromine slowly and quantitatively yields BrF_3 and xenon [15].

BiF_5 fluorinates bromine with formation of BrF_3 and BiF_3 at 450 K in a moderately fast reaction with a BrF_3 yield of about 50% [16]. The reaction of gaseous bromine and MF_6 (M = Np, Pu) yields BrF_3 and MF_4 at 350 K; see "Transurane" C, 1972, p. 113. BrF_3 is formed by fluorination of NaBr with a twofold molar excess of XeF_2 at 520 K in an autoclave. Additional products are $NaBrF_4 \cdot 2\,NaF$ and Br_2 [17].

References:

[1] Haszeldine, R. N. (J. Chem. Soc. **1950** 3037/41).
[2] Lange, G.; Dehnicke, K. (Naturwissenschaften **53** [1966] 38).

[3] Mit'kin, V. N.; Zemskov, S. V. (Zh. Prikl. Khim. **54** [1981] 2180/6; J. Appl. Chem. [USSR] **54** [1981] 1913/8).
[4] Nikolaev, N. S.; Sukhoverkhov, V. F. (Bul. Inst. Politeh. Iasi [2] **3** [1957] 61/6).
[5] Musgrave, W. K. R. (Advan. Fluorine Chem. **1** [1960] 1/28, 11/2).
[6] Davis, R. A.; Larsen, E. R. (J. Org. Chem. **32** [1967] 3478/81).
[7] Ruff, O.; Ascher, E. (Z. Anorg. Allgem. Chem. **176** [1928] 258/70, 267).
[8] Ruff, O.; Braida, A. (Z. Anorg. Allgem. Chem. **214** [1933] 81/90).
[9] Stein, L. (J. Am. Chem. Soc. **81** [1959] 1273/6).
[10] Chernyaev, I. I.; Nikolaev, N. S.; Ippolitov, E. G. (Dokl. Akad. Nauk SSSR **130** [1960] 1041/3; Proc. Acad. Sci. USSR Chem. Sect. **130/135** [1960] 167/9).

[11] Irsa, A. P.; Friedman, L. (J. Inorg. Nucl. Chem. **6** [1958] 77/90).
[12] Tattershall, B. W.; Cady, G. H. (J. Inorg. Nucl. Chem. **29** [1967] 3003/5).
[13] Seppelt, K. (Chem. Ber. **106** [1972] 157/64).
[14] Nikitin, I. V.; Rosolovskii, V. Ya. (Zh. Neorg. Khim. **20** [1975] 263/4; Russ. J. Inorg. Chem. **20** [1975] 143/4).
[15] Meinert, H.; Groß, U. (Z. Chem. [Leipzig] **8** [1968] 343/4).
[16] Fischer, J.; Rudzitis, E. (J. Am. Chem. Soc. **81** [1959] 6375/7).
[17] Popov, A. I.; Kiselev, Yu. M.; Sukhoverkhov, V. F.; Chumaevskii, N. A.; Krasnyanskaya, O. A.; Sadikova, A. T. (Zh. Neorg. Khim. **32** [1987] 1007/12; Russ. J. Inorg. Chem. **32** [1987] 619/22).

10.10.1.3 Disproportionation of Bromine

Formation of BrF_3 by bromine disproportionation occurs in excess solution of AgF in HF at 370 K. The formed BrF_3 was separated from HF by adding NaF and KF and evaporating the BrF_3. The overall reaction is described by $2\,Br_2 + 3\,AgF \rightarrow BrF_3 + 3\,AgBr$. The disproportionation probably involves initial formation of BrF as an unstable intermediate.

Reference:

Russell, J. L.; Jache, A. W. (Inorg. Nucl. Chem. H.H. Hyman Mem. Vol. **1976** 81/3).

10.10.1.4 Reduction of BrF_5

Electrolytical reduction of BrF_5 yields BrF_3 and minor amounts of other products [1]. Partial hydrolysis of BrF_5 yields BrF_3, HF, and O_2 in bulk reactions [2, 3], in HF solution at 210 K [4], and probably also during heating with alkali fluorides containing water at temperatures as high as 670 K [5]. Likely intermediates of the hydrolysis are $BrOF_3$, BrO_2F [4], and BrO_2F_3 [3]. Nearly equimolar amounts of BrF_5 and $KBrO_3$ yield BrF_3, $KBrO_2F_2$, and oxygen at ambient temperature [6].

References:

[1] Meinert, H.; Groß, U. (Z. Chem. [Leipzig] **12** [1972] 150/1).
[2] Ustinov, V. I.; Sukhoverkhov, V. F.; Podzolko, L. G. (Zh. Fiz. Khim. **52** [1978] 610/4; Russ. J. Phys. Chem. **52** [1978] 344/7).
[3] Sloth, E. N.; Stein, L.; Williams, C. W. (J. Phys. Chem. **73** [1969] 278/80).
[4] Gillespie, R. J.; Spekkens, P. H. (J. Chem. Soc. Dalton Trans. **1977** 1539/46).
[5] Green, G. L.; Hunt, J. B.; Sutula, R. A. (J. Inorg. Nucl. Chem. **35** [1973] 4305/7).
[6] Tantot, G.; Bougon, R. (Compt. Rend. C **281** [1975] 271/3).

10.10.1.5 Thermal Decomposition of Bromine-Fluorine Compounds

Compounds yielding BrF_3 by thermal decomposition are compiled in Table 6. They are mainly of the general type $BrF_3 \cdot MF_n$ and are formulated with the ions BrF_2^+ or BrF_4^-, if they are identified as salts. None of the reactions was used preparatively. Decomposition of BrF_5 into BrF_3 and F_2 above 820 K is indicated by the corresponding free energy of formation [1].

Table 6
Formation of BrF_3 by Thermal Decomposition of Bromine-Fluorine Compounds.

compound	products other than BrF_3	remarks	Ref.
$BrF_3 \cdot BF_3$	BF_3	dissociation noticeable at 242 K, extensive dissociation at 298 K	[2, 3]
$BrF_3 \cdot 2\,BF_3$	BF_3	dissociation pressure $\geq$4 Torr at 193 K	[2]
$(BrF_2)_2GeF_6$	GeF_4	dissociation pressure $\sim$4 Torr at room temperature	[4]
BrF_2AsF_6	AsF_5	dissociation pressure $\sim$2 Torr at room temperature	[4]
BrF_2SbF_6	SbF_5	dissociation pressure 40 and 650 Torr at 498 and 623 K	[5]
BrF_2MF_6 (M = Bi, Nb, Ta)	MF_5	educt thermal stability decreases in the order Ta > Nb > Bi	[6]
BrF_2RuF_6	RuF_5	decomposition in vacuum at 390 K	[7]
BrF_2PdF_4	PdF_3	decomposition at 450 K	[8]
$(BrF_2)_2PtF_6$	PtF_4	decomposition at 470 K	[8, 9]
BrF_2AuF_4	AuF_3	rapid decomposition at 450 K	[10]
$MBrF_4$ (M = Na, K, Rb, Cs)	MF	rapid evolution of BrF_3 at 590, 750, 870, and 890 K for M = Na, K, Rb, and Cs; the results agree with the qualitative observations in [12, 13]; lower decomposition temperatures [5, 14] are probably due to pyrohydrolysis	[11]
NO_2BrF_4	NO_2F	dissociation pressure $\sim$2 Torr at room temperature	[15]
MBr_3F_{10} (M = Rb, Cs)	MBr_2F_7	by pumping on the melt at 388 to 403 K; the product contains $MBrF_4$ for M = Rb	[16]

Table 6 (continued)

compound	products other than BrF_3	remarks	Ref.
$BrF_3 \cdot 3\,NaF$		decomposition at 510 to 540 K	[11]
$BrOF_3$	O_2	decomposition at room temperature	[17]
BrO_2F	Br_2, O_2	slow decomposition at room temperature, accelerated in HF solution	[19]
		violent decomposition at 329 K	[18]
$BrO_2Sb_2F_{11} \cdot 0.24\,SbF_5$ and excess NaF	Br_2, O_2, $NaSbF_4$	vacuum pyrolysis at 520 K	[20]

References:

[1] Stein, L. (J. Phys. Chem. **66** [1962] 288/91).
[2] Christe, K. O. (J. Phys. Chem. **73** [1969] 2792/3).
[3] Cyr, T.; Brownstein, S. (J. Inorg. Nucl. Chem. **39** [1977] 2143/5).
[4] Christe, K. O.; Schack, C. J. (Inorg. Chem. **9** [1970] 2296/9).
[5] Sheft, I.; Martin, A. F.; Katz, J. J. (J. Am. Chem. Soc. **78** [1956] 1557/9).
[6] Gutmann, V.; Emeléus, H. J. (J. Chem. Soc. **1950** 1046/50).
[7] Hepworth, M. A.; Peacock, R. D.; Robinson, P. L. (J. Chem. Soc. **1954** 1197/201).
[8] Sharpe, A. G. (J. Chem. Soc. **1950** 3444/50).
[9] Chernyaev, I. I.; Nikolaev, N. S.; Ippolitov, E. G. (Dokl. Akad. Nauk SSSR **130** [1960] 1041/3; Proc. Acad. Sci. USSR Chem. Sect. **130/135** [1960] 167/9).
[10] Sharpe, A. G. (J. Chem. Soc. **1949** 2901/2).

[11] Popov, A. I.; Kiselev, Yu. M.; Sukhoverkhov, V. F.; Chumaevskii, N. A.; Krasnyanskaya, O. A.; Sadikova, A. T. (Zh. Neorg. Khim. **32** [1987] 1007/12; Russ. J. Inorg. Chem. **32** [1987] 619/22).
[12] Asprey, L. B.; Margrave, J. L.; Silverthorn, M. E. (J. Am. Chem. Soc. **83** [1961] 2955/6).
[13] Goldberg, G.; Meyer, A. S., Jr.; White, J. C. (Anal. Chem. **32** [1960] 314/7).
[14] Sharpe, A. G.; Emeléus, H. J. (J. Chem. Soc. **1948** 2135/8).
[15] Christe, K. O.; Schack, C. J. (Inorg. Chem. **9** [1970] 1852/8).
[16] Stein, L. (J. Fluorine Chem. **27** [1985] 249/56).
[17] Christe, K. O.; Curtis, E. C.; Bougon, R. (Inorg. Chem. **17** [1978] 1533/9).
[18] Schmeisser, M.; Pammer, E. (Angew. Chem. **69** [1957] 781).
[19] Gillespie, R. J.; Spekkens, P. H. (J. Chem. Soc. Dalton Trans. **1977** 1539/46).
[20] Jacob, E. (Angew. Chem. **88** [1976] 189/90).

10.10.1.6 "Neutralization" Reaction

BrF_3 is formed in "neutralization" reactions of the Lewis acid BrF_2^+ with "bases", or of the Lewis base BrF_4^- with "acids" or "acid anhydrides", e.g.,

$$BrF_2SbF_6 + KBrF_4 \rightarrow 2\,BrF_3 + KSbF_6$$

Individual reactions are given in Table 7.

Reactions of the strong Lewis acid BrF_2SbF_6 and the strong Lewis base $KBrF_4$ or $AgBrF_4$ proceed with a heat of neutralization of about −19 kJ/mol. Lower values are found in reactions of weaker acids or bases, e.g., BrF_2TaF_6, $(BrF_2)_2SnF_6$, $NaBrF_4$, and $Ba(BrF_4)_2$. Details are given in [1].

Table 7
Formation of BrF_3 by Neutralization Reactions.

reactants	products other than BrF_3	remarks	Ref.
BrF_2AuF_4, $AgBrF_4$	$AgAuF_4$	in BrF_3 solution	[2]
$BrF_2M'F_6$, $MBrF_4$	$MM'F_6$	in BrF_3 solution with M=K, Ag; M'=Bi, Nb, Ta	[3]
BrF_2^+, MBr	Br_2	equimolar amounts of BrF_3 and Br_2, fast reaction for M=Na, K and slow reaction for M=Rb, Cs	[4]
$(BrF_2)_2GeF_6$, SeF_4			[5]
BrF_2PdF_4, KF	Br_2, K_2PdF_6		[6]
BrF_2PdF_4, SeF_4	Br_2, $PdF_4 \cdot 2\,SeF_4$	with Pd oxidation in boiling SeF_4	[7]
$(BrF_2)_2PtF_6$, SeF_4		during heating	[8]
BrF_2AuF_4, SeF_4		during heating	[8]
BrF_2AuF_4, XeF_6	XeF_5AuF_4	during melting	[9]
$CsBrF_4$, HF	HF_2^-	equilibrium $BrF_4^- + HF \rightleftharpoons BrF_3 + HF_2^-$ with an equilibrium constant of 0.45 (at room temperature?)	[10]
NO_2BrF_4, SiF_4		during warming to room temperature	[11]
$NOBrF_4$, MF_n	$(NO)_2MF_6$	with n=4 for M=Si, Ge, Sn, Ti and n=3 for M=Mn; reactant ratio 2:1 in BrF_3 solution (Ti, Mn, Sn) or in the gas phase (Si, Ge)	[12]
$MBrF_4$, UF_4		in BrF_3 solution, the composition of the formed fluorouranates varies with M=Na, K, Rb, or Cs	[4, 13]

References:

[1] Richards, G. W.; Woolf, A. A. (J. Fluorine Chem. **1** [1971/72] 129/39).
[2] Sharpe, A. G. (J. Chem. Soc. **1949** 2901/2).
[3] Gutmann, V.; Emeléus, H. J. (J. Chem. Soc. **1950** 1046/50).
[4] Martin, D. (Rev. Chim. Miner. **4** [1967] 367/97).
[5] Woolf, A. A. (Advan. Inorg. Chem. Radiochem. **9** [1966] 217/314, 264).
[6] Bartlett, N.; Lohmann, D. H. (J. Chem. Soc. **1964** 619/26).
[7] Bartlett, N.; Quail, J. W. (J. Chem. Soc. **1961** 3728/32).
[8] Bartlett, N.; Robinson, P. L. (J. Chem. Soc. **1961** 3417).
[9] Lutar, K.; Jesih, A.; Leban, I.; Zemva, B.; Bartlett, N. (Inorg. Chem. **28** [1989] 3467/71).
[10] Surles, T.; Quarterman, L. A.; Hyman, H. H. (J. Fluorine Chem. **3** [1973/74] 293/306).
[11] Christe, K. O.; Schack, C. J. (Inorg. Chem. **9** [1970] 1852/8).
[12] Bouy, P. (Ann. Chim. [Paris] [13] **4** [1959] 853/90).
[13] Chretien, A.; Martin, P. (Compt. Rend. C **263** [1966] 235/8).

10.10.2 Purification

Principal impurities in freshly prepared or commercial samples of BrF_3 include the starting materials fluorine and bromine (possibly containing chlorine), BrF and BrF_5, and HF and bromine oxyfluorides formed by contact of BrF_3 with moisture. All of these impurities with the exception of the bromine oxyfluorides are more volatile than BrF_3 [1].

The distillative purification of BrF_3 at temperatures below 303 K can be carried out in quartz [2], Pyrex [3], or Vycor glass stills [4]. At higher temperatures, a nickel still with a column packed with nickel helices [5], a steel apparatus [6], an aluminium still [7], or a Fluorothene still and a column with Monel packing were used [8].

Separation of BrF_3 from HF by absorption of HF by anhydrous NaF and KF was described [9]. However, the removal of HF from BrF_3 by contact with NaF is questionable because of a possible reaction of BrF_3 and NaF [10]. Pure solid BrF_3 is colorless [11, 12]; the liquid is colorless [11] to yellow [12].

References:

[1] Leech, H. R. (in: Mellor, J. W.; A Comprehensive Treatise on Inorganic and Theoretical Chemistry, Suppl. Vol. II, Pt. I, Longmans, London 1956, pp. 147/81, 160/1).
[2] Ruff, O.; Braida, A. (Z. Anorg. Allgem. Chem. **206** [1932] 59/64).
[3] Slivnik, J.; Zemljic, A. (Vestn. Sloven. Kem. Drustva **9** [1962] 57/9).
[4] Popov, A. I.; Glockler, G. (J. Am. Chem. Soc. **74** [1952] 1357/8).
[5] Stein, L.; Vogel, R. C.; Ludewig, W. H. (J. Am. Chem. Soc. **76** [1954] 4287/9).
[6] Sheft, I.; Hyman, H. H.; Katz, J. J. (J. Am. Chem. Soc. **75** [1953] 5221/3).
[7] Yosim, S. J. (J. Phys. Chem. **62** [1958] 1596/7).
[8] Rogers, M. T.; Garver, E. E. (J. Phys. Chem. **62** [1958] 952/4).
[9] Russell, J. L.; Jache, A. W. (Inorg. Nucl. Chem. H. H. Hyman Mem. Vol. **1976** 81/3).
[10] Muetterties, E. L.; Tullock, C. W. (Prep. Inorg. React. **2** [1965] 237/99, 270/1).
[11] Kwasnik, W. (in: Brauer, G.; Handbuch der Präparativen Anorganischen Chemie, Vol. 1, Enke, Stuttgart 1975, pp. 169/70).
[12] Christe, K. O.; Schack, C. J. (Inorg. Chem. **9** [1970] 1852/8).

10.10.3 Toxicity

BrF_3 is very toxic and corrosive and reacts violently with many organic and inorganic compounds, e.g., water [1, 2]. No toxicity (LD_{50}) data on BrF_3 have been published, but its high reactivity makes it highly toxic and damaging to biological tissues [3, 4]. The treshold limiting value (TLV) for BrF_3 in air is 0.1 ppm. Reactions of BrF_3 liberate very toxic fumes [1]. First aid requirements are those used for HF, which forms by hydrolysis [5].

References:

[1] Sax, N. J. (Dangerous Properties of Industrial Materials, 6th Ed., Van Nostrand Reinhold, New York 1984, pp. 519/20).
[2] Gerhartz, W. (Ullmann's Encycl. Ind. Chem. 5th Ed. A **4** [1985] 425).
[3] Stokinger, H. E. (Patty's Ind. Hyg. Toxicol. 3rd Revis. Ed. B **2** [1981] 2937/3043, 2971).
[4] Kwasnik, W. (in: Brauer, G.; Handbuch der Präparativen Anorganischen Chemie, Vol. 1, Enke, Stuttgart 1975, pp. 169/70).

[5] Kessie, R. W.; Lawroski, S.; Levenson, M.; Lumatainen, R. C.; Mecham, W. J.; Rodger, W. A.; Seefeldt, W. B.; Vogel, G. J.; Goring, G. E. (TID-7534 [1957] 576/613, 607/9; C.A. **1958** 907).

10.10.4 Handling. Storage

Personnel who are exposed to unknown concentrations should be adequately protected because of the toxicity and corrosivity of BrF_3 [1], see above. Only CO_2 fire extinguishers should be used in fires involving BrF_3 [2]. "Oxidizer" and "Poison" labels are required for rail shipment [3] in specified cylinders [1], see also [4].

Laboratory techniques to manipulate BrF_3 and other fluorides in spectroscopic investigations are described in [5]. Equipment fabrication, pilot plant techniques, BrF_3 handling and sampling are discussed in [6].

At moderate temperatures, BrF_3 can be handled in vessels made of Teflon, Kel-F, or Hostaflon [7, 8]. CaF_2 can be used as transparent containment material [7]. Reactions above 400 K require equipment made of appropriate metals. Nickel equipment is quite inert to BrF_3 [9, 10] at temperatures as high as 1000 K [11]. Slow reaction of purified liquid BrF_3 with preconditioned nickel results in a reddish impurity which may be BrF or bromine [12]. The maximum corrosion rate of nickel, Monel, and Inconel in BrF_3 at 398 K is less than 1.3×10^{-3} mm/year [13]. Handling of BrF_3 below 670 K is possible in copper vessels [11]. Copper alloys can also be used [14]. Iron [9] and iron-based alloys [14] resist boiling BrF_3. Mild steel resists BrF_3 up to about 520 K [11]; the rate of BrF_3 attack is several times that of nickel, but mild steel is an attractive material for economic reasons. Stainless steel is more resistant to BrF_3 than mild steel [7]. Aluminium corrodes slightly faster in BrF_3 at 398 K than nickel and its alloys [15]. Diverging results on the resistance of platinum to BrF_3 [10, 14] at moderate temperatures may be due to HF, since attack by BrF_3-HF is appreciable [16], see also p. 80. Pure BrF_3 attacks quartz very slowly at 298 K [17] and noticeably at 303 K [18]. Glass is not attacked appreciably within a short time when moisture is excluded rigorously [19]. Borosilicate glass also resists BrF_3 [20]. Compatibility data for storing and handling of BrF_3 with metals and nonmetals are summarized in [21].

The use of vacuum grease must be limited to the absolute minimum, because silicon grease [22] and even Kel-F grease react with liquid BrF_3. Gaseous BrF_3 does not seem to attack Kel-F grease [23]. A tapless dropping funnel especially suited for the addition of BrF_3 was described in [24]. The high reactivity of BrF_3 also limits the range of materials in spectroscopic investigations. An IR cell with diamond windows was described in [25]. Other window materials are sapphire [26], CaF_2 [27] and Irtran-2 [26], and prefluorinated polyethylene [28]. An UV spectrum was measured in a cell with windows made of polychlorotrifluoroethane [29].

Gas chromatographic separation of BrF_3 from nonmetal elements and compounds was carried out at 350 to 370 K in polytetrafluoroethene columns [30, 31] with a packing of the same material [31]. A gas-liquid chromatographic (GLC) separation of BrF_3 from BrF_5, UF_6, and bromine succeeded with a perfluoroalkane oil [32].

Purified BrF_3 can be stored for a limited time only, because it gradually decomposes [33]. Nickel [34] and Monel vessels [35] were occasionally prefluorinated before use [36]. Teflon vessels [37] and steel bottles with screw-on steel caps were also used [38]. Larger quantities of BrF_3 must be stored in a cool, ventilated area out of the direct sun exposure and away from combustible materials [1]. Smaller quantities were stored in a dry atmosphere [37] or under helium [36].

References:

[1] Sax, N. J. (Dangerous Properties of Industrial Materials, 6th Ed., Van Nostrand Reinhold, New York 1984, pp. 519/20).
[2] Kessie, R. W.; Lawroski, S.; Levenson, M.; Lumatainen, R. C.; Mecham, W. J.; Rodger, W. A.; Seefeldt, W. B.; Vogel, G. J.; Goring, G. E. (TID-7534 [1957] 576/613, 608; C.A. **1958** 907).
[3] Gerhartz, W. (Ullmann's Encycl. Ind. Chem. 5th Ed. A **4** [1985] 425).
[4] U.S. Dept. of Transportation (Fed. Regist. **50** [1985] 41092/7; C.A. **104** [1986] No. 115277).
[5] Canterford, J. H.; O'Donnell, T. A. (in: Jonassen, H. B.; Weissberger, A.; Technique of Inorganic Chemistry, Vol. 7, Wiley-Interscience, New York 1968, pp. 273/306).
[6] Kessie, R. W.; et al. (from [2, pp. 592/601]).
[7] Kessie, R. W.; et al. (from [2, pp. 583/5]).
[8] Dixon, K. R.; Sharp, D. W. A.; Sharpe, A. G. (Inorg. Syn. **12** [1970] 232/7).
[9] I.G. Farbenindustrie A.-G. (Fr. 803855 [1936] 2 pp.; C.A. **1937** 2760).
[10] Toy, M. S.; Cannon, W. A. (Electrochem. Technol. **4** [1966] 520/3).

[11] Swinehart, C. F. (private communication, ref. 33 in Booth, H. S.; Pinkston, J. T.; Fluorine Chem. **1** [1950] 189/200, 195/6).
[12] Claassen, H. H.; Weinstock, B.; Malm, J. G. (J. Chem. Phys. **28** [1958] 285/9).
[13] Kessie, R. W.; et al. (from [2, pp. 576/7]).
[14] Hise, D. R. (Process Ind. Corros. **1975** 240/6).
[15] Kessie, R. W.; et al. (from [2, pp. 576/7, 582/3]).
[16] Surles, T.; Hyman, H. H.; Quarterman, L. A.; Popov, A. I. (Inorg. Chem. **10** [1971] 611/3).
[17] Muetterties, E. L.; Phillips, W. D. (J. Am. Chem. Soc. **79** [1957] 322/6).
[18] Ruff, O.; Braida, A. (Z. Anorg. Allgem. Chem. **206** [1932] 59/64).
[19] Toy, M. S.; Cannon, W. A. (J. Phys. Chem. **70** [1966] 2241/4).
[20] Toy, M. S.; Cannon, W. A. (Advan. Chem. Ser. **54** [1965] 237/44).

[21] Boyd, W. K.; Berry, W. E.; White, E. L. (AD-613553 [1965] 45 pp.; C.A. **67** [1967] No. 23666).
[22] Gutmann, V. (Angew. Chem. **62** [1950] 312/5).
[23] Cleaver, B.; Condlyffe, D. H. (J. Chem. Soc. Faraday Trans. I **1985** 2453/64).
[24] Zakharov, L. N. (Khim. Zhizn **1983** No. 10, p. 60).
[25] Hyman, H. H.; Surles, T.; Quarterman, L. A.; Popov, A. I. (Appl. Spectrosc. **24** [1970] 464/5).
[26] Surles, T.; Hyman, H. H.; Quarterman, L. A.; Popov, A. I. (Inorg. Chem. **9** [1970] 2776/30).
[27] Haendler, H. M.; Bukata, S. W.; Millard, B.; Goodman, E. I.; Littman, J. (J. Chem. Phys. **22** [1954] 1939).
[28] Stein, L. (J. Am. Chem. Soc. **81** [1959] 1273/6).
[29] Katz, J. J.; Hyman, H. H. (Rev. Sci. Instrum. **24** [1953] 1066/7).
[30] Sukhoverkhov, V. F.; Podzolko, L. G. (Zh. Analit. Khim. **38** [1983] 715/21; J. Anal. Chem. [USSR] **38** [1983] 551/7).

[31] Pervov, V. S.; Sukhoverkhov, V. F.; Podzolko, L. G. (Zh. Analit. Khim. **34** [1979] 2369/73; J. Anal. Chem. [USSR] **34** [1979] 1840/4).
[32] Pappas, W. S.; Million, J. G. (Anal. Chem. **40** [1968] 2176/80).
[33] Sakurai, T.; Kobayashi, Y.; Iwasaki, M. (J. Nucl. Sci. Technol. **3** [1966] 10/3).
[34] Sheft, I.; Hyman, H. H.; Katz, J. J. (J. Am. Chem. Soc. **75** [1953] 5221/3).
[35] Shoolery, J. N.; Goodman, E. I.; Littman, J. (BNL-2382 [1955] 6 pp.).
[36] Long, R. D.; Martin, J. J.; Vogel, R. C. (Chem. Eng. Data Ser. **3** [1958] 28/34).

[37] Sukhoverkhov, V. F.; Takanova, N. D.; Uskova, A. A. (Zh. Neorg. Khim. **21** [1976] 2245/9; Russ. J. Inorg. Chem. **21** [1976] 1234/70).
[38] Sharpe, A. G.; Emeléus, H. J. (J. Chem. Soc. **1948** 2135/38).

10.10.5 Disposal

Decomposition of BrF_3 by slow addition to a large volume of CCl_4 is vigorous, but not dangerous. Very small quantities of BrF_3 left in reaction vessels may be destroyed by washing with CCl_4 [1]. Pure gaseous BrF_3 decomposes to bromine and SiF_4 when passed through a steel trap filled with silicon lumps [2].

Air is freed from BrF_3 vapor by passage through a trap filled with sodium lime or activated aluminia and a secondary scrubber unit containing an aqueous wash solution [3]. Hydrolysis of BrF_3 in air by aqueous KOH in a spray tower was also described [4]. The exit gas of uranium fluorination is scrubbed of excess BrF_3 by contact with anhydrous K_2CO_3 at ambient temperatures [5].

References:

[1] Dixon, K. R.; Sharp, D. W. A.; Sharpe, A. G. (Inorg. Syn. **12** [1970] 232/7).
[2] Czerepinski, R. G.; Margrave, J. L. (Inorg. Chem. **2** [1963] 875/6).
[3] Stein, L. (J. Inorg. Nucl. Chem. **35** [1973] 39/43).
[4] Kessie, R. W.; Lawroski, S.; Levenson, M.; Lumatainen, R. L.; Mecham, W. J.; Rodger, W. A.; Seefeldt, W. B.; Vogel, G. J.; Goring, G. E. (TID-7534 [1957] 576/631, 606/7; C.A. **1958** 907).
[5] Rosen, F. D. (NAA-SR-213 [1957] 34 pp.; N.S.A. **11** [1957] No. 9637).

10.10.6 Uses

An industrial use of BrF_3 does not seem to have been described in the literature. BrF_3 was proposed as an electrolyte in batteries [1 to 5], as a gaseous etching agent in semiconductor devices fabrication [6 to 8], as a hypergolic propellant in rocket motors [9 to 11] and explosive charges [12, 13], and as a reactant in underground cutting of pipes [14].

Treatment with BrF_3 removes Rn [15 to 17] or moisture from air [18]. Fluorination of U by BrF_3 yields UF_6 and is applied in the reprocessing of nuclear fuels; see "Uranium" Suppl. Vol. A 4, 1982, pp. 333/4. Reaction of BrF_3 with a perfluorinated solid copolymer yields a wax-type product of reduced molecular weight [19, 20]. Traces of BrF_3 catalyze the fluorination of coke to CF_4 [21] and prevent decomposition of UF_6 during isotopic enrichment [22].

Decomposition by BrF_3 was used to analyze normally unreactive compounds of Ru, Rh, Pd, Ir, Pt, and Au [23 to 26]. Quantitative liberation of O_2 from many metal and several nonmetal oxides by BrF_3 was described [27 to 29]. Traces of O_2 and N_2 in metals were also determined by this method [30, 31].

References:

[1] Park, K. H.; Miles, M. H.; Bliss, D. E.; Stilwell, P.; Hollins, R. A.; Rhein, R. A. (J. Electrochem. Soc. **135** [1988] 2901/2).

[2] Mellors, G. W.; Union Carbide Corp. (U.S. 4186248 [1980] 4 pp.; C.A. **92** [1980] No. 150128).
[3] Goodson, F. R.; Shipman, W. H.; McCartney, J. F.; U.S. Dept. of the Navy (U.S. Appl. 871877 [1978] 7 pp. from C.A. **90** [1979] No. 124642).
[4] Toy, M. S.; Cannon, W. A. (Electrochem. Technol. **4** [1966] 520/3).
[5] Toy, M. S.; Cannon, W. A.; Berger, C.; U.S. Dept. of the Army (U.S. 3427207 [1969] 4 pp.; C.A. **70** [1969] No. 83640).
[6] Ibbotson, D. E.; Mucha, J. A.; Flamm, D. L.; Cook, J. M. (J. Appl. Phys. **56** [1984] 2939/42).
[7] Cook, J. M.; Donnelly, V. M.; Flamm, D. L.; Ibbotson, D. E.; Mucha, J. A.; American Telephone and Telegraph Co. (Ger. Offen. 3427599 [1985] 18 pp.; C.A. **102** [1985] No. 177633).
[8] Hitachi, Ltd. (Japan. Tokkyo Koho 60-12779 [1985] from C.A. **103** [1985] No. 133431).
[9] Ciepluch, C. C.; Allen, H., Jr.; Fletcher, E. A. (Am. Rocket Soc. J. **31** [1961] 514/8).
[10] Tarpley, W. B., Jr.; Aeroprojects Inc. (U.S. 3449178 [1969] 4 pp.; C.A. **71** [1969] No. 51823).

[11] Lewis, B.; Ethyl Corp. (U.S. 3177652 [1965] 5 pp.; C.A. **63** [1965] 5442).
[12] von Elbe, G.; McHale, E.T. (AD-A 069415 [1980] 27 pp.; C.A. **95** [1981] No. 206318).
[13] Tulis, A. J. (Proc. Intern. Pyrotech. Semin. **6** [1978] 615/33 from C.A. **93** [1980] No. 49736).
[14] Anonymous (Chem. Eng. News **33** [1955] 3395).
[15] Stein, L. (J. Inorg. Nucl. Chem. **35** [1973] 39/43).
[16] Stein, L. (Science **175** [1972] 1463/5).
[17] Stein, L. (Radiochim. Acta **32** [1983] 163/71).
[18] Schneider, H.; Staudt, A.; Siemens A.-G. (Ger. Offen. 3701977 [1987] 6 pp.; C.A. **110** [1989] No. 43667).
[19] Aramaki, M.; Sakayuchi, H.; Central Glass Co., Ltd. (Ger. Offen. 3540280 [1986] 14 pp.; C.A. **106** [1987] No. 177/23).
[20] Aramaki, M.; Nakano, H.; Kubo, M.; Central Glass Co., Ltd. (Ger. Offen. 3740565 [1988] 8 pp.; C.A. **109** [1988] No. 129903).

[21] Ranto Denka Kogyo Co., Ltd. (Japan. Kokai Tokkyo Koho 58-162536 [1983] 6 pp. from C.A. **100** [1984] No. 51075).
[22] Bacher, W.; Bier, W.; Stober, S. (KfK-4031 [1986] 40 pp., pp. 16/20, 27; C.A. **104** [1986] No. 214873).
[23] Benediktov, A. B.; Mit'kin, V. N.; Belyaev, A. V.; Zemskov, S. V. (Zh. Analit. Khim. **40** [1985] 1286/9; J. Anal. Chem. [USSR] **40** [1985] 1018/21).
[24] Mit'kin, V. N.; Morozova, N. B.; Isakova, V. G.; Zemskov, S. V. (Izv. Sibirsk. Otd. Akad. Nauk SSSR Ser. Khim. Nauk **1988** No. 3, pp. 78/81).
[25] Zemskov, S. V.; Mit'kin, V. N.; Torgov, V. G.; Glinskaya, A. N. (Zh. Analit. Khim. **38** [1983] 38/41; J. Anal. Chem. [USSR] **38** [1983] 30/3).
[26] Mit'kin, V. N.; Vasil'eva, A. A.; Korda, T. M.; Zemskov, S. V.; Torgov, V. G.; Tatarchuk, A. N. (Zh. Analit. Khim. **44** [1989] 1589/93).
[27] Emeléus, H. J.; Woolf, A. A. (J. Chem. Soc. **1950** 164/8).
[28] Hoekstra, H. R.; Katz, J. J. (Anal. Chem. **25** [1953] 1608/12).
[29] Woolf, A. A. (J. Labelled Compounds Radiopharm. **27** [1989] 17/22).
[30] Dupraw, W. A.; O'Neill, H. J. (Anal. Chem. **31** [1959] 1104/7).

[31] Kirtchik, H.; Culp, S. (NASA-CR-57951 [1965] 81 pp., pp. 48/60, 67/9; C.A. **65** [1966] 11339).

10.10.7 Molecular Properties and Spectra

Electronic Structure

The BrF_3 molecule is usually described as a trigonal bipyramid with the F atoms at two axial (ax) and one equatorial (eq) positions and the two lone electron pairs of bromine at the remaining two equatorial positions. Repulsion by the lone electron pairs of bromine leads to a narrowing of the expected $F_{ax}BrF_{eq}$ angle of 90° and to a shortening of the equatorial BrF bond compared to the axial BrF bonds [1, 2], as was confirmed by experiment (see below). Semiempirical quantum chemical calculations indicate considerably weaker axial BrF bonds and a more ionic character of the equatorial BrF bond [3]. An ionic character of the bonds in BrF_3 (49%) also results from the analysis of the nuclear quadrupole coupling constant according to Townes and Dailey's theory [4]. The interpretation due to a modified theory indicates a significant d-orbital participation in the hybridization of the Br atom [5]. For bond polarities determined from atomic constants, see [6].

Molecular Structure

The nearly T-shaped BrF_3 molecule has C_{2v} symmetry in the gas phase and solid state. The results of an electron-diffraction investigation in the gas phase are in excellent agreement [7, 8] with earlier results of a microwave investigation [9]. Given are also X-ray results on solid BrF_3 [10]. Internuclear distances are in Å and angles in degrees:

method	r(Br-F_{eq})	r(Br-F_{ax})	∢(F_{ax}-Br-F_{eq})	Ref.
microwave study	1.721	1.810	86° 12.6'	[9]
electron diffraction	1.728(15)	1.809(17)	85(2)°	[7, 8]
X-ray investigation	1.72	1.84, 1.85	82.0°, 88.4°	[10]

Barrier to Intramolecular Fluorine Exchange

The barrier to intramolecular fluorine exchange by pseudorotation is small in liquid BrF_3. A value of about 16 kJ/mol was estimated from NMR spectra. Small anharmonic effects due to fluorine exchange in electron-diffraction studies were taken as an indication for a high barrier in the gas phase [7]. BrF_3 can be expected to have a smaller barrier to pseudorotation than ClF_3, where the calculated ab initio value is about 146 kJ/mol [11].

Ionization Potentials

The He I photoelectron (PE) spectrum of BrF_3 yields eight vertical ionization potentials E_i(vert) in eV:

E_i(vert)	12.38(5)	13.94(2)	14.60(4)	15.05(3)	15.61(3)	16.67(3)	17.59(2)	18.76(4)	[12]
assignment ..	$6b_2$	$15a_1$	$8b_1$, $14a_1$	$2a_2$	$5b_2$	$7b_1$	$4b_2$, $6b_1$	$13a_1$	[3,13]

The PE bands were assigned using the discrete-variational Xα (DMV Xα) method with a HF basis including 4d functions [13] and the self-consistent charge Xα (SCC Xα) method. Both calculations are in excellent agreement [3]. They suggest a correction of the original assignment which had been obtained in analogy to ClF_3 [12].

The adiabatic ionization potential of 12.15(4) eV from PE spectra agrees fairly well [12] with the value of 12.9(3) eV from electron impact mass spectrometry [14]. Additional adiabatic ionization potentials from PE spectra are 13.58(1) (2nd E_i) and 16.26(3) eV (6th E_i) [12].

Dipole Moment. Polarizability

A dipole moment of 1.0 D was estimated from the Stark effect in two transitions of the microwave spectrum [9]. A value of 1.19(12) D was calculated from the dielectric constants measured for gaseous BrF_3 [15].

The δ-function potential model yields an average polarizability $\bar{\alpha}=5.189\times10^{-24}$ cm^3 for BrF_3 [16].

Nuclear Quadrupole Coupling

Measurement of the fluorine quadrupole coupling constants in crystalline BrF_3 by the time-differential, perturbed-angular-distribution method after F excitation to the J=5/2 state yielded the modulation amplitude A_{22} and the resonance frequency n_Q [17]:

T in K	A_{22} in %	n_Q in MHz
10	2.0(2)	71.5(3)
80	1.1(3)	71.8(5)

An earlier publication lists two interaction frequencies obtained by the same method [18].

Bromine quadrupole coupling coefficients were determined from the microwave spectrum; values in MHz are [9]:

molecule	eQV_{aa}	eQV_{bb}	eQV_{cc}
$^{79}BrF_3$	607.57	501.78	−1109.35
$^{81}BrF_3$	506.13	419.21	−925.34

Constants of Molecular Rotation and Vibration

Rotational Constants. Moments of Inertia. Rotational transitions within the BrF_3 vibrational ground state were analyzed. BrF_3 is an asymmetric top with the asymmetry parameters $\kappa=-0.71609$ for $^{79}BrF_3$ and $\kappa=-0.71438$ for $^{81}BrF_3$. Rotational constants A_o, B_o, C_o in MHz, moments of inertia I and the inertial defect $\Delta=I_c-I_a-I_b$ in amu·Å^2 are [9]:

molecule	A_o	B_o	C_o	I_a	I_b	I_c	Δ
$^{79}BrF_3$	10841.25	4077.57	2958.59	46.626	123.966	170.852	0.260
$^{81}BrF_3$	10806.99	4077.21	2956.01	46.744	123.977	171.001	0.250

An inertial defect of $\Delta=0.22\pm0.05$ amu·Å^2, which was calculated from the fundamental vibrations in [19] by a general formula, agrees with the experimental value and confirms the planar structure of BrF_3 [20].

Centrifugal Distortion. Centrifugal distortion constants were calculated from structural parameters and force constants by Cyvin's method. Values in kHz are [21]:

D_J	D_K	D_{JK}	$-R_5$	$-R_6$	δ_J
0.665	4.502	9.557	2.034	0.018	0.158

Fundamental Vibrations. There are six normal modes of vibration for the T-shaped BrF_3 molecule of C_{2v} symmetry: $\nu_1(A_1)$, $\nu_2(A_1)$, $\nu_3(A_1)$, $\nu_4(B_1)$, $\nu_5(B_1)$, and $\nu_6(B_2)$. All of them are infrared and Raman active. IR and Raman wave numbers are listed in Table 8, p. 62.

Coriolis Coupling Constants. Coriolis coupling constants calculated from force constants are [21]:

ζ^x_{16}	ζ^x_{26}	ζ^x_{36}	ζ^y_{14}	ζ^y_{15}	ζ^y_{24}	ζ^y_{25}	ζ^y_{34}	ζ^y_{35}	ζ^z_{46}	ζ^z_{56}
0.165	0.075	0.983	0.251	0.814	0.019	0.545	0.968	0.201	0.995	0.104

Mean Amplitudes of Vibration. Experimental mean square amplitudes of vibration are known from an electron-diffraction study of gaseous BrF_3 at 293 K [7, 8]; values calculated for a temperature of 298 K from force constants [21] are given for comparison (u in pm):

method	$u(Br\text{-}F_{eq})$	$u(Br\text{-}F_{ax})$	$u(F_{ax}\cdots F_{eq})$	$u(F_{ax}\cdots F_{ax})$
electron diffraction	4.8(1)	5.0(1)	9.2(1)	9.1(1)
calculation	4.197	4.600	8.345	6.132

Force Constants

The most recent normal coordinate analysis [21] of BrF_3 was performed by the kinetic constants method using the vibrational frequencies of [25]. The indices D, d, a, β, and γ refer to the $Br\text{-}F_{ax}$ bond, the $Br\text{-}F_{eq}$ bond, the $F_{ax}\text{-}Br\text{-}F_{ax}$ angle, the $F_{eq}\text{-}Br\text{-}F_{ax}$ angle, and the torsion. The potential constants in 10^2 N/m at the general quadratic force field are:

f_D	f_d	$-f_{Dd}$	f_{dd}	$f_{d\alpha}$	f_β	$-f_{D\alpha}$	$-f_{d\alpha}$	$-f_{\alpha\beta}$	f_γ
4.149	3.139	0.031	0.266	0.105	0.232	0.061	0.004	0.062	0.004

Earlier publications report calculations of force constants [22, 23, 29 to 31], mean amplitudes of vibration [29 to 34], Coriolis coupling constants [30, 31], and centrifuged distortion constants [31].

Bond Energy. Atomization Energy

An average Br-F bond energy in BrF_3 of 220.1 kJ/mol or more at 300 K was obtained, probably from thermodynamic data [35]. A value of 201.7 kJ/mol was also cited [36]. The heat of BrF_3 atomization of 768.2 kJ/mol was calculated from force constants [37] and is considerably higher than the experimental value of 604.2 kJ/mol cited here in [37], for example, in [38]. A value of 661 kJ/mol was cited in [39].

Quantum Chemical Calculations

An ab initio calculation within the Hartree-Fock approximation using small basis sets (3-21G, 3-21G$^{(*)}$) deals with the equilibrium geometry, the electric dipole moment, and the hydrogenation energy [40].

Table 8
Fundamental Vibrations of BrF_3.

phase	spectrum	fundamental vibrations in cm^{-1}						Ref.
		$\nu_1(A_1)$ $\nu_s(Br{-}F_{eq})$	$\nu_2(A_1)$ $\nu_s(F_{ax}{-}Br{-}F_{ax})$	$\nu_3(A_1)$ $\delta_s(F_{ax}{-}Br{-}F_{ax})$	$\nu_4(B_1)$ $\nu_{as}(F_{ax}{-}Br{-}F_{ax})$	$\nu_5(B_1)$ $\delta_{as}(F_{eq}{-}Br{-}F_{ax})$	$\nu_6(B_2)$ $\tau(F_{ax}{-}Br{-}F_{ax})$ out of plane	
Ar matrix,	IR	678.1, 676.5	547	240	599	347	251.5	[22]
5 to 25 K [a]		672.0 [b], 670.4 [b]		235 [b]	597 [b], 594 [b]			
Ar matrix, 4 K	IR	672 s	545 mw	235 mw	592 vs	346 mw	250 m	[23]
solid	IR	670	480 [c]	235	590, 570	335, 260	—	[22]
liquid	IR	674 s	528 vw	—	614 vs	344 w	269 m	[24]
vapor	IR	(682, 668) s	(557, 547) w	242 s	(621, 614, 604) vs	(359, 350, 342) vw	242 s	[25]
liquid	Raman	672 p, s	576.5 p, m	237 s	—	—	338	[26]
liquid	Raman	673	531	236	—	341	265	[24]
vapor, 398 K	Raman	675 p, s	552 p, vs	233 p, w [d]	612 vvw	—	—	[25]

[a] Isotopic splitting of $^{79}BrF_3$ and $^{81}BrF_3$ is 1.6 cm^{-1} in the ν_1 and ν_4 modes. Absorptions in Ne and N_2 matrices are also listed [22]. — [b] Second trapping site. — [c] Probably assignable to Br–F–Br stretching vibration of the dimer [27]. — [d] From liquid BrF_3 [28].

A number of semiempirical methods (SCC Xα [3], DVM Xα [13], AM1 [41], CNDO/2, INDO [42], NDDO-2(a, b) [43], HMO [44, 45]) was employed to calculate some molecular properties. Simple models were applied to interpret structural details [46, 47].

Spectra

Nuclear Magnetic Resonance. The ^{19}F NMR spectrum of BrF_3 consists of a singlet [48 to 51]. It arises from rapid fluorine exchange [49], which is possibly favored by the approximately equivalent environments of all Br atoms in the associated liquid with a coordination number close to four [52]. Reported chemical shifts (external references, negative high field shifts) are:

δ in ppm	$-23^{a)}$	$-16.5^{b)}$	$+54.3$	-46.1 ± 0.1	$+80$
reference	CCl_3F	CCl_3F	CF_3COOH	F_2	cyclo-C_4F_8
Ref.	[51]	[50]	[53, 54]	[55]	[48]

[a] At ambient temperature. — [b] In SO_2ClF solution below 253 K. The sharp signal broadens from 253 to 223 K, the chemical shift is quite independent of concentration and temperature [50].

The longitudinal relaxation $T_1 \approx 10^2$ s of solid BrF_3 depends only slightly on the temperature. The relaxation of the dipolar energy depends approximately on T^{-2} as usually found in the relaxation of quadrupolar nuclei. This indicates that the observed effect possibly results from relaxation of the central Br atom in BrF_3 [56].

Microwave Spectrum. Thirteen lines for each of the isotopic molecules $^{79}BrF_3$ and $^{81}BrF_3$ with J from 1 to 3 are known between 7.6 and 19.8 GHz [9].

Infrared and Raman Spectra. Recently measured fundamental vibrations of BrF_3 are given on p. 62. Earlier IR data for the gas phase are given in [19, 57, 58], an earlier Raman spectrum is described in [19].

Bands in the IR spectrum of liquid BrF_3 from 1154 to 1337 cm^{-1} [59] belong to an overtone region [24]. Weak bands of overtones and combinations were found between 795 and 1345 cm^{-1} in matrix-isolated BrF_3 [23]. Similar bands in BrF_3 vapor were observed from 1116 to 1340 cm^{-1} [25]. Raman spectra of solid BrF_3 retain their general appearance from 281 to 170 K. However, bands of external vibrations were observed and in some cases assigned tentatively [26].

Bands of dimeric and polymeric BrF_3 were identified from about 200 to 700 cm^{-1} in vibrational spectra of matrix-isolated [22, 23], liquid [19, 23, 24, 26], and gaseous samples [25] by dilution experiments and by comparison of BrF_3 spectra in different phases. The assignment to distinct associated species is not unequivocal. A tentative assignment of dimer bands based on C_{2h} symmetry is given in [22]. The dimerization probably involves the two equivalent Br-F_{ax} bands, while the unique Br-F_{eq} band is essentially unaltered [22, 23, 26]. In the liquid, the intensity of polymer bands decreases with increasing sample temperature [24, 28], while the intensity of bands assigned to dimers increases [24, 60].

Ultraviolet Spectra. The maximum of the UV absorption of gaseous BrF_3 shifts from 215 nm at 293±2 K [61] to 240 nm at 348 K [62]. Early measurements were made on the pressure dependence of the head of absorption [63].

The UV spectrum of BrF_3 in SO_2ClF has a weak maximum with a molar extinction coefficient of $\varepsilon = 10$ at 370 nm [50].

References:

[1] Pauling, L. (Die Natur der Chemischen Bindung, 3rd Ed., Verlag Chemie, Weinheim 1962, p. 174).
[2] Gillespie, R. J. (Can. J. Chem. **39** [1961] 318/23).
[3] Grodzicki, M.; Männing, V.; Trautwein, A. X.; Friedt, J. M. (J. Phys. B **20** [1987] 5595/625, 5606/12).
[4] Gupta, L. C. (Indian J. Pure Appl. Phys. **5** [1967] 437/8).
[5] Narain, P.; Chandra, S. (Indian J. Pure Appl. Phys. **13** [1975] 496/7).
[6] Lakatos, B. (Z. Elektrochem. **61** [1957] 944/9).
[7] Ishchenko, A. A.; Myakshin, I. N.; Romanov, G. V.; Spiridonov, V. P.; Sukhoverkhov, V. F. (Dokl. Akad. Nauk SSSR **267** [1982] 1143/6; Dokl. Chem. Proc. Acad. Sci. USSR **262/267** [1982] 994/6).
[8] Sukhoverkhov, V. F.; Myakshin, I. N.; Romanov, G. V.; Spiridonov, V. P.; Ishchenko, A. A. (Tez. Dokl. 6th Vses. Simp. Khim. Neorg. Ftoridov, Novosibirsk 1981, p. 58).
[9] Magnuson, D. W. (J. Chem. Phys. **27** [1957] 223/6).
[10] Burbank, R. D.; Bensey, F. N., Jr. (J. Chem. Phys. **27** [1957] 982/3).

[11] Charkin, O. P.; Bold'irev, A. I. (Itogi Nauki Tekh. Ser. Neorg. Khim. **8** [1980] 61).
[12] DeKock, R. L.; Higginson, B. R.; Lloyd, D. R.; Breeze, A.; Cruickshank, D. W. J.; Armstrong, D. R. (Mol. Phys. **24** [1972] 1059/72).
[13] Gutzev, G. L.; Smolyar, A. E. (Chem. Phys. Letters **71** [1980] 296/9).
[14] Irsa, A. P.; Friedman, L. (J. Inorg. Nucl. Chem. **6** [1958] 77/90).
[15] Rogers, M. T.; Pruett, R. D.; Speirs, J. L. (J. Am. Chem. Soc. **77** [1955] 5280/2).
[16] Lippincott, E. R.; Nagarajan, G.; Stutman, J. M. (J. Phys. Chem. **70** [1966] 78/84).
[17] Barfuβ H.; Gubitz, F.; Ittner, W.; Kreische, W.; Lanzendorfer, G.; Röseler, B. (Congr. AMPERE Magn. Resonance Relat. Phenom. 22nd Proc., Zürich 1984, pp. 285/6).
[18] Barfuß, H.; Böhnlein, G.; Gubitz, F.; Kreische, W.; Röseler, B. (Hyperfine Interact. **15/16** [1983] 911/4).
[19] Claassen, H. H.; Weinstock, B.; Malm, J. G. (J. Chem. Phys. **28** [1958] 285/9).
[20] Oka, T.; Morino, Y. (J. Mol. Spectrosc. **11** [1963] 349/67).

[21] Mohan, S.; Gunasekaran, S. (Bull. Soc. Chim. France **1983** I 173/6).
[22] Frey, R. A.; Redington, R. L.; Aljibury, A. L. K. (J. Chem. Phys. **54** [1971] 344/55).
[23] Christe, K. O.; Curtis, E. L.; Pilipovich, D. (Spectrochim. Acta A **27** [1971] 931/6).
[24] Surles, T.; Hyman, H. H.; Quarterman, L. A.; Popov, A. I. (Inorg. Chem. **9** [1970] 2726/30).
[25] Selig, H.; Claassen, H. H.; Holloway, J. H. (J. Chem. Phys. **52** [1970] 3517/21).
[26] Drifford, M.; Martin, D.; Bougon, R. (Rev. Chim. Miner. **7** [1970] 1069/86).
[27] Surles, T.; Quarterman, L. A.; Hyman, H. H. (J. Fluorine Chem. **3** [1973] 453/6).
[28] Selig, H.; Holzman, A. (private communication, ref. 12 in [25]).
[29] Müller, A.; Krebs, B.; Fadini, A.; Glemser, O.; Cyvin, S. J.; Brunvoll, J.; Cyvin, B. N.; Elvebredd, J.; Hagen, G.; Vizi, B. (Z. Naturforsch. **23a** [1968] 1656/60).
[30] Ramaswamy, K.; Muthusubramanian, P. (J. Mol. Struct. **9** [1971] 193/6).

[31] Thirugnanasambandam, P.; Karunanidhi, N. (J. Annamalai Univ. B **31** [1977] 73/84 from C.A. **90** [1979] No. 112246).
[32] Müller, A.; Nagarajan, G. (Z. Physik. Chem. [Leipzig] **235** [1967] 113/26).
[33] Cyvin, S. J.; Brunvoll, J.; Robiette, A. G. (J. Mol. Struct. **3** [1969] 259/61).
[34] Baran, E. J. (Z. Naturforsch. **25a** [1970] 1292/5).
[35] Slutsky, L.; Bauer, S. H. (J. Am. Chem. Soc. **76** [1954] 270/5).
[36] Downs, A. J.; Adams, C. J. (in: Bailar, J. C., Jr.; Emeléus, H. J.; Nyholm, R.; Trotman-Dickenson, A. F.; Comprehensive Inorganic Chemistry, Vol. 2, Pergamon, Oxford 1973, pp. 1476/1534, 1491).

[37] Thyagarajan, G.; Subhedar, M. K. (Indian J. Pure Appl. Phys. **18** [1980] 392/5).
[38] Meinert, H. (Z. Chem. [Leipzig] **7** [1967] 1/57, 42).
[39] Gruzu, V. V. (Zap. Vses. Mineral. Obshch. **107** [1978] 711/20; C.A. **90** [1979] No. 93310).
[40] Dobbs, K. D.; Hehre, W. J. (J. Computat. Chem. **7** [1986] 359/78).

[41] Dewar, M. J. S.; Zoebisch, E. G. (J. Mol. Struct. **180** [1988] 1/21).
[42] Deb, B. M.; Coulson, C. A. (J. Chem. Soc. A **1971** 958/70).
[43] Smolyar, A. E.; Charkin, O. P.; Klimenko, N. M. (Zh. Strukt. Khim. **15** [1974] 993/1003; J. Struct. Chem. [USSR] **15** [1974] 885/93).
[44] Wiebenga, E. H. (Stereochim. Inorg. Acad. Nazl. Lincei 9th Corso Estivo Chim., Rome 1965 [1967], pp. 319/31).
[45] Wiebenga, E. H.; Kracht, D. (Inorg. Chem. **8** [1969] 738/46).
[46] Havinga, E. E.; Wiebenga, E. H. (Recl. Trav. Chim. **78** [1959] 724/38).
[47] Searcy, A. W. (J. Chem. Phys. **31** [1959] 1/4).
[48] Shoolery, J. N.; Goodman, E. I.; Littman, J. (BNL-2382 [1955] 6 pp.).
[49] Muetterties, E. L.; Phillips, W. D. (J. Am. Chem. Soc. **79** [1957] 322/6).
[50] Cyr, T.; Brownstein, S. (J. Inorg. Nucl. Chem. **39** [1977] 2143/5).

[51] Gilbreath, W. P.; Cady, G. H. (Inorg. Chem. **2** [1963] 496/9).
[52] Mit'kin, V. N.; Yur'ev, G. S.; Zemskov, S. V.; Kazakova, V. I. (Zh. Strukt. Khim. **28** [1987] 73/9; J. Struct. Chem. [USSR] **28** [1987] 60/7).
[53] Muetterties, E. L.; Phillips, W. D. (J. Am. Chem. Soc. **81** [1959] 1084/8).
[54] Selig, H.; Sander, W. A.; Vasile, M. J.; Stevie, F. A.; Gallagher, P. K.; Ebert, L. B. (J. Fluorine Chem. **12** [1978] 397/412).
[55] Gutowsky, H. S.; Hoffman, C. J. (J. Chem. Phys. **19** [1951] 1259/67).
[56] Weulersse, J. M. (CEA-R-4868 [1977] 213 pp., pp. 97/8; C.A. **89** [1978] No. 206985).
[57] Stein, L. (J. Am. Chem. Soc. **81** [1959] 1273/6).
[58] Csejka, D. A. (AD-617964 [1965] 39 pp.; C.A. **64** [1966] 267).
[59] Haendler, H. M.; Bukata, S. W.; Millard, B.; Goodman, E. I.; Littman, J. (J. Chem. Phys. **22** [1954] 1939).
[60] Selig, H. (private communication, ref. 20 in [24]).

[61] Buben, S. N.; Chaikin, A. M. (Kinet. Katal. **21** [1980] 1591/2).
[62] Steunenberg, R. K.; Vogel, R. C.; Fischer, J. (J. Am. Chem. Soc. **79** [1957] 1320/3).
[63] White, C. F.; Goodeve, C. F. (Trans. Faraday Soc. **30** [1934] 1049/51).

10.10.8 Crystallographic Properties

Large, shiny needles of BrF_3 are obtained by freezing the liquid and small prisms by condensing the vapor [1]. Solid BrF_3 crystallizes at 148 K in the orthorhombic space group $Cmc2_1$-C_{2v}^{12} (No. 36) with the lattice parameters a=5.34, b=7.35, and c=6.61 Å; Z=4. The structure was refined to R=0.21; atomic positions were given. The molecule has a distorted T-shaped geometry (point group C_{2v}) similar to that in the gaseous state. For bond distances and angles see p. 59 [2]. Molecular movements in solid BrF_3 were found to be extremely slow in relaxation measurements [3].

References:

[1] Ruff, O.; Braida, A. (Z. Anorg. Allgem. Chem. **214** [1933] 91/3).
[2] Burbank, R. D.; Bensey, F. N., Jr. (J. Chem. Phys. **27** [1957] 982/3).
[3] Weulersse, J. M. (CEA-R-4868 [1977] 1/213, 170; C.A. **89** [1978] No. 206985).

10.10.9 Constitution of the Liquid and Gaseous Phases

Only slight thermal disproportionation in liquid BrF_3 with formation of BrF and BrF_5 takes place at moderate temperatures; see pp. 73/74 for details. Pure BrF_3 contains dimeric and polymeric molecules [1 to 5]. An analysis of the experimental X-ray diffraction curve of liquid BrF_3 at 298 K showed the presence of free BrF_3, its dimer, trimer, and tetramer, and of fragments of quasi-crystallites of solid BrF_3 [6]. An increasing concentration of the dimer and decreasing concentrations of the polymers upon heating were deduced from intensity changes of Raman bands [1, 7]. The manner of BrF_3 association is uncertain. Raman spectra suggest the formation of fluorine bridges involving the two equivalent fluorine atoms, but not the unique one [4, 5, 8]. In contrast, X-ray diffraction indicates four nearly equivalent F atoms surrounding each Br atom, and the nonequivalence of the Br-F bands disappears [6]. In earlier papers association was deduced from the high value of the Trouton constant [9, 10].

Association in liquid BrF_3 can be related to rapid fluorine exchange and readily permits formation of ionic BrF_2^+ and BrF_4^- [6] by donor-acceptor reactions via the equilibrium $2\,BrF_3 \rightleftharpoons BrF_2^+ + BrF_4^-$. Among the halogen fluorides the autoionization of BrF_3 is most noteable [2]. The ion concentration decreases with increasing temperature at constant pressure as determined by measuring the intensity of Raman bands [11]:

concentration $[BrF_2^+]=[BrF_4^-]$ in mol/L	0.90 ± 0.1	0.84 ± 0.1	0.82 ± 0.1	0.74 ± 0.1
T in K	303	318	324	335

The decreasing ion concentrations and the associated shift of the equilibrium to the left lowers the electrical conductivity (cf. p. 72) and reduces the overall volume of the system. This is consistent with a negative value for the "activation volume" $\Delta V_\kappa = -RT(\partial \ln\kappa/\partial p)_T$ which was determined from the slope of the pressure dependence of the electrical conductivity κ of BrF_3 at various temperatures [12]:

T in K	$10^6\Delta V_\kappa$ in m^3/mol	E_p in J/mol
301	−8.1	−920
315	−10.3	−1550

The decreasing ion concentration towards higher temperatures at constant pressure results in an increasingly negative value of the "activation energy" $E_p = -R[\partial \ln\kappa/\partial(l/T)]_p$ [12].

Association in gaseous BrF_3 is suggested from Raman spectra; the dimer mole fraction at ambient temperature is about 0.03 and at 400 K about twice as much [13]. Gas-phase association of BrF_3 was also inferred from the experimental molar weight of 140.5 (calculated 136.9) g/mol [14]. The equilibrium constant for the dissociation $(BrF_3)_2 \rightleftharpoons 2\,BrF_3$ was estimated to be given by $\log K_p = -973/T + 3.04$ from measurements of the molecular weight of BrF_3 vapor at 348 and 373 K [3].

References:

[1] Surles, T.; Hyman, H. H.; Quarterman, L. A.; Popov, A. I. (Inorg. Chem. **9** [1970] 2726/30).
[2] Meinert, H.; Groß, U. (J. Fluorine Chem. **2** [1973] 381/6).
[3] Claassen, H. H.; Weinstock, B.; Malm, J. G. (J. Chem. Phys. **28** [1958] 285/9).
[4] Christe, K. O.; Curtis, E. C.; Pilipovich, D. (Spectrochim. Acta A **27** [1971] 931/6).

[5] Drifford, M.; Martin, D.; Bougon, R. (Rev. Chim. Miner. **7** [1970] 1069/86).
[6] Mit'kin, V. N.; Yur'ev, G. S.; Zemskov, S. V.; Kazakova, V. I. (Zh. Strukt. Khim. **28** [1987] 73/9; J. Struct. Chem. [USSR] **28** [1987] 60/7).
[7] Selig, H. (private communication, ref. 20 in [1]).
[8] Frey, R. A.; Redington, R. L.; Aljibury, A. L. K. (J. Chem. Phys. **54** [1971] 344/55).
[9] Rogers, M. T.; Garver, E. E. (J. Phys. Chem. **62** [1958] 952/4).
[10] Ruff, O.; Braida, A. (Z. Anorg. Allgem. Chem. **214** [1933] 91/3).

[11] Surles, T.; Quarterman, L. A.; Hyman, H. H. (J. Fluorine Chem. **5** [1973] 453/6).
[12] Cleaver, B.; Condlyffe, D. H. (J. Chem. Soc. Faraday Trans. I **85** [1989] 2453/64).
[13] Selig, H.; Claassen, H. H.; Holloway, J. H. (J. Chem. Phys. **52** [1970] 3517/21).
[14] Steunenberg, R. K.; Vogel, R. C.; Fischer, J. (J. Am. Chem. Soc. **79** [1957] 1320/3).

10.10.10 Mechanical and Thermal Properties

Density

Solid BrF_3 at 282 K has a pycnometrically measured density of at least 3.23 g/cm^3 [1]. Reported values of the pycnometrically measured density of liquid BrF_3 are:

T in K	282 (m.p.)	298	323	~399 (b.p.)
density in g/cm^3	2.843	2.8030 ± 0.0005	2.7351 ± 0.0005	2.49
Ref.	[1]	[2]	[2]	[3]

The density is represented by the equation $d(g/cm^3) = 3.6137 - 0.00272 \cdot T$ from 298 to 323 K [2]; an earlier equation for liquid BrF_3 is $d = 3.623 - 0.00277 \cdot T$ [1].

Surface Tension

The relatively high surface tension of liquid BrF_3 indicates strong intermolecular forces. Experimental values are [4]:

γ in dyn/cm	37.1	36.4	35.6	33.8
T in K	285.2	292.1	300.3	318.2

Viscosity

The dynamic viscosity of liquid BrF_3 measured between 286.4 and 312.8 K is described by $\eta(P) = 3.41 \times 10^{-5} \exp(1944/T)$; measured values are [4]:

η in cP	3.036	2.553	2.219	1.775
T in K	286.4	293.7	298.2	312.8

By applying the theory of absolute reaction rate to viscosity, thermodynamic parameters of activation for viscous flow were calculated: $\Delta G^* = 14.0$ kJ/mol, $\Delta H^* = 16.86$ kJ/mol, and $\Delta S^* = 7.24\ J \cdot mol^{-1} \cdot K^{-1}$. The ratio $\Delta_{vap}H : \Delta H^* = 2.50$ (normally 3 to 4) and the positive value of ΔS^* confirm molecular association in liquid BrF_3 [4].

The viscosity of gaseous BrF_3 was measured from 403 to 473 K [5] and from 338 to 422 K [6]. The deviation of both sets of values is about 2% [6]. Average values are:

η in μP	173.8	181.6	197.6	219.4	233.7
T in K	338.7	363	403	443	473
Ref.	[6]	[6]	[5]	[5]	[5]

The temperature relationship of the BrF_3 viscosity between 403 and 473 K is given by $\eta(\mu P) = 21.33\ (\pm 0.61) \cdot T^{1/2} - 230.2\ (\pm 12.8)$ with a 95% confidence limit [5].

These viscosity data were used to compute the parameters of the Stockmayer (3-6-12) and the Lennard-Jones (6-12) potentials: $\sigma = 4.588$ Å, $\varepsilon/k = 339$ K [6] and $\sigma = 4.04$ Å, $\varepsilon/k = 595$ K [5]. (From other physical data: $\sigma = 4.366$ Å, $\varepsilon/k = 481.7$ K for the Lennard-Jones (6-12) potential [7].)

Melting Point. Triple Point. Boiling Point

Melting points of BrF_3 were observed at 281.9 K [8], 282.0 K [1, 2, 9], and 282.1 K [10]. The triple point temperature of 281.93 ± 0.05 K was determined by the fractional melting point method [11]. The boiling temperature of BrF_3 was reported to be 398.90 ± 0.05 [2, 11, 12] and 399.2 K [9].

Heats of Transition

The heat of BrF_3 fusion is 12.0273 ± 0.0013 kJ/mol. The heat of vaporization is given by $\Delta_{vap}H(kJ/mol) = 32.28\ [T/(T-52.58)]^2$. The value at the boiling point is 42.823 kJ/mol [11]. A value of 48.95 kJ/mol was obtained from vapor pressure investigations at lower temperatures [13]. The entropy of vaporization, the Trouton constant, has a value of 107.1 $J \cdot K^{-1} \cdot mol^{-1}$ which is unusually large and indicates association in liquid BrF_3 [4].

A heat of sublimation of 69.5 kJ/mol was obtained for BrF_3 from vapor pressure measurements at low temperatures [13]. A value of $\Delta_{subl}H^{\circ}_{298} = 53.1$ kJ/mol was cited and found to fit closely a curve relating $\Delta_{subl}H^{\circ}_{298}$ values with melting points of halides [14].

Critical Constants

The critical temperature of BrF_3 is estimated to be 600 [15] to 612 K [16]. The estimated critical pressure is 69 atm [16].

Vapor Pressure

The vapor pressure data of BrF_3 were fitted to the Antoine equation $\log(p/Torr) = 7.65757 - 1627.5(T-58)^{-1}$ for the temperature range 336.64 to 453.89 K [17] and $\log(p/Torr) = 7.74853 - 1685.8(T-52.58)^{-1}$ for the temperature range 311.9 to 428.0 K [11]. Both sets of measured data excellently agree with each other [17]. Higher pressure values were obtained earlier by a static method [15] and might have been caused by volatile impurities [17].

References:

[1] Ruff, O.; Braida, A. (Z. Anorg. Allgem. Chem. **206** [1932] 59/64).
[2] Stein, L.; Vogel, R. C.; Ludewig, W. H. (J. Am. Chem. Soc. **76** [1954] 4287/9).
[3] Ruff, O.; Ebert, F.; Menzel, W. (Z. Anorg. Allgem. Chem. **207** [1932] 46/60).
[4] Rogers, M. T.; Garver, E. E. (J. Phys. Chem. **62** [1958] 952/4).
[5] Selby, T. W.; Smith, H. A. (J. Inorg. Nucl. Chem. **30** [1968] 1183/6).
[6] Ostorero, J. (CEA-N-1293 [1070] 36 pp.; C.A. **74** [1971] No. 130644).

[7] Svehla, R. A. (NASA-TR-R 132 [1962] 140 pp., pp. 34, 37, 129; C.A. **57** [1962] 79).
[8] Fischer, J.; Steunenberg, R. K.; Vogel, R. C. (J. Am. Chem. Soc. **76** [1954] 1497/8).
[9] Mit'kin, V. N.; Zemskov, S. V. (Zh. Prikl. Khim. **54** [1981] 2180/6; J. Appl. Chem. [USSR] **54** [1981] 1913/8).
[10] Muetterties, E. L.; Phillips, W. D. (J. Am. Chem. Soc. **81** [1959] 1084/8).

[11] Oliver, G. D.; Grisard, J. W. (J. Am. Chem. Soc. **74** [1952] 2705/7).
[12] Clark, H. C.; Sadana, Y. N. (Can. J. Chem. **42** [1964] 50/6).
[13] Schumacher, H. J. (Anales Asoc. Quim. Argent. **38** [1950] 209/24).
[14] Gruza, V. V. (Kap. Vses. Mineral. Obshch. **107** [1978] 711/20; C.A. **90** [1979] No. 93310).
[15] Ruff, O.; Braida, A. (Z. Anorg. Allgem. Chem. **214** [1933] 91/3).
[16] Long, R. D. (Diss. Univ. Michigan 1955, 132 pp.; Diss. Abstr. **15** [1955] 1581).
[17] Long, R. D.; Martin, J. J.; Vogel, R. C. (Chem. Eng. Data Ser. **3** [1959] 28/34).

10.10.11 Thermodynamic Data of Formation. Heat Capacity. Thermodynamic Functions

The standard enthalpy and free energy of formation of gaseous BrF_3 from the gaseous elements at 298.15 K are $\Delta_fH° = -271.04 \pm 2.93$ kJ/mol and $\Delta_fG° = -231.0$ kJ/mol, respectively, measured calorimetrically. The values are uncorrected for the heat of mixing, which is probably small, and the partial association of BrF_3, where data are lacking [1]. A value of $\Delta_fH° = -255.59 \pm 2.9$ kJ/mol results, if the calculation is based on liquid bromine [2 to 4]. A value of $\Delta_fH° = -300.8$ kJ/mol was obtained for the formation of liquid BrF_3 [4]. The values of $\Delta_fH° = -314$ kJ/mol and $\Delta_fG° = -238$ kJ/mol for the formation of liquid BrF_3 from liquid bromine and gaseous fluorine are based on data for the formation of BrF from BrF_3 and bromine [5]. An average value of $\Delta_fH° = -321.7 \pm 3.6$ kJ/mol of liquid BrF_3 in a BrF_3-Br_2 solution was calculated from reactions of the solution with molybdenum, KBr, KIO_3, K_2SO_4, and $K_2S_2O_3$ [6]. The value agrees with the result of [5] and approaches that of [1], if the removal of bromine from the solution is taken into consideration [6].

The heat capacity of solid BrF_3 was measured over the temperature range 14.5 to 273 K; excerpted values are [7]:

T in K	15.32	47.45	85.18	108.69	149.96	185.93	235.62	273.51
C_p in $J \cdot K^{-1} \cdot mol^{-1}$	5.351	30.05	47.74	57.11	69.83	78.24	88.07	100.00

Heat capacities of liquid BrF_3 were measured from 285 to 316 K; excerpted values are [7]:

T in K	285.55	287.65	293.53	300.56	308.14	316.27
C_p in $J \cdot K^{-1} \cdot mol^{-1}$	124.06	124.10	124.56	124.73	125.19	125.52

The standard Gibbs free energy of liquid BrF_3 at 298.15 K is $\Delta_fG° = -240.6$ kJ/mol [4]. A standard entropy of 178.11 ± 0.4 $J \cdot K^{-1} \cdot mol^{-1}$ at 298.15 K was obtained from heat capacity measurements [7], a value of 178.2 $J \cdot K^{-1} \cdot mol^{-1}$ is listed in [4].

Thermodynamic data of BrF_3 in the ideal gas state are given in Table 9. The calculations are based on the rigid-rotor harmonic-oscillator approximation [8]. Thermodynamic data in the JANAF Tables [2, 3] and in [9] partly used estimated (erroneous) vibrational frequencies and have not yet been revised. Calculated thermodynamic data using unknown experimental data are also given in [10].

Table 9
Thermodynamic Properties of BrF_3 in the Ideal Gas State at 1 atm Pressure [8].

T in K	$H^\circ - H^\circ_0$ in kJ/mol	$-(G^\circ - H^\circ_0)/T$	S°	C°_p
		in $J \cdot mol^{-1} \cdot K^{-1}$		
0	0	0	0	0
100	3.544	201.229	236.651	41.622
200	8.540	227.957	270.663	57.534
298.15	14.719	246.295	295.662[a)]	67.434
500	29.388	274.265	333.046	76.283
700	45.007	294.993	359.288	79.417
1000	69.153	318.825	387.978	81.253
1500	110.089	347.766	421.157	82.291
2000	151.134	369.221	444.893	82.663

a) A value of 299.45 $J \cdot K^{-1} \cdot mol^{-1}$ was obtained from heat capacity measurements [7].

References:

[1] Stein, L. (J. Phys. Chem. **66** [1962] 288/91).
[2] Chase, M. W., Jr.; Davies, C. A.; Downey, J. R., Jr.; Frurip, D. J.; McDonald, R. A.; Syverud, A. N. (JANAF Thermochemical Tables, 3rd Ed., Pt. I [1985] 427).
[3] Stull, D. R.; Prophet, H. (JANAF Thermochemical Tables, 2nd Ed., NSRDS-NBS-37 [1971]).
[4] Wagman, D. D.; Evans, W. H.; Halow, J.; Parker, V. B.; Bailey, S. M.; Schumm, R. H. (NBS-TN-270-1 [1965] 124 pp., p. 34).
[5] Steunenberg, R. K.; Vogel, R. C.; Fischer, J. (J. Am. Chem. Soc. **79** [1957] 1320/3).
[6] Richards, G. W.; Woolf, A. A. (J. Chem. Soc. A **1969** 1072/6).
[7] Oliver, G. D.; Grisard, J. W. (J. Am. Chem. Soc. **74** [1952] 2705/7).
[8] Christe, K. O.; Curtis, E. C.; Pilipovich, D. (Spectrochim. Acta A **27** [1971] 931/6).
[9] Claassen, H. H.; Weinstock, B.; Malm, J. G. (J. Chem. Phys. **28** [1958] 285/9).
[10] Glushko, V. P.; Gurvich, L. V.; Bergman, G. A. et al. (Thermodynamic Properties of Individual Substances, Vol. 1, Book 2, Nauka, Moscow 1978, p.119).

10.10.12 Optical and Magnetic Properties

Refractive Index

Refractive indexes of liquid BrF_3 are $n_D^{25} = 1.4536 \pm 0.0002$ and $n_D^{70} = 1.4302 \pm 0.002$; the approximate temperature coefficient is $-0.00052\ K^{-1}$ [1]. The refractive index of gaseous BrF_3 at the Hg green line is $(n-1) \times 10^6 = 725$ at 326 K and 1 atm [2]. The molar refraction of liquid BrF_3 is 13.32 cm^3 [1] and agrees within the limit of error with the value of $12.92 \pm 0.3\ cm^3$ obtained for gaseous BrF_3 [2].

Diamagnetic Susceptibility

Liquid BrF_3 is diamagnetic at room temperature with an average molar diamagnetic susceptibility of $\chi_M = -33.9 \times 10^{-6}$ [3].

References:

[1] Stein, L.; Vogel, R. C.; Ludewig, W. H. (J. Am. Chem. Soc. **76** [1954] 4287/9).
[2] Rogers, M. T.; Malik, J. G.; Speirs, J. L. (J. Am. Chem. Soc. **78** [1956] 46/7).
[3] Rogers, M. T.; Panish, M. B.; Speirs, J. L. (J. Am. Chem. Soc. **77** [1955] 5292/3).

10.10.13 Electric Properties

Dielectric Constant

Liquid BrF_3 has a dielectric constant of $\varepsilon = 106.8 \pm 0.4$ at 298 K. The dielectric loss factor ε'' is zero at 100 kHz. Values obtained at lower frequencies are unreliable. The high value of ε shows BrF_3 to be one of the most polar liquids, comparable to $HCONH_2$ and HCN [1].

The dielectric constant of gaseous BrF_3 ranges from $\varepsilon = 1.003748$ at 415.5 K to $\varepsilon = 1.002964$ at 448.2 K [2].

Electric Conductivity

The conductivity of solid BrF_3 has a positive temperature coefficient [3 to 5], see **Fig. 2**. A marked discontinuity of the conductivity occurs at 253 K [4]. The accuracy of other published data on conductivity is uncertain because imperfect contact with the electrodes could

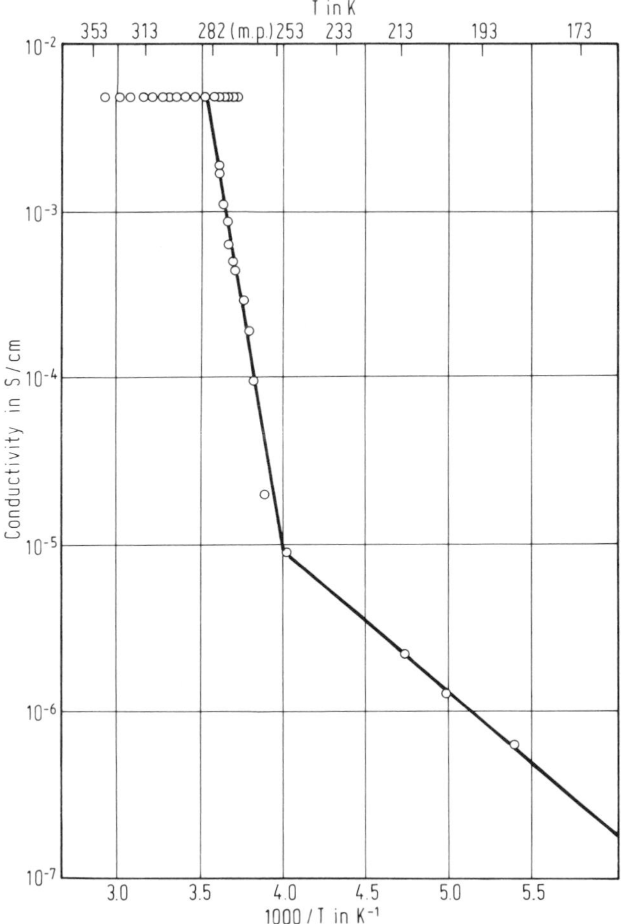

Fig. 2. Electric conductivity of BrF_3 as a function of temperature (taken from [4]).

not be excluded [3]. Data in [5] are erroneous, because they depend on the applied ac voltage due to interfering reactance [6]. The conductivity of solid BrF_3 suggests an ionic lattice and ionic conductivity [3].

The conductivity of liquid BrF_3 at 293 to 298 K is in the range of 7×10^{-3} to 8×10^{-3} S/cm and varies with sample purity [7 to 9]. Substantially lower conductivity data indicate the presence of impurities [7]. The temperature coefficient of the conductivity is negative; it is small below 293 K and more marked at higher temperature [3]. Selected data are:

T in K	277.5[a)]	283	293	298	303	313	323	333	Ref.
conductivity in 10^{-3} S/cm	8.12	8.14	8.09	–	7.87	7.58	7.19	6.75	[3]
	–	–	8.20[b)]	8.03	7.97	7.71	7.31	–	[7]

[a)] Supercooled liquid. – [b)] At 291.2 K.

Differing conductivities in [5] are erroneous because they depend on the applied ac voltage due to interfering reactance [6].

The high conductivity of liquid BrF_3 is thought to result from a high mobility of the parent ions BrF_2^+ and BrF_4^-, perhaps by a chain-conducting mechanism [7] via fluoride transfer in the solvate sphere of the ions [10]. The negative temperature coefficient of the conductivity possibly results from decreasing stability of the ions [3], which might result from increasing decomposition of the solvate spheres of the ions on heating [10]. An explanation by slow evaporation of nonremovable, complexed impurities [1] is in contrast to reproducible low-temperature measurements after completing heating cycles [3, 5]. A conductivity of 10^{-6} to 10^{-9} S/cm of pure BrF_3 was proposed to agree with the high reactivity of the substance [11]. This proposal was rejected, because the observed conductivity correlates with a low value of the heat of neutralization as observed generally [12].

The logarithm of the conductivity of liquid BrF_3 is a linear function of the pressure; read-off values from the curve in [13] are 0.8 S/m at 10 MPa and 1.1 S/m at 100 MPa for 301 K. Bromine addition initially reduces the conductivity of BrF_3 to a temperature-dependent minimum which shifts from 4 to 6 mol% of Br_2 with temperature increasing from 298 to 323 K. Continued Br_2 addition results in a conductivity increase [14]:

	298 K		323 K	
added Br_2 in mol%	4	16.4	6	16.4
conductivity in 10^{-3} S/m	7.21	12.52	5.95	7.64

Similar conductivities were reported earlier [3]. The initial decrease of the conductivity was attributed to the formation of BrF from BrF_3 and Br_2 which decreases the BrF_3 concentration [14].

The conductivity of BrF_3-HF mixtures was described in "Fluorine" Suppl. Vol. 3, 1982, p. 192. The conductivity of BrF_3 containing 0.0241 mol of HSO_3F per kg of BrF_3 is 11.25×10^{-3} S/cm and increases to 15.56×10^{-3} S/cm at a HSO_3F concentration of 0.0892 mol per kg of BrF_3 at 298 K [15].

Addition of a small quantity of ClF_3 to pure BrF_3 produces a sharp drop in the conductivity which becomes more gradual in solutions containing about 85 mol% of BrF_3. Conductivities at 298 K are [16]:

concentration of BrF_3 in mol%	100	94.6	83.85	63.17	14.24
conductivity in 10^{-3} S/cm	8.01	5.09	2.52	0.717	0.00403

The conductivities of the BrF_3-ClF_3 solutions have very slight negative temperature coefficients, see figure 2 in [17].

Addition of BF_3 increases the conductivity of BrF_3 and BrF_3-ClF_3 mixtures. The conductivities have slightly positive temperature coefficients. Dilute solutions of BF_3 in BrF_3 resemble strong electrolytes in water in accordance with the high dielectric constant of liquid BrF_3 [17].

Dissolution of XeF_2 or XeF_4 in BrF_3 slightly decreases the conductivity of BrF_3 which indicates the absence of xenon fluoride ionization and the unimportance of solvation. Dissolution of increasing amounts of MF_5 (M = P, As, Sb) in BrF_3 linearly increases the conductivity at molalities below about 0.05 to 0.09 and demonstrates MF_5 ionization [18].

The conductivity of BrF_3 is unchanged when the insoluble salts LiF, BeF_2, and CaF_2 are added. In contrast, addition of MF (M = Na, K, Rb, Cs, NH_4) increases the conductivity of BrF_3 about linearly [8, 19] at molalities of about 0.1 or less and begins to level at higher concentrations [8]. The conductivity curve of KF agrees well with the curve given in [20] for a $KBrF_4$ solution [8].

References:

[1] Martin, D.; Tantot, G. (J. Fluorine Chem. **6** [1975] 477/9).
[2] Rogers, M. T.; Pruett, R. D.; Speirs, J. L. (J. Am. Chem. Soc. **77** [1955] 5280/2).
[3] Banks, A. A.; Emeléus, H. J.; Woolf, A. A. (J. Chem. Soc. **1949** 2861/5).
[4] Toy, M. S.; Cannon, W. A. (Advan. Chem. Ser. No. 54 [1965] 237/44).
[5] Christe, K. O. (J. Phys. Chem. **73** [1969] 2792/3).
[6] Christe, K. O. (J. Phys. Chem. **74** [1969] 2038/9).
[7] Hyman, H. H.; Surles, T.; Quarterman, L. A.; Popov, A. (J. Phys. Chem. **74** [1970] 2038/9).
[8] Martin, D. (Rev. Chim. Miner. **4** [1967] 367/97).
[9] Bouy, P. (Ann. Chim. [Paris] [13] **4** [1959] 853/90).
[10] Meinert, H.; Gross, K. (J. Fluorine Chem. **2** [1973] 381/6).
[11] Muetterties, E. L.; Tullock, C. W. (Prepar. Inorg. React. **2** [1965] 237/99, 247).
[12] Richards, G. W.; Woolf, A. A. (J. Fluorine Chem. **1** [1971] 129/39).
[13] Cleaver, B.; Condlyffe, D. H. (J. Chem. Soc. Faraday Trans. I **85** [1989] 2453/64).
[14] Quarterman, L. A.; Hyman, H. H.; Katz, J. J. (J. Phys. Chem. **61** [1957] 912/7).
[15] Woolf, A. A. (J. Chem. Soc. **1955** 433/43).
[16] Surles, T.; Hyman, H. H.; Quarterman, L. A.; Popov, A. I. (Inorg. Chem. **10** [1971] 913/6).
[17] Toy, M. S.; Cannon, W. A. (J. Phys. Chem. **70** [1966] 2241/4).
[18] Martin, D. (Compt. Rend. C **265** [1967] 919/22).
[19] Chrétien, A.; Martin, D. (Compt. Rend. C **263** [1966] 235/8).
[20] Woolf, A. A.; Emeléus, H. J. (J. Chem. Soc. **1949** 2865/71).

10.10.14 Chemical Behavior

10.10.14.1 Thermal Decomposition. Radiolysis

Beginning thermal decomposition of BrF_3 at about 370 K was noticed in vapor pressure investigations [1]. Extensive thermal decomposition of BrF_3 occurs at about 450 K [2]. Thermal disproportionation proceeds via the reaction $2\ BrF_3 \rightleftharpoons BrF + BrF_5$. About 2% of the initial amount of BrF_3 disproportionates at 420 to 470 K in the gas phase, the equilibrium constant being in the range of 10^{-4}. The disproportionation takes place only to a small extent at 298 K in the gas and liquid states [3]. The reaction is catalyzed in CCl_3F solution at 230 K by small amounts of RbF or CsF [4]. Standard free energies of $\Delta_rG^\circ_{298}=+33$ and $\Delta_rG^\circ_{500}=+40$ kJ/mol with respect to BrF_3 were given [5].

Disproportionation of BrF_3 into Br_2 and BrF_5 to a very small extent is known from the detection of a very low concentration of Br_2 in gaseous BrF_3 at 498 K [6]. In the presence of bromine, the concentration of BrF_5 is reduced in accordance with the equilibrium $3\,BrF_5+Br_2 \rightleftharpoons 5\,BrF_3$ [7]. A standard free energy of $\Delta_rG^\circ_{298}=-28.9$ kJ/mol with respect to formed BrF_5 was reported [8]. The considerable stability of BrF_3 with respect to dissociation into BrF and F_2 at moderate temperature is reflected by $\Delta G^\circ_{298}=+161.9$ kJ/mol. A positive ΔH value shows that this dissociation will increase at higher temperature [5, 8].

Irradiation of liquid BrF_3 with 1 MeV electrons and beam intensities from 5 to 25 μA at 298 K is believed to induce the overall reaction $5\,BrF_3 \rightarrow Br_2+3\,BrF_5$. The decomposition is accompanied by a pressure rise, and its rate increases with the beam intensity. The fraction y of BrF_3 decomposed at a dosage D in μA·h is given by $y=0.107(1-\exp(-0.0470\,D))$. A steady state seems to be reached when about 10% of the initial amount of BrF_3 has reacted [9]. Bands of BrF become observable when BrF_3 vapor is irradiated by UV [10] or discharged in a hollow-cathode lamp [11]. Some radiative decomposition of BrF_3 by dissolved oxidized radon and its short-lived daughters reduces BrF_3 to Br_2 and BrF [12]. No new radicals resulted from γ irradiation of ~5 mol% of BrF_3 in solid Xe or SF_6 [13].

References:

[1] Oliver, G. D.; Grisard, J. W. (J. Am. Chem. Soc. **74** [1952] 2705/7).
[2] Long, R. D.; Martin, J. J.; Vogel, R. C. (Chem. Eng. Data Ser. **3** [1958] 28/34).
[3] Stein, L. (J. Am. Chem. Soc. **81** [1959] 1273/6).
[4] Naumann, D.; Lehmann, E. (J. Fluorine Chem. **5** [1975] 307/21).
[5] Stein, L. (Halogen Chem. **1** [1967] 133/224, 156).
[6] Selig, H.; Claassen, H. H.; Holloway, J. H. (J. Chem. Phys. **52** [1970] 3517/21).
[7] Richards, G. W.; Woolf, A. A. (J. Chem. Soc. A **1969** 1072/6).
[8] Wiebenga, E. H.; Havinga, E. E.; Boswijk, K. H. (Advan. Inorg. Chem. Radiochem. **3** [1961] 133/69, 138).
[9] Yosim, S. J. (J. Phys. Chem. **62** [1958] 1596/7).
[10] Smith, D. F.; Tidwell, M.; Williams, D. V. P. (Phys. Rev. [2] **77** [1950] 420/1).
[11] Brodersen, P. H.; Mayo, S. (Z. Physik **143** [1955] 477/8).
[12] Stein, L. (J. Am. Chem. Soc. **91** [1969] 5396/7).
[13] Mamantov, G.; Smyrl, N. R. (AD-A041439 [1977] 1/58, 15; C.A. **87** [1977] No. 210448).

10.10.14.2 Fluorine Exchange Reactions

Fluorine exchange reactions of BrF_3 and HF in the gaseous phase were described in "Fluorine" Suppl. Vol. 3, 1982, p. 167. Rapid fluorine exchange between BrF_3 and ClF or ClF_3 on the NMR time scale was mentioned in [1]. The exchange reaction with $Cl^{18}F_3$ in solution at 300 K is complete in 10 ± 3 min; an ionic mechanism was proposed [2]. Fluorine exchange between BrF_3 and BrF_5 is slow on the NMR time scale [1, 3], but becomes observable after adding UF_6 [3].

19% of the fluorine in BrF_3 chemically absorbed on $Na^{18}F$ at 383 K are exchanged after desorption at 413 K. The exchange probably takes place in the adduct formed between BrF_3 and NaF [4]. Fluorine exchange between BrF_3 and dissolved UF_6 is rapid; the mean lifetime in one or the other molecular species is shorter than 10^{-6} s (at ambient temperature probably) [3].

References:

[1] Hamer, A. N. (J. Inorg. Nucl. Chem. **9** [1959] 98/9).
[2] Rogers, M. T.; Katz, J. J. (J. Am. Chem. Soc. **74** [1952] 1375/7).
[3] Shoolery, J. N.; Goodman, E. J.; Littman, J. (BNL-2382 [1955] 6 pp.; N.S.A. **13** [1959] No. 10839).
[4] Sakurai, T. (Inorg. Chem. **11** [1972] 3110/2).

10.10.14.3 Electrolysis

Electrolysis of liquid BrF_3 at 298 K at 25 V produces a brown color in the catholyte which is separated from the clear, yellow rest of the liquid by a sharp boundary. No gas is evolved. Possible reactions are $2\,BrF_2^+ + 2\,e^- \rightarrow BrF_3 + BrF$ (brown) at the cathode, and $2\,BrF_4^- \rightarrow BrF_3 + 2\,e^- + BrF_5$ (colorless) at the anode [1].

Ohm's law is obeyed during electrolysis at voltages from 0.0336 to 0.875 V, and there are no measurable polarization effects. Similar results were obtained in solid BrF_3 at 273 K, the current being of the order of 10^{-3} of that passing through the liquid [1]. The validity of Ohm's law and the absence of polarization were attributed to reformation of BrF_3 from BrF and BrF_5 at the boundary of the catholyte with an accompanying low concentration of electrolysis products close to the electrodes, as the equilibrium $2\,BrF_3 \rightleftharpoons BrF + BrF_5$ lies far to the left [2].

Electrolysis of BrF_3-HSO_3F leads to a brown layer in the catholyte, supposedly resulting from the accumulation of bromine [3]. The electrolysis of the mixtures BrF_3-$KBrF_4$, BrF_3-K_2ClO_3F, BrF_3-$KBrF_4$-ClO_3F, and BrF_3-$KBrF_4$-N_2F_4 was reported [4].

References:

[1] Banks, A. A.; Emeléus, H. J.; Woolf, A. A. (J. Chem. Soc. **1949** 2861/5).
[2] Richards, G. W.; Woolf, A. A. (J. Chem. Soc. A **1969** 1072/6).
[3] Woolf, A. A. (J. Chem. Soc. **1955** 433/43).
[4] Didchenko, R.; Toeniskoetter, R. H. (AD-433910 [1964] 17 pp. from C.A. **62** [1965] 8673).

10.10.14.4 Reactions with Electrons

The intensity of ions in the 50 eV mass spectrum of BrF_3 relative to the intensity of the molecular ion and appearance potentials (AP) are as follows:

ion	Br_2^+	BrF_3^+	BrF_2^+	BrF_2^-	BrF^+	Br^+	Br^-	F_2^-	F^-
m/z	158	136	117	117	98	79	79	38	19
intensity	48.7	100.0	144	1.9	202	26.6	1.5	0.6	4.7
AP in eV	10.7±0.3	12.9±0.3	13.5±0.3		11.8±0.2	10.3±0.3			

The dissociation can be formulated by $BrF_3 \rightarrow BrF_2^+ + F + e^-$. Rate of effusion data at ambient temperature suggest that BrF^+ forms primarily from BrF_3. Roughly equal proportions of Br^+ are formed from Br_2 and BrF, while very little Br^+ originates directly from BrF_3. Ions containing oxygen and resulting from impurities were also identified in the spectrum.

Reference:

Irsa, A. P.; Friedman, L. (J. Inorg. Nucl. Chem. **6** [1958] 77/90).

10.10.14.5 Reactions of BrF_3 with Elements and Compounds

BrF_3 fluorinates almost everything which dissolves in it [1]. The resulting fluorides are usually in the highest valency state [2]. Exceptions are reactions where BrF_3 is oxidized, and which are described separately on pp. 96/7, and solutions of inert fluorides, e.g., MoF_6, see Table 11. Incomplete fluorination was observed mainly in the case of some transition metal oxides and complexes; see Table 11 for examples.

Several of the formed fluorinated compounds can react with additional BrF_3. The resulting adducts or fluoro complexes with the cation BrF_2^+ or the anion BrF_4^- are frequently isolable [3, 4]. Mixtures of both types of salts are obtained by the reaction of ternary fluorides with the solvent BrF_3, e.g., $4\,BrF_3 + K_2PtF_6 \rightleftharpoons 2\,KBrF_4 + (BrF_2)_2PtF_6$ [5, 6].

Most reactions of BrF_3 are carried out by direct combination of the reactants and boiling when necessary. The reactions are often violent. The presence of Br_2 or the use of solvents, like $Cl_2FC\text{-}CClF_2$, moderates the reactions, especially as the use of an appropriate solvent allows cooling to temperatures far below 273 K; see Table 11 for examples. The products can frequently be isolated by evaporation of the volatiles and drying at an appropriate temperature; see for example [7].

Reactions of BrF_3 with mixtures containing elements are described in Table 12. Earlier results on reactions of BrF_3 were described in "Brom" 1931, p. 338.

References:

[1] Downs, A. J.; Adams, C. J. (in: Bailar, J. C., Jr.; Emeléus, H. J.; Nyholm, R.; Trotman-Dickenson, A. F.; Comprehensive Inorganic Chemistry, Vol. 2, Pergamon, Oxford 1973, pp. 1476/1534, 1521).
[2] Leech, H. R. (in: Mellor's Comprehensive Treatise on Inorganic and Theoretical Chemistry, Suppl. II, Pt. I, Longmans Green, London 1956, pp. 147/81, 164).
[3] Gutmann, V. (Quart. Rev. [London] **10** [1956] 451/62, 457).
[4] Weidlein, J. (Z. Chem. [Leipzig] **9** [1969] 134/40, 138).
[5] Sharpe, A. G. (J. Chem. Soc. **1950** 3444/50).
[6] Hepworth, M. A.; Peacock, R. D.; Robinson, P. L. (J. Chem. Soc. **1954** 1197/201).
[7] Emeléus, H. J.; Woolf, A. A. (J. Chem. Soc. **1950** 164/8).

10.10.14.5.1 Reactions with Nonmetal Elements

Rn. Oxidation of Rn by BrF_3 occurs when the frozen mixture of the pure components or solutions in Br_2, HF, or IF_5 are warmed to ambient temperature [1, 2]. Oxidized Rn is involatile even after evaporation of the solvent. Electrolysis of the BrF_3 solution indicates a cationic Rn species, while in BrF_3-HF solutions, the presence of anionic Rn ions is possible and may result from the high concentration of the fluoride ion [1].

H_2. Ignition of a BrF_3-H_2 mixture leads to the formation of HF in an explosive reaction [3]. The enthalpy of the reaction $BrF_3 + 2\ H_2 \rightarrow HBr + 3\ HF$ is −619 kJ/mol at 0 K from literature data. The reaction energy is poorly described in MO calculations using 3-21G basis sets [4].

O_3. Cocondensation of BrF_3 and O_3 in an IR cell at 83 K yields a faint yellow glass which changes to a white solid at 173 to 178 K. IR absorptions due to the BrF_3-O_3 product were recorded, but an identification of the product was not possible [5].

F_2. Br_2. The reaction of BrF_3 and F_2 yields BrF_5; see p. 116 for details. The system BrF_3-Br_2 is described on pp. 103/5.

C. The exothermal reaction of BrF_3 with graphite leads to its incomplete oxidation with formation of C-F bonds, simultaneous intercalation of BrF_3 into the fluorinated graphite matrix, and release of Br_2 [6 to 9]. An X-ray investigation of $C_{1.8}F \cdot 0.1\ BrF_3$ showed that liquid BrF_3 at room temperature enters through graphite pores with an average diameter of about 20 Å followed by partial fluorination to $C_{1.8}F$ leaving residual liquid BrF_3 in the pores [10]. Additional fluorination results during heating [7, 10]. Graphite swells by a factor of 2 to 4 during the reaction with BrF_3 [11, 12]. The reaction is formally described by $4\ BrF_3 + 3\ e^- \rightarrow 3\ BrF_4^- + 1/2\ Br_2$ [13]. However, the intercalated species is molecular BrF_3 rather than an ionic species [8, 11, 13]. Interactions of BrF_3 with the fluorinated graphite matrix are mainly of the van der Waals type [8]. Intercalation of formed Br_2 into the matrix can be neglected [14].

Single-phase intercalation products usually exist only in equilibrium with the appropriate vapor pressure of BrF_3, isolated products are frequently inhomogeneous [14]. Products of identical composition may be quite different chemically as demonstrated by the different properties of nominally identical samples of stoichiometry $C_{8.9}BrF_3$ [9, 11]. Such differences are probably due to variations in the experimental conditions, including the type of graphite used [9] and the size of the graphite particles [15].

The reaction of solid BrF_3 and graphite begins at 263 to 278 K and accelerates with the melting of BrF_3 at 282 K [6]. Graphite is preferably added to excess liquid BrF_3, because addition of BrF_3 to graphite leads to occasional explosions [7]. By allowing more time for liquid BrF_3 to react at ambient temperature, the ratio of carbon-bonded F to F in BrF_3 is raised from 1.65 in weakly fluorinated dark specimens to 3.5 for strongly fluorinated, light-colored samples [6]. Reaction of liquid BrF_3 with graphite at ambient temperature and removing excess BrF_3 by reaction with CCl_4 yields C_2F_x. The value of x is determined by the reaction time and varies from 0.84 to a limiting value of 1.02 [16].

Reactions with gaseous BrF_3 are carried out by a two-zone vapor transport method. The degree of graphite intercalation by gaseous BrF_3 at 298 K is pressure-dependent [14]. The ordered intercalation compound of the first stage, $C_2F \cdot 0.33\ BrF_3$, is formed at 70% or more BrF_3 saturation, while below 30% saturation an inhomogeneous product, $C_2F \cdot 0.13\ BrF_3$, results. Intermediate partial pressures lead to continuously changing product compositions [8]. The amount of absorbed, gaseous BrF_3 drops continuously with increasing temperature from 81.9 wt% BrF_3 in $C_{2.16}BrF_3$ at a graphite temperature of 322 K to 56.8 wt% BrF_3 in $C_{8.9}BrF_3$ at 425 K, which is the limiting composition obtained between 425 and 446 K [11]. Additional heating decomposes the intercalate either at 425 K [11] or at 623 K [9]. The preparation of an intercalate with higher F concentration using gaseous BrF_3 at 323 to 400 K was described in a patent [17]. The reaction of expanded graphite and gaseous BrF_3 in Ar yields $(CF_{0.88})_n$ at 283 K and $(CF_{1.04})_n$ at 400 K. The extent of the reaction can be controlled easily by the applied temperature [18].

Individual reactions are described in the cited literature and in [19 to 21]. Earlier results on the reaction of BrF_3 and graphite are given in "Kohlenstoff" B, 1968, p. 816.

The reaction of BrF_3 in HF and graphite at 295 K yields intercalation products consisting of bright, blue plates after evaporating the liquid. Chemical analyses of the obtained samples are given; a typical product composition is $C_{17.6}F_{7.2} \cdot BrF_3 \cdot 2.1\ HF$. The presence of BrF_3 and HF in the products was determined by IR and ^{19}F NMR spectroscopy. The degree of graphite fluorination by BrF_3 in the presence of HF is similar to that reached in the absence of HF [22].

An excess of a 1:1 mixture of BrF_3 and IF_5 with graphite yields an intercalate of the chemical composition $C_xF \cdot y\ BrF_3 \cdot z\ IF_5$ with $y:z \approx 1:2$ at 295 K and containing intercalated BrF_3 and IF_5 according to ^{19}F NMR spectra [23].

Gaseous BrF_3 at 3.5 Torr reacts with carbon electrodes under condition of an electric arc. The products $CBrF_3$ (40 mol%), CF_4 (45 mol%), C_2F_4 (10 to 15 mol%), and some C_2F_6 and CBr_2F_2 remain after freezing out unreacted BrF_3 and formed Br_2 [24].

Diamond is resistant to BrF_3 [25].

Si. Silicon is spontaneously etched by BrF_3 with a pressure near 1 Torr. The etch rate decreases with increasing temperature up to about 370 K. Near this temperature a minimum and then an increase of the etch rate with temperature was expected. The etch rate as a function of substrate temperature in the range studied (up to 370 K) was given in the general form of the Arrhenius equation: $R = 1.16 \times 10^{-18}\ n\ T^{1/2} \exp(6.4/kT)$ Å/min with the BrF_3 density n in cm^{-3} [26].

Ge. Sb. The very vigorous reaction of BrF_3 and Ge yields GeF_4 [27]. The heat of reaction of a gaseous BrF_3-Br_2 mixture (mole ratio 7.04:1) and gaseous Sb at 298 K was calculated to be -932.1 ± 4.9 kJ/mol based on experiments in solution [28].

References:

[1] Stein, L. (Science **168** [1970] 362/4).
[2] Stein, L. (J. Am. Chem. Soc. **91** [1969] 5396/7).
[3] Gregg, D. W.; Pearson, R. K. (U.S. 3928821 [1975] 7 pp.; C.A. **84** [1976] No. 67727).
[4] Dobbs, K. D.; Hehre, W. J. (J. Computat. Chem. **7** [1986] 359/78).
[5] Csejka, D. A. (AD-617964 [1965] 39 pp.; C.A. **64** [1966] 267).
[6] Nikonorov, Yu. I. (Izv. Akad. Nauk SSSR Neorg. Mater. **18** [1982] 130/4; Inorg. Mater. [USSR] **18** [1982] 110/4).
[7] Nikonorov, Yu. I.; Gornostaev, L. L. (Izv. Sibirsk. Otd. Akad. Nauk SSSR Ser. Khim. Nauk **1979** No. 4, pp. 55/9).
[8] Panich, A. M.; Danilenko, A. M.; Nazarov, A. S.; Gabuda, S. P.; Yakovlev, I. I. (Zh. Strukt. Khim. **29** No. 2 [1988] 55/61; J. Struct. Chem. [USSR] **29** [1988] 211/6).
[9] Selig, H.; Sunder, W. A.; Vasile, M. J.; Stevie, F. A.; Gallagher, P. K.; Ebert, L. B. (J. Fluorine Chem. **12** [1978] 397/412).
[10] Gornostaev, L. L.; Zemskov, S. V.; Yur'ev, G. S.; Yakovlev, I. I.; Tkachev, S. V. (Izv. Akad. Nauk SSSR Neorg. Mater. **16** [1980] 2075/6).

[11] Opalovskii, A. A.; Nazarov, A. S.; Uminskii, A. A.; Chichagov, Yu. V. (Zh. Neorg. Khim. **17** [1972] 2350/3; Russ. J. Inorg. Chem. **17** [1972] 1227/9).
[12] Rosen, F. D. (NAA-SR-213 [1957] 34 pp.; N.S.A. **11** [1957] No. 9637).
[13] McQuillan, B. W. (LBL-12228 [1981] 210 pp.; C.A. **95** [1981] No. 143377 see also Diss. Abstr. Intern. B **42** [1982] 2830).

[14] Danilenko, A. M.; Nazarov, A. S.; Yakovlev, I. I. (Zh. Neorg. Khim. **31** [1986] 1953/6; Russ. J. Inorg. Chem. **31** [1986] 1124/6).
[15] Danilenko, A. M.; Nazarov, A. S.; Yakovlev, I. I.; Potapova, O. G.; Fadeeva, V. P. (Izv. Akad. Nauk SSSR Neorg. Mater. **26** [1990] 1441/5 from C.A. **113** [1990] No. 203711).
[16] Yudanov, N. F.; Boguslavskii, E. G.; Yakovlev, I. I.; Gabuda, S. P. (Izv. Akad. Nauk SSSR Ser. Khim. **1988** 272/6; Bull. Acad. Sci. USSR Div. Chem. Sci. **37** [1988] 200/3).
[17] Nikonorov, Yu. I. (U.S.S.R. 710930 [1980] from C.A. **92** [1980] No. 165862).
[18] Yakovlev, P. V. (U.S.S.R. 1308550 [1987] 2 pp.; C.A. **107** [1987] No. 80507).
[19] Nikonorov, Yu. I.; Khainovskii, N. G.; Khairetdinov, E. F. (Izv. Akad. Nauk SSSR Neorg. Mater. **21** [1985] 1952/4; Inorg. Mater. [USSR] **21** [1985] 1701/3).
[20] Nikonorov, Yu. I.; Marusin, V. V. (Zh. Neorg. Khim. **26** [1981] 2662/6; Russ. J. Inorg. Chem. **26** [1981] 1426/8).

[21] Nikonorov, Yu. I.; Khairetdinov, E. F. (Izv. Akad. Nauk SSSR Neorg. Mater. **15** [1979] 1593/7; Inorg. Mater. [USSR] **15** [1979] 1254/7).
[22] Danilenko, A. M.; Nazarov, A. S.; Yakovlev, I. I.; Fadeeva, V. P. (Izv. Akad. Nauk SSSR Neorg. Mater. **25** [1989] 1303/6; Inorg. Mater. [USSR] **25** [1989] 1100/3).
[23] Danilenko, A. M.; Nazarov, A. S.; Yakovlev, I. I. (Zh. Neorg. Khim. **34** [1989] 1693/6; Russ. J. Inorg. Chem. **34** [1989] 958/61).
[24] Farlow, M. W.; Muetterties, E. L.; E. I. du Pont de Nemours and Co. (U.S. 2732410 [1956] 3 pp.; C.A. **1956** 15574).
[25] Hyman, H. H.; Surles, T.; Quarterman, L. A.; Popov, A. I. (Appl. Spectrosc. **24** [1970] 464/5).
[26] Ibbotson, D. E.; Mucha, J. A.; Flamm, D. L.; Cook, J. M. (J. Appl. Phys. **56** [1984] 2939/42).
[27] Gutmann, V. (Svensk Kem. Tidskr. **68** [1956] 1/25, 9).
[28] Richards, G. W.; Woolf, A. A. (J. Fluorine Chem. **1** [1971/72] 129/39).

10.10.14.5.2 Reactions with Metals

K. Mg. Ba. K reacts violently with BrF_3 to yield $KBrF_4$ [1]. The corrosion of Mg by BrF_3 is extensive at 298 K [2]; Ba yields $Ba(BrF_4)_2$ in a violent reaction [1].

Al. In. Fluorination of Al (see also p. 55) by BrF_3 is very slow under normal conditions, even in the presence of KF. Complete fluorination can be achieved by BrF_3-HF-KF at 600 K under an autogenous pressure of about 75 atm. Anodic dissolution of Al in BrF_3 requires the presence of KF and a voltage of at least 85 V, and yields a nonadherent precipitate at the anode [3]. The reaction of In and BrF_3 is incomplete [1].

Sn. Pb. The noticeable to violent reaction of Sn with BrF_3 yields a solution of $(BrF_2)_2SnF_6$ [4, 5], from which $SnF_4 \cdot 1.76\,BrF_3$ was isolated at ambient temperature and SnF_4 at 460 K [4]. The specific dissolution rate of Sn in BrF_3 at 293 K is 5.84×10^{-5} $mol \cdot s^{-1} \cdot cm^{-2}$ and does not depend on the concentration of added Br_2 [5]. The heat of reaction of a gaseous BrF_3-Br_2 mixture (mole ratio 7.04:1) and gaseous Sn at 298 K of -762.7 ± 2.2 kJ/mol was calculated from experiments in solution [6]. The attack of BrF_3 on Pb is slight at 298 K [2].

Bi. Po. The initial reaction with Bi is vigorous yielding BiF_3 but does not go to completion [1]. A tracer experiment with Po gave no volatile product [7].

Ti. Zr. Metallic Ti and its alloys dissolve in BrF_3 with the formation of TiF_4 and Br_2 [8]. Corrosion [9] and even ignition of Zr by BrF_3 were observed [10]. Zr and Zr-U alloy react very slowly with BrF_3 and BrF_3-KF under normal conditions. Anodic dissolution of

Zr requires the presence of SbF_5 at about 50 V and does not occur in the presence of KF. Anodic dissolution of a Zr–U alloy is vigorous in the presence of KF at 25 V and takes place in BrF_3-SbF_5 at about 50 V, while no reaction was observed in a BrF_3 or BrF_3-HF solution. Fluorination of Zr by BrF_3-HF-KF at 600 K under an autogenous pressure of about 75 atm is complete [3].

V. Nb. Ta. Powdered V forms VF_5 with BrF_3 at 570 K with incandescence; there is no reaction at 430 K [11]. However, formation of VF_5 at 343 K was described in [12]. Incandescence of Nb and Ta upon contact with BrF_3 was prevented by reacting frozen Br_2 suspensions with the powdered metals followed by boiling. Evaporation of the solvents leaves BrF_2MF_6 with M = Nb, Ta [13].

Mo. W. Powdered Mo or W and BrF_3 yield MoF_6 or WF_6 and Br_2 [14]; Mo reacts with incandescence [11]. The enthalpy of the reaction $Mo + 2\,BrF_3(sln) \rightarrow MoF_6(g) + Br_2(sln)$ is -894.9 ± 1.5 kJ/mol in BrF_5-Br_2 (mole ratio 7.04:1) at 298 K [15].

Mn. Re. The reaction of Mn and BrF_3 is incomplete and yields a mixture of MnF_2 and MnF_3 [16], Re reacts to ReF_6 with incandescence [1].

Fe. Ni. Fe and Ni are quite inert in BrF_3; see p. 55 for details.

Ru. The violent reaction of Ru with BrF_3 at 283 to 288 K yields $RuBrF_8$ and Br_2 [17], whereas a mixture of RuF_5 and $RuOF_4$ results at 293 K with a mixture of BrF_3 and Br_2 after drying the products at 390 K [18].

Os. Ir. The extent and ease of the reaction of Os and BrF_3 depends markedly on the state of the metal and ranges from complete reaction at about 300 K to incomplete reaction during boiling. Ir is not oxidized at ambient temperature; the reaction is slow and incomplete after boiling [19]. Powdered Ir dissolves in boiling BrF_3 [20]. Evaporation of the volatiles at ~300 K yields BrF_2IrF_6 [21].

Pt. Reports on the reaction of Pt and BrF_3 are contradictory. Dissolution of Pt foil in pure BrF_3 was observed at 323 K [22] or during boiling [23]. No attack on Pt by pure BrF_3 was reported [5], even in powdered form [24] or during boiling [25]. However, finely divided Pt was removed from the electrodes during conductivity investigations in BrF_3 at 295 and 315 K, while smooth Pt was not attacked [26]. Solutions of BrF_3 in Br_2 attack Pt at appreciable rates at 298 K, whereas BrF_3-rich solutions are much less corrosive [23]. The rate of Pt dissolution in BrF_3 containing 0.5 to 2.8 mol/L of Br_2 is 2.6×10^{-9} $mol \cdot s^{-1} \cdot cm^{-2}$ at 353 K under kinetic conditions, there is no noticeable reaction below 323 K [5].

Cu. Ag. Cu is quite inert in BrF_3. The reaction of Ag and BrF_3 yields a solution of $AgBrF_4$. A dissolution rate of $(2.28 \pm 0.1) \times 10^{-6}$ $mol \cdot s^{-1} \cdot cm^{-2}$ was measured at 293 K under kinetic conditions. The rate increases linearly with the Br_2 concentration. The rate law is $w = w^{\circ} + K[Br_2]^n$ where w° is the reaction rate in the absence of Br_2, $K = (3.2 \pm 0.1) \times 10^{-6}$ $L \cdot mol \cdot (mol \cdot s \cdot cm^2)^{-1}$, and $n = 0.52 \pm 0.15$. The rate-determining step is the reaction of BrF_3 with an AgBr surface layer [5].

Au. Pure BrF_3 does not attack Au [5]; dissolution with evolvement of Br_2 requires gentle warming. Evaporation of the volatiles leaves $AuBrF_6$ at 323 K [27] and AuF_3 at 570 K [28]. The dissolution rate $w = 0.95 \times 10^{-8}$ $mol \cdot s^{-1} \cdot cm^{-2}$ was measured at 293 K in a BrF_3 solution containing 0.009 mol/L of Br_2 under kinetic conditions. The dissolution rate of Au has a reaction order of 1 ± 0.12 with respect to the Br_2 concentration. The rate-determining step is the formation of AuBr or AuF from Au and Br_2 or BrF [5].

Zn. Cd. Zn corrodes extensively in BrF_3 at 298 K, while Cd is attacked only slightly [2].

Actinides. The fluorination of Th by BrF_3 or BrF_3-KF is very slow under normal conditions. Complete fluorination is achieved by BrF_3-HF-KF at 600 K under an autogenous pressure of about 75 atm [3]. Tracer experiments with Pa did not yield volatile products [7]. Fluorination of U by BrF_3 yields UF_6; see "Uran" Suppl. Vol. A 3, 1981, pp. 133/4, and "Uran" Suppl. Vol. C 8, 1980, p. 76. The reactions of Np and Pu with BrF_3 yields NpF_6 and PuF_4, respectively; see "Transurane" A 1, II, 1974, pp. 252 and 259/60.

References:

[1] Gutmann, V. (Svensk Kem. Tidskr. **68** [1956] 1/25, 9).
[2] Toy, M. S.; Cannon, W. A. (Electrochem. Technol. **4** [1966] 520/3).
[3] Osborne, D. W. (ANL-4469 [1957] 63 pp., 49, 53/5; N.S.A. **12** [1958] No. 12215).
[4] Woolf, A. A.; Emeléus, H. J. (J. Chem. Soc. **1949** 2865/71).
[5] Mit'kin, V. N.; Zemskov, S. V. (Zh. Prikl. Khim. **54** [1981] 2180/6; J. Appl. Chem. [USSR] **54** [1981] 1913/8).
[6] Richards, G. W.; Woolf, A. A. (J. Fluorine Chem. **1** [1971] 129/39).
[7] Emeléus, H. J.; Maddock, A. G.; Miles, G. L.; Sharpe, A. G. (J. Chem. Soc. **1948** 1991).
[8] Dupraw, W. A.; O'Neill, H. J. (Anal. Chem. **31** [1959] 1104/7).
[9] Hendel, F. J. (Chem. Eng. **68** No. 7 [1961] 131/48).
[10] Gutmann, V. (unpublished, ref. 31 in: Gutmann, V.; Angew. Chem. **62** [1950] 312/5).

[11] Emeléus, H. J.; Gutmann, V. (J. Chem. Soc. **1949** 2979/82).
[12] Popov, A. I.; Sukhoverkhov, V. F.; Chumaevskii, N. A. (Zh. Neorg. Khim. **35** [1990] 1111/22; Russ. J. Inorg. Chem. **35** [1990] 626/32).
[13] Gutmann, V.; Emeléus, H. J. (J. Chem. Soc. **1950** 1046/50).
[14] Cox, B.; Sharp, D. W. A.; Sharpe, A. G. (J. Chem. Soc. **1956** 1242/4).
[15] Richards, G. W.; Woolf, A. A. (J. Chem. Soc. A **1969** 1072/6).
[16] Gutmann, V.; Emeléus, H. J. (Monatsh. Chem. **81** [1950] 1157/9).
[17] Hepworth, M. A.; Peacock, R. D.; Robinson, P. L. (J. Chem. Soc. **1954** 1197/1201).
[18] Holloway, J. H.; Peacock, R. D. (J. Chem. Soc. **1963** 527/30).
[19] Hepworth, M. A.; Robinson, P. L.; Westland, G. J. (J. Chem. Soc. **1954** 4269/75).
[20] Zemskov, S. V.; Mit'kin, V. N.; Gornostaev, L. L.; Shipachev, V. A.; Isakova, V. G. (U.S.S.R. 1203021 [1986] 2 pp.; C.A. **104** [1986] No. 112268).

[21] Mit'kin, V. N.; Mironov, Yu. I.; Zemskov, S. V.; Zil'berman, B. D.; Gabuda, S. P. (Koord. Khim. **9** [1983] 20/5; Soviet J. Coord. Chem. **9** [1983] 18/23).
[22] Popov, A. I.; Glockler, G. (J. Am. Chem. Soc. **74** [1952] 1357/8).
[23] Quarterman, L. A.; Hyman, H. H.; Katz, J. J. (J. Phys. Chem. **61** [1957] 912/7).
[24] Sharpe, A. G. (J. Chem. Soc. **1950** 3444/50).
[25] Sharpe, A. G.; Emeléus, H. J. (J. Chem. Soc. **1948** 2135/8).
[26] Cleaver, B.; Condlyffe, P. H. (J. Chem. Soc. Faraday Trans I **85** [1989] 2453/64).
[27] Sharpe, A. G. (J. Chem. Soc. **1949** 2901/2).
[28] Asprey, L. B.; Kruse, F. H.; Jack, K. H.; Maitland, R. (Inorg. Chem. **3** [1964] 602/4).

10.10.14.5.3 Reactions with Nonmetal Compounds

Reactions of BrF_3 with nonmetal compounds are listed in Table 10. More details on specific reactions are given in the following text.

The hydrolysis of BrF_3 by liquid or frozen water is violent [1]. Hydrolysis in a ferrous sulfate solution is less violent than in water [2]. Reaction with a 2.5% NaOH solution yields fluoride, while bromine is found as bromide (20 to 30%), hypobromite (BrO^-, 40 to 50%), and bromate (BrO_3^-, about 30%). The initially formed bromite (BrO_2^-) is too unstable to be detected [3]. Hydrolysis in an NaOH solution and subsequent bromine reduction to bromide by hydrazine can be used for the quantitative analysis of BrF_3 [3, 4]. Partial hydrolysis of frozen BrF_3 by cocondensed water (molar ratio about 1.3:1) proceeds violently during thawing. Additional species, such as BrO_x (x = 1, 2), BrO_2F, HBrO, HBr, and Br_2 were identified by mass spectrometry [5].

Products formed from BrF_3 and BF_3 are stable below 193 K [6, 7]. Formation of BrF_2BF_4 in SO_2ClF was observed at 120 K and higher temperatures. The equilibrium constants of the product formation are 450 at 246 K, 55 at 259 K, and 14 at 273 K [8]. Isolation of BrF_2BF_4 at ambient temperature was reported [9, 10], see also "Borverbindungen" 8 New Suppl. Ser. Vol. 34 Pt. 10, 1976, p. 223. This substance most likely is an impurity or a hydrolysis product [6].

Excess BrF_3 and SbF_5 yield solid BrF_2SbF_6 by sublimation after evaporation of the solvent [21, 32, 53 to 55]. The heat of solution of SbF_5 in BrF_3-Br_2 (molar ratio 7.04:1) at 298 K is -92.38 ± 0.33 kJ/mol [12, 49] and includes a component of the heat of reaction. The system BrF_3-SbF_5 contains the compounds $3\,BrF_3 \cdot SbF_5$ (approximate composition), $3\,BrF_3 \cdot 2\,SbF_5$, BrF_2SbF_6, and $BrF_3 \cdot 3\,SbF_5$ [56]. The titration of BrF_3 with SbF_5 in a HSO_3F solution has an end point at a reactant ratio of about 2:1 [57].

The reaction of SO_3 with excess BrF_3 yields a golden yellow liquid, BrF_2SO_3F [11, 12], contradicting an earlier report of $Br_2(SO_3F_2)_3$ formation [13]. The reaction requires moderate temperatures as heating induces decomposition [12, 14]. Polymerized solid SO_3 is formed occasionally and is favored by excess SO_3 and traces of moisture [12].

The reaction between BrF_3 and ClF_3 proceeds via the equilibrium $BrF_3 + ClF_3 \rightleftharpoons ClF_2BrF_4$ with an equilibrium constant of $(1 \pm 0.4) \times 10^{-4}$ probably at or slightly below room temperature. ClF_3 is probably a much weaker fluoride ion acceptor than BrF_3 and behaves as a weak base in BrF_3 solution [15].

Table 10
Reactions of BrF_3 with Nonmetal Compounds.

reactants	products	conditions, remarks
H_2O	HF, O_2, HBrO, $HBrO_3$, HBr	violent reaction [3], see text
B_2O_3	BF_3, O_2	quantitative reaction [16, 17]
H_3BO_3	–	rapid reaction with evolvement of much gas [1]
BF_3	BrF_2BF_4	stable at 193 K or less [6, 7], see text
	$BrF_2B_2F_7$	stable at 153 K or less [6 to 8]
CO	COBrF, COF_2	by bubbling CO through BrF_3 at 281 to 303 K in about quantitative yields [18]
$(CF)_n$	–	only physical absorption of BrF_3 [19]
$C_{10.1}F_{4.3} \cdot BrF_5$	$C_{17.1}F_{7.2} \cdot BrF_3$	exchange reaction with excess BrF_3 at 295 K [20]
SiO_2, GeO_2	SiF_4, GeF_4	quantitative reaction [16]
GeF_4	$GeF_4 \cdot 2\,BrF_3$	exothermal reaction at room temperature [7, 21]
NH_4F, NH_4Cl	–	explosive reaction [22]
NH_4F	NH_4BrF_4 (?)	white solid (not analyzed) when warming the reactants from 77 K to ambient temperature [23]

Table 10 (continued)

reactants	products	conditions, remarks
NO	$NOBrF_4$, Br_2	60 to 70% yield in a vigorous reaction [24, 25]
N_2O_5	$NO_2[Br(ONO_2)_2]$, NO_2F	in CCl_3F at 243 K [52], wrong product composition in [26]
		the frozen pure reactants explode when melted [27]
NOF, NOF·3 HF	$NOBrF_4$	at ambient temperature [24, 28, 29]
NO_2F	NO_2BrF_4	formation at ambient temperature, isolation at 228 K [29, 30]
PF_5	$BrF_3 \cdot n\ PF_5$ (n = 1, 2)	in SO_2ClF at low temperature [8]
PBr_5	PF_5	quantitative yield [31]
As_2O_3	AsF_5	quantitative reaction [16, 31]
AsF_5	$AsF_5 \cdot n\ BrF_3$	n = 1, white solid at 273 K, of marginal stability at 293 K [21, 32]
		product mixture with n = 1, 2 in SO_2ClF [8]
Sb_2O_3	BrF_2SbF_6	violent reaction, quantitative yield [16, 33]
		Sb_2O_3 volatilization with gaseous BrF_3 sets in at 420 K [34]
Sb(O)Cl	BrF_2SbF_6	very violent reaction [33]
SbF_3	BrF_2SbF_6, Br_2	mild reaction [33]
SbF_5	BrF_2SbF_6	and other products, see text
ClO_2SbF_6	BrF_2SbF_6	[35]
SO_3	BrF_2SO_3F	[11, 12, 14], see text
H_2SO_4	(HSO_3F ?)	decomposition of H_2SO_4, volatile products [36]
$(NO)_2S_2O_7$	$NOSO_3F$	[14]
$SOCl_2$	SOClF	the reaction is of explosive violence [37]
Cl_3CSO_2Cl	SOF_2	at high initial concentration of BrF_3 [38]
SF_4	SF_5Br	with BrF_3-Br_2 (1:1) at 363 K in the presence of CsF [39, 40]
		no reaction at 300 and 570 K in the absence of CsF [41]
SeO_2	SeF_6	quantitative reaction [16]
SeF_4	—	no reaction in the liquid phase [42]
F_5SeOH	$Br(OSeF_5)_3$	71% yield, separated from HF by distillation [43]
TeO_2	—	volatilization by gaseous BrF_3 sets in at 370 K [34]
TeF_6	—	no involatile product obtained at 298 K in the liquid and at 348 and 373 K in the gas phase [44]
HF	—	completely miscible, see "Fluorine" Suppl. Vol. 3, 1982, p. 192
ClO_2	ClO_2F	at 303 K [45]
ClO_2SbF_6	ClO_2F	[46]
ClF_3	ClF_2BrF_4	[15], see text
BrCl, $BrCl_3$	—	no reaction at room temperature [5, 47]
I_2O_5	IF_5	quantitative; reaction at 233 K, initially [16, 48]
IF_5	—	no reaction [49]
IO_2F	IF_5, O_2, Br_2	quantitative reaction [50]
XeF_6	—	no stable adduct at room temperature [51]

References:

[1] Rosen, F. D. (NAA-SR-213 [1957] 34 pp.; N.S.A. **11** [1957] No. 9637).
[2] Schnizlein, J. G. (personal communication, ref. 11 in: Dupraw, W. A.; O'Neill, H. J.; Anal. Chem. **31** [1959] 1104/7).
[3] Sakurai, T.; Kobayashi, Y.; Iwasaki, M. (J. Nucl. Sci. Technol. [Tokyo] **3** [1966] 10/3).
[4] Fischer, J.; Steunenberg, R. K.; Vogel, R. C. (J. Am. Chem. Soc. **76** [1954] 1497/8).
[5] Sloth, E. N.; Stein, L.; Williams, C. W. (J. Phys. Chem. **73** [1969] 278/80).
[6] Christe, K. O. (J. Phys. Chem. **73** [1969] 2792/3).
[7] Brown, D. H.; Dixon, K. R.; Sharp, D. W. A. (Chem. Commun. **1966** 654/5).
[8] Cyr, T.; Brownstein, S. (J. Inorg. Nucl. Chem. **39** [1977] 2143/5).
[9] Toy, M. S.; Cannon, W. A. (J. Phys. Chem. **70** [1966] 2241/4).
[10] Toy, M. S.; Cannon, W. A.; U.S. Dept. of the Army (U.S. 3645702 [1972] 2 pp.; C.A. **76** [1972] No. 143022).

[11] Groß, U.; Meinert, H.; Grimmer, A. R. (Z. Chem. [Leipzig] **10** [1970] 441/3).
[12] Woolf, A. A. (J. Fluorine Chem. **1** [1971] 127/8).
[13] Gilbreath, W. P.; Cady, G. H. (Inorg. Chem. **2** [1963] 496/9).
[14] Woolf, A. A. (J. Chem. Soc. **1950** 1053/6).
[15] Surles, T.; Hyman, H. H.; Quarterman, L. A.; Popov, A. I. (Inorg. Chem. **10** [1971] 913/6).
[16] Emeléus, H. J.; Woolf, A. A. (J. Chem. Soc. **1950** 164/8).
[17] Woolf, A. A. (Diss. Cambridge 1950, ref. 21 in: Gutmann, V.; Angew. Chem. **62** [1950] 312/5).
[18] Kwasnik, W. (unpublished results, ref. 5 in: Rüdorff, W.; FIAT Rev. Ger. Sci. 1939/46 **23** I [1949] 239/55, 241/2).
[19] Opalovskii, A. A.; Nazarov, A. S.; Uminskii, A. A. (Izv. Sibirsk. Otd. Akad. Nauk SSSR Ser. Khim. Nauk **1972** No. 2, pp. 52/5).
[20] Danilenko, A. M.; Nazarov, A. S.; Yakovlev, I. I. (Zh. Neorg. Khim. **34** [1989] 1693/6; Russ. J. Inorg. Chem. **34** [1989] 958/61).

[21] Christe, K. O.; Schack, C. J. (Inorg. Chem. **9** [1970] 2296/9).
[22] Sharpe, A. G.; Emeléus, H. J. (J. Chem. Soc. **1948** 2135/8).
[23] Whitney, E. D.; McLaren, R. O.; Fogle, C. E.; Hurley, T. J. (J. Am. Chem. Soc. **86** [1964] 2583/6).
[24] Chrétien, A.; Bouy, P. (Compt. Rend. **246** [1958] 2493/5).
[25] Bouy, P. (Ann. Chim. [Paris] [13] **4** [1959] 853/90).
[26] Schmeisser, M.; Taglinger, L. (Angew. Chem. **71** [1959] 523).
[27] Schmeisser, M.; Taglinger, L. (Chem. Ber. **94** [1961] 1533/9).
[28] Seel, F.; Birnkraut, W.; Werner, D. (Angew. Chem. **73** [1961] 806).
[29] Christe, K. O.; Schack, C. J. (Inorg. Chem. **9** [1970] 1852/8).
[30] Aynsley, E. E.; Hetherington, G.; Robinson, P. L. (J. Chem. Soc. **1954** 1119/24).

[31] Martin, D. (Compt. Rend. C **265** [1967] 919/22).
[32] Surles, T.; Hyman, H. H.; Quarterman, L. A.; Popov, A. I. (Inorg. Chem. **9** [1970] 2726/30).
[33] Woolf, A. A.; Emeléus, H. J. (J. Chem. Soc. **1949** 2865/71).
[34] Sakurai, T. (J. Nucl. Sci. Technol. [Tokyo] **10** [1973] 130/1).
[35] Woolf, A. A. (J. Chem. Soc. **1954** 4113/6).
[36] Ruff, O.; Braida, A. (Z. Anorg. Allgem. Chem. **214** [1933] 91/3).
[37] Jonas, H. (Z. Anorg. Allgem. Chem. **265** [1951] 273/83).
[38] Jonas, H. (Ger. Appl. R 99870 [1937], ref. 13 in [37]).
[39] Christe, K. O.; Curtis, E. C.; Schack, C. J.; Roland, A. (Spectrochim. Acta A **33** [1977] 69/73).

[40] Wessel, J.; Kleemann, G.; Seppelt, K. (Chem. Ber. **116** [1983] 2399/407).

[41] Cotton, F. A.; George, J. W. (J. Inorg. Nucl. Chem. **7** [1958] 397/403).
[42] Bartlett, N.; Robinson, P. L. (J. Chem. Soc. **1961** 3417).
[43] Seppelt, K. (Chem. Ber. **106** [1972] 157/64).
[44] Fischer, J.; Steunenberg, R. K. (ANL-5593 [1956] 19 pp.; C.A. **1956** 16449).
[45] Schmeisser, M.; Fink, W. (Angew. Chem. **69** [1957] 780).
[46] Woolf, A. A. (Chem. Ind. [London] **1954** 346).
[47] Irsa, A. P.; Friedman, L. (J. Inorg. Nucl. Chem. **6** [1958] 77/90).
[48] Olah, G.; Pavláth, A.; Kuhn, I. (J. Inorg. Nucl. Chem. **7** [1958] 301/2).
[49] Richards, G. W.; Woolf, A. A. (J. Fluorine Chem. **1** [1971] 129/39).
[50] Aynsley, E. E.; Sampath, S. (J. Chem. Soc. **1959** 3099).

[51] Pullen, K. E. (Diss. Seattle 1967, 75 pp., 51; C.A. **69** [1968] No. 24083).
[52] Wilson, W. W.; Christe, K. O. (Inorg. Chem. **26** [1987] 1573/8).
[53] Edwards, A. J.; Jones, G. R. (J. Chem. Soc. A **1969** 1467/70).
[54] Sheft, J.; Katz, J. J. (Anal. Chem. **29** [1957] 1322/5).
[55] Sheft, J.; Martin, A.; Katz, J. (J. Am. Chem. Soc. **78** [1956] 1557/9).
[56] Fischer, J.; Liimatainen, R.; Bingle, J. (J. Am. Chem. Soc. **77** [1955] 5848/9).
[57] Woolf, A. A. (J. Chem. Soc. **1955** 433/43).

10.10.14.5.4 Reactions with Metal Compounds

The reactions of BrF_3 with metal compounds are listed in Table 11. More details for some reactions are given in the following text.

Enthalpies of reaction were determined in BrF_3-Br_2 (mole ratio 7.04:1) solution at 298 K for the following reactions [1]:

reaction	Δ_rH in kJ/mol
$3\ KBr + BrF_3(sln) \rightarrow 3\ KF(sln) + 2\ Br_2(sln)$	-161.4 ± 1.2
$2\ KIO_3 + 4\ BrF_3(sln) \rightarrow 2\ KF(sln) + 2\ IF_5(sln) + 2\ Br_2(sln) + 3\ O_2$	-395.2 ± 0.8
$6\ K_2SO_4 + 4\ BrF_3(sln) \rightarrow 6\ KF(sln) + 6\ KSO_3F(sln) + 2\ Br_2(sln) + 3\ O_2$	-190.0 ± 1.5
$6\ K_2S_2O_8 + 4\ BrF_3(sln) \rightarrow 12\ KSO_3F(sln) + 2\ Br_2(sln) + 3\ O_2$	-231.1 ± 2.1

An average cryoscopic constant of 7.53 [2] to 7.62 mol/kg [3] was measured for BrF_3 with alkali fluorides; see also [4].

Sorption of gaseous BrF_3 on powdered NaF at 380 K proceeds by diffusion of sorbed BrF_3 into the crystals and reaches a ratio of NaF:BrF_3 = 3, which probably is also the composition of the formed intermediate complex. There is no sorption beyond 450 K. Desorption sets in after heating to 410 K. The sorption-desorption process increases the surface area of the NaF sample [5].

Catalytic amounts of RbF or CsF induce disproportionation of BrF_3 in a CCl_3F solution at 233 K with formation of BrF and BrF_5. The reaction of BrF_3 with CsF (mole ratio 2:1) in CCl_3F at 195 K also leads to disproportionation, but the expected formation of $CsBrF_4$ does not take place [6]. Mixtures of BrF_3, HF(l), and RbF [74] or CsF [75] at 293 K yielded the solids $MF \cdot 2\ BrF_3$ (M = Rb, Cs), $MBrF_4$, $RbF \cdot 3.5\ HF$, and $CsF \cdot 3\ HF$ upon varying the proportions of the reactants.

The reaction of excess BrF_3 and SnF_4 yields $(BrF_2)_2SnF_6$ [76, 77]. The products $SnF_4 \cdot n\ BrF_3$ with n=0.9, 1.0, 1.7 were synthesized by decomposing $(BrF_2)_2SnF_6$ or by reacting BrF_3 with SnF_4 at various molar ratios over varying lengths of time. Preparative details were not given [77]. The isolation of $SnF_4 \cdot m\ BrF_3$ with m=1, 1.7, 2 at about 313 K from mixtures of BrF_3 and SnF_4 was mentioned [78, 79]. However, later it was stated that stoichiometric $SnF_4 \cdot 2\ BrF_3$ at 298 K is impossible to isolate [16]. The heat of solution of SnF_4 in BrF_3-Br_2 (molar ratio 7.04:1) at 298 K is -58.8 ± 0.7 kJ/mol [16] and probably includes a component for the heat of reaction.

Chlorine substitution in M_2PtCl_6 (M = alkali metal) by BrF_3 is a stepwise process. The composition of the resulting product mixtures of $M_2PtCl_{6-n}F_n$ (n=1 to 6) depends on the reaction time, temperature, and the use of a moderator, e.g., $Cl_2FCCClF_2$. Long reaction times and high temperature favor the formation of $[PtF_6]^{2-}$. Separation of the mixtures is described in [7 to 9]. The isolated $M_2PtCl_3F_3$ described in earlier papers [10, 11] is actually a product mixture [9].

BrF_3 reacts rapidly and quantitatively at temperatures below its boiling point with most uranium compounds and forms the volatile UF_6 [12]. Individual results are given in "Uran" Suppl. Vol. A 3, 1981, pp. 133/4, Suppl. Vol. C 2, 1978, pp. 121, 124, 235, and Suppl. Vol. C 8, 1980, pp. 78/80. The solid-liquid phase diagram of BrF_3-UF_6 is displayed in "Uran" Suppl. Vol. C 8, 1980, p. 268. A more recent investigation of the kinetics of UO_2 conversion to UF_6 by BrF_3 is described in [13].

Table 11
Reactions of BrF_3 with Metal Compounds.

reactant	products	conditions, remarks
LiCl	LiF	product quite insoluble in BrF_3, see p. 102 [14]
LiBr	–	reactant insoluble in BrF_3 [4]
Li_2CO_3	LiF	evolvement of 1/3 of the bonded O_2 [15]
NaX (X = F, Cl, Br, I)	$NaBrF_4$	impure products with initial formation of NaF for X ≠ F [14], see also pp. 109/10
		the observed heat upon NaF dissolution is −50.5(5) kJ/mol at 298 K [16]
$Na_2B_4O_7$	$NaBF_4$	[17]
Na_2CO_3	$NaBrF_4$	[15]
Na_3PO_4	$NaBrF_4$, $NaPF_6$	sodium and potassium metaphosphate yield insoluble MPF_6 after boiling [15]
$Na_2S_2O_4$, Na_2SO_3	$NaBrF_4$	the isolated solid is free of sulfur and its compounds [15]
$Na_2S_2O_8$	$NaSO_3F$	[18]
KOH, $KF \cdot 2\ H_2O$		violent reaction [19]
KF, KNO_3	$KBrF_4$	reactions with good yields [17, 20 to 23]
		the observed heat of KF dissolution is −91.9(6) to −94.8(5) kJ/mol [1, 16]
KX (X = Cl, Br, I)	$KBrF_4$, Cl_2 or Br_2	the violence of the reaction increases from KCl to KI, iodine is fluorinated [14]
$KClO_3$	$KBrF_4$, ClO_2F, O_2	[15, 24, 25]
$KBrO_3$, KIO_3	$KBrF_4$, O_2	quantitative liberation of O_2 [15, 26]; see also p. 85
$KBrO_2F_2$	$KBrF_4$	[26]

Table 11 (continued)

reactant	products	conditions, remarks
K_2SO_4, $K_2S_2O_8$	KSO_3F	additional $KBrF_4$ from K_2SO_4 [1, 15, 18]; see also p. 85
$K_2S_2O_3$, $K_2S_2O_5$	$KBrF_4$, KSO_3F	[15]
$K_2(NO)_2SO_3$	$KBrF_4$, KSO_3F	equimolar mixture of the products [27]
RbCl	$RbBrF_4$	impure product [14]
RbF	$RbF \cdot 2\ BrF_3$	reaction with the double molar quantity of BrF_3 containing 4% HF at 323 K, 78% yield [28]; see also p. 105
$RbBrF_4$	$RbF \cdot n\ BrF_3$	heating with appropriate amounts of BrF_3 yields an impure product for n=2 and a pure product for n=3 [29]; see also p. 105
CsCl	$CsBrF_4$	impure product [14]
CsF	$CsBrF_4$	at 353 K or from the vapor over a Br_2 solution of BrF_3 at room temperature [20, 30]
	$CsBrF_2$	31% in CsF from the vapor over a Br_2 solution of BrF_3 at 273 K [30]; see also p. 105
	$CsF \cdot 2\ BrF_3$	99% yield after the reaction with the double molar quantity of BrF_3 containing 10% HF at 323 K [28]; see p. 105
$CsBrF_4$	$CsF \cdot n\ BrF_3$	n=2, 3 with appropriate quantities of BrF_3 in about quantitative yields [29]; see p. 105
$CsClF_4$, Cs_2XeF_6	$CsBrF_4$	and ClF_3 or XeF_4 at increased temperature [31]
BeO	BeF_2	about 20% yield after boiling [15, 32]
		complete fluorination at 590 K in a Ni bomb tube [23]
$BeCO_3$	BeF_2	with evolvement of a third of the bonded O_2 [15]
MgO	MgF_2	about 30% yield after boiling, no reaction at 350 K [15, 32, 33]
		incomplete fluorination at 590 K in a Ni bomb tube [23]
CaO, $Ca(OH)_2$	CaF_2	violent and incomplete reaction [15, 19]
CaF_2, $CaCl_2$		ill-defined products of varying Br content [14]
$CaCO_3$	CaF_2	quantitative with evolvement of CO_2 and O_2 (2:1) [34]
SrF_2		ill-defined products of varying Br content [14]
BaF_2	$Ba(BrF_4)_2$	about 50 mol% in unreacted BaF_2, the heat of BaF_2 dissolution is −88.75(5) kJ/mol at 298 K [16]
$BaCl_2$	$Ba(BrF_4)_2$, Cl_2	99% yield [14]
$BaSO_4$	$Ba(BrF_4)_2$	with formation of SO_3F^- [33]
Al_2O_3	AlF_3	7% yield after boiling [15, 32]
		incomplete reaction at 590 K in a Ni bomb tube [23]
$AlCl_3$	AlF_3	very slow and incomplete reaction [14]
Tl_2O_3	TlF_3	quantitative conversion [15]

Table 11 (continued)

reactant	products	conditions, remarks
$TlCl$	TlF, TlF_3	insoluble product mixture, 80 to 90% TlF_3 [14]
$TlIO_3$	TlF_3	with quantitative evolvement of O_2 [15]
SnO_2	SnF_4	in 20% yield after boiling [15, 32]
SnF_4	$SnF_4 \cdot n\ BrF_3$	n=0.9 to 2, see p. 86
$SnCl_2$, $SnCl_4$	$SnF_4 \cdot 1.76\ BrF_3$	$SnF_4 \cdot 2\ BrF_3$ is stable in solution only; $SnCl_2$ reacts with incandescence, dilution with Br_2 is required [35]
		formation of an adduct $SnCl_2 \cdot n\ BrF_3$ was mentioned [16]
$(NO)_2SnCl_4$	$(NO)_2SnF_6$	the product contains about 12% of SnF_4 [27]
PbO, PbO_2	PbF_4	incomplete reactions after boiling [15, 32]
PbX_2 (X=F, Cl, Br, IO_3)	PbF_2, PbF_4	product mixture with up to 20% PbF_4 [12, 14]
Bi_2O_3	BiF_3	41 to 87% yield in boiling BrF_3 [12, 32]
		complete fluorination at 590 K in a Ni bomb tube [23]
Bi_2O_3	BiF_3, $BiOF_3$	[36]
BiF_3	$BiF_3 \cdot n\ BrF_3$	the adduct is relatively stable [37]
BiF_5	BrF_2BiF_6	the only compound in the BrF_3–BiF_5 system, precipitates in HF solution [38]
$Bi(O)Cl$	BiF_3	[15]
$NOBiCl_4$	BiF_3	a NO^+ salt does not form [27]
$BiPO_4$	BiF_3, PF_5	with quantitative evolvement of O_2 [15, 39]
TiO_2	TiF_4	with quantitative evolvement of O_2 [15]
$TiCl_4$	TiF_4	vigorous reaction with incandescence, impure product [2]
$TiBr_4$	$TiBrF_4$	exothermal reaction in CCl_4 at 263 K [40]
K_2TiF_6	$K_2TiF_6 \cdot 0.95\ BrF_3$	product contains $KBrF_4$ due to solvolysis or incomplete reaction [41]
ZrO_2	ZrF_4	13% yield in boiling BrF_3 [15, 32]
		complete conversion at 590 K in a Ni bomb tube [23]
VO_2	VF_5	exclusive product at 293 to 343 K [31]
V_2O_5, $VOCl_3$	VOF_3	main product [12, 42]
$NaVO_3$	$NaVOF_4$	prolonged boiling yields $NaVF_6$ [43]
VF_5		dissolves without formation of VF_6^- in appreciable amounts [44]
VCl_3	VF_5	[42]
M_2O_5 (M=Nb, Ta)	BrF_2MF_6	[36], volatilization of Nb_2O_3 by BrF_3 starts at 370 K [45]
$LiMO_3$ (M=Nb, Ta)	$LiMF_6$	in the case of Nb the product contains $LiNbOF_4$ in the reaction at 293 and at 399 K [46]
TaF_5	BrF_2TaF_6	the heat of TaF_5 dissolution is −45.2(8) kJ/mol at 298 K [16]

Table 11 (continued)

reactant	products	conditions, remarks
XMF_6	BrF_2MF_6, XF	with X = XeF, ClF_2 and M = Nb, Ta in HF solution [31]
Cr_2O_3	$CrOF_n$, O_2	n is unknown, the product contains BrF_3 [43]
		no quantitative fluorination at 590 K in a Ni bomb tube [23]
CrO_3	$CrOF_3 \cdot n\ BrF_3$	n = 0.25 to 0.3 after drying at 370 K [47, 48]
	$CrF_3 \cdot BrF_3$	after drying at 390 K [47]
M_2CrO_4 (M = K, Cs)	$MCrOF_4$, $MBrF_4$	[43, 48]
$M_2Cr_2O_7$ (M = K, Cs, Ag)	$MCrOF_4$	contains BrF_3 for M = K after drying at 425 K [43, 47, 48]
Cs_2CrOCl_4	$CsCrOF_4$, $CsBrF_4$	[48]
CrF_4	$CrF_3 \cdot 0.5\ BrF_3$	forms during boiling, no reaction at 370 K [49]
MoO_3	$MoOF_4$	at 350 K [33]
	MoF_6	[32], the apparent activation energy of the reaction with gaseous BrF_3 is 15.5 kJ/mol at 370 K [45]
$MoBr_2$	MoF_6	main product [42]
$M_2M'F_6$ (M = K, Rb, Cs; M' = Mo, W)	$MBrF_4$	residue after evaporation of volatile $M'F_6$ [50]
MnO_2		incomplete reaction when boiled, no reaction at 350 K [15, 33]
	MnF_n	n = 3 or 4, complete reaction at 590 K in a Ni bomb tube [23]
$KMnO_4$	$KMnF_5$	[43]
MnF_4	$BrF_2Mn_2F_9$	at 291 to 298 K with liquid or gaseous BrF_3 [51]
$MnCl_2$	MnF_3, Cl_2, Br_2	product of low solubility [2]
$Mn(IO_3)_2$	MnF_3, O_2, IF_5	the product solution contains Mn^{IV} [43]
	MnF_2, MnF_3	the products contain BrF_3 [15]
Re_2O_7	ReF_3O_2	about 20% of the volatile product forms in the presence of one equivalent of KBr and contains ~5% of $ReFO_3$ [52]
$MReO_4$ (M = Na to Cs, Ag)	$MReF_4O_2$	quiet, exothermic reactions, $NaReF_4O_2$ contains $NaBrF_4$ [53]
$M'(ReO_4)_2$ (M' = Ca, Sr, Ba)	$M'(ReF_4O_2)_2$	quiet, exothermic reactions, incomplete for M' = Ca, Sr [53]
ReF_3N, $ReF_5(NCO)$	$ReF_5(NBr)$	also from ReF_6 and 1) $[(CH_3)_3Si]_2NBr$, 2) BrF_3 [54]
Fe_2O_3	FeF_3	incomplete reaction in a Ni bomb tube at 590 K [15, 23]
$FeCl_3$	FeF_3	slow and incomplete reaction at 390 K [14]
Co_2O_3	CoF_3	complete fluorination at 590 K in a Ni bomb tube [23]
$CoCl_2$, $Co(IO_3)_2$	CoF_2, CoF_3	[14, 15, 55]
NiO	NiF_2	complete fluorination at 590 K in a Ni bomb tube [23]

Table 11 (continued)

reactant	products	conditions, remarks
RuO_2	RuF_6	formation of volatile RuF_6 starts at 425 K with gaseous BrF_3 [45]
$KRuO_4$	$KRuF_6$	[56]
RhF_4		a stable product could not be isolated [57]
Na_3RhCl_6	Na_3RhF_7	[57], product composition Na_2RhF_6 given in [50]
$RhBr_3$	RhF_4	the product contains Br [57]
$PdCl_2$, $PdBr_2$	$PdF_3 \cdot BrF_3$	approximate composition after drying at 293 K [57]
K_2PdCl_4	K_2PdF_5	after drying at 550 K, with a low BrF_3 content at lower temperature [57, 58]
M_2PdCl_6 (M=K, Rb, Cs)	M_2PdF_6	after drying at 420 K, 10% BrF_3 content for M=K [58]
OsO_4	OsO_3F_2	forms in presence of KBr, isolated by sublimation [56]
$OsBr_4$	OsO_3F_2	and a compound of unknown composition [56, 59]
M_2OsCl_6 (M=K, $N(C_4H_9)_4$)	$M_2OsF_nCl_{6-n}$	n=1 to 6, composition of the mixture depends on the reaction conditions; in $ClFCCClF_2$ [60 to 62]
$MOs(NO)Cl_5$ (M=K, Rb, Cs)	$MOs(NO)F_5$	[63]
K_2OsBr_4, K_2OsI_4	K_2OsF_6	exclusive product [61]
Ir_2Br_7		two products of unknown composition [59]
M_2IrCl_6 (M=K, Rb, $N(C_4H_9)_4$)	$M_2IrF_nCl_{6-n}$	in $Cl_2FCCClF_2$, n=1 to 6; composition of the mixture depends on the reaction conditions [9, 64]
$[IrF_2Cl_4]^{2-}$	$[IrF_nCl_{6-n}]^{2-}$	n=3, 4, in equal amounts [64]
$[N(C_2H_5)_4]_2IrF_nCl_{6-n}$ (n=3 to 6)	$[N(C_2H_5)_4]$-IrF_nCl_{6-n}, $N(C_2H_5)_4F$, Br_2	oxidation of Ir in CH_2Cl_2 at 243 K [65]
Cs_2IrF_5Cl	$CsIrF_5Cl$, CsF	in HF solution [66]
PtF_4	$PtF_4 \cdot 2\,BrF_3$	[80]; and unisolable $PtF_4 \cdot 7\,BrF_3$ [81, 82]
$Pt(O_2)F_6$	$PtF_4 \cdot 2\,BrF_3$	[67]
$PtCl_4$, $PtBr_4$	$PtF_4 \cdot 2\,BrF_3$	by drying at 293 K; PtF_4 by drying at 450 K [57]
K_2PtF_6, K_2PtBr_6	$K_2PtF_6 \cdot 1.1\,BrF_3$	by drying at 293 K [57]
K_2PtCl_4	$K_2PtF_{6-n}Cl_n$	n=1 to 3 [68], an earlier report of n=3 [10] is erroneous
M_2PtCl_6 (M=Na to Cs, NO)	M_2PtF_6	stepwise substitution of Cl, see p. 86 [10, 11, 58, 69]
M_2PtX_6 (M=K, Rb, Cs; X=Br, I)	M_2PtF_6	more vigorous reaction than for X=Cl, the products contain small amounts of Br or I [10]
cis-, trans-$[Pt(NH_3)_2Cl_2]$	$[Pt(NH_3)_2Cl_2F(OH)] \cdot 6\,H_2O$	in Freon-113, reactions of the other Pt-amine complexes were also described [70]

Table 11 (continued)

reactant	products	conditions, remarks
$Na_2PtCl_6 \cdot 6\,H_2O$	$Na_2[Pt(OH)F_5]$	same product from moist Na_2PtCl_6 [68]
$Na_2[Pt(OH)_2Cl_4]$	$Na_2[Pt(OH)_2F_4]$	[68]
CuO	CuF_2	with quantitative evolvement of O_2 [15] incomplete reaction at 590 K in a Ni bomb tube [23]
Cu_2Cl_2	Cu_2F_2, CuF_2	[14]
AgF, AgCl	$AgBrF_4$	98 to 99% yield in mild reactions [14]
AgBr	$AgBrF_4$	with initial formation of AgF; the heat of AgBr dissolution is −36.15(25) kJ/mol at 298 K [16]
Ag_3PO_4	$AgBrF_4$, $AgPF_6$	in a molar ratio of 2:1 [15]
AuF_3, $AuCl_3$	$AuF_3 \cdot BrF_3$	[71]; exclusive product with excess BrF_3 [82]
$NaAuCl_4$	$NaAuF_4$	the product contains BrF_3 [71]
MO (M = Zn, Cd, Hg)	MF_2	in yields of 23% (M = Zn), 30% (M = Cd), 7% (M = Hg) from boiling BrF_3 [15, 32] complete fluorination at 590 K in a Ni bomb tube only for M = Hg [23]
CdI_2	CdF_2	[14]
$Hg_2(IO_3)_2$	HgF_2	with quantitative evolvement of O_2 [15]
La_2O_3	LaF_3	incomplete reaction at 590 K in a Ni bomb tube [23]
CeO_2	CeF_4	incomplete reaction in boiling BrF_3 [15] quantitative reaction at 590 K in a Ni bomb tube [23]
Pr_2O_3, Nd_2O_3		the dry compounds do not react with BrF_3 [72]
$M_2(C_2O_4)_3 \cdot n\,H_2O$ (M = Pr, Nd)	MF_3	[72]
ThO_2	ThF_4	incomplete reaction in boiling BrF_3 [15]
UO_2	UF_6	complete reaction, see p. 86
$NOUF_6$	UF_6	relatively slow and incomplete reaction [73]
NpO_2, NpF_4	NpF_6	see "Transuranium Elements" C, 1972, pp. 109, 113
PuO_2	PuF_4	readily separable from the volatile UF_6 [13]

References:

[1] Richards, G. W.; Woolf, A. A. (J. Chem. Soc. A **1969** 1072/6).
[2] Bouy, P. (Ann. Chim. [Paris] [13] **4** [1959] 853/90).
[3] Martin D. (Rev. Chim. Miner. **4** [1967] 367/97).
[4] Chretien, A.; Martin, D. (Compt. Rend. C **263** [1966] 235/8).
[5] Sakurai, T. (Inorg. Chem. **11** [1972] 3110/2).
[6] Naumann, D.; Lehmann, E. (J. Fluorine Chem. **5** [1975] 307/21).

[7] Preetz, W.; Erlhöfer, P. (Z. Naturforsch. **44b** [1989] 412/8).
[8] Erlhöfer, P.; Preetz, W. (Z. Naturforsch. **44b** [1989] 619/26).
[9] Preetz, W.; Kühl, H. (Z. Anorg. Allgem. Chem. **425** [1976] 97/103).
[10] Brown, D. H.; Dixon, K. R.; Sharp, D. W. A. (J. Chem. Soc. A **1966** 1244/6).

[11] Dixon, K. R.; Sharp, D. W. A.; Sharpe, A. G. (Inorg. Synth. **12** [1970] 232/7).
[12] Emeléus, H. J.; Maddock, A. G.; Miles, G. L.; Sharpe, A. G. (J. Chem. Soc. **1948** 1991).
[13] Galkin, N. P.; Veryatin, U. D.; Zuev, V. A. (Radiokhimiya **1980** 750/3; Soviet Radiochem. **22** [1980] 578/81).
[14] Sharpe, A. G.; Emeléus, H. J. (J. Chem. Soc. **1948** 2135/8).
[15] Emeléus, H. J.; Woolf, A. A. (J. Chem. Soc. **1950** 164/8).
[16] Richards, G. W.; Woolf, A. A. (J. Fluorine Chem. **1** [1971] 129/39).
[17] Woolf, A. A.; Emeléus, H. J. (J. Chem. Soc. **1950** 1050).
[18] Richards, G. W.; Woolf, A. A. (J. Chem. Soc. A **1968** 470/6).
[19] Rosen, F. D. (NAA-SR-213 [1975] 34 pp.; N.S.A. **11** [1957] No. 9637).
[20] Christe, K. O.; Schack, C. J. (Inorg. Chem. **9** [1970] 2296/9).

[21] Goldberg, G.; Meyer, A. S., Jr.; White, J. C. (Anal. Chem. **32** [1960] 314/7).
[22] Surles, T.; Hyman, H. H.; Quarterman, L. A.; Popov, A. I. (Inorg. Chem. **9** [1970] 2726/30).
[23] Sheft, I.; Martin, A. F.; Katz, J. J. (J. Am. Chem. Soc. **78** [1956] 1557/9).
[24] Woolf, A. A. (J. Chem. Soc. **1954** 4113/6).
[25] Woolf, A. A. (Chem. Ind. [London] **1954** 346).
[26] Tantot, G.; Bougon, R. (Compt. Rend. C **281** [1975] 271/3).
[27] Woolf, A. A. (J. Chem. Soc. **1950** 1053/6).
[28] Sukhoverkhov, V. F.; Tanakova, N. D. (U.S.S.R. 559897 [1977] 2 pp.; C.A. **87** [1977] No. 119976).
[29] Stein, L. (J. Fluorine Chem. **27** [1985] 249/56).
[30] Surles, T.; Quarterman, L. A.; Hyman, H. H. (J. Inorg. Nucl. Chem. **35** [1973] 668/70).

[31] Popov, A. J.; Sukhoverkhov, V. F.; Chumaevskii, N. A. (Zh. Neorg. Khim. **35** [1990] 1111/22; Russ. J. Inorg. Chem. **35** [1990] 626/32).
[32] Woolf, A. A. (Diss. Cambridge 1950; ref. 21 in: Gutmann, V.; Angew. Chem. **62** [1950] 312/5).
[33] Hoekstra, H. R.; Katz, J. J. (Anal. Chem. **25** [1953] 1608/12).
[34] Clayton, R. N. (J. Chem. Phys. **34** [1961] 724/6).
[35] Woolf, A. A.; Emeléus, H. J. (J. Chem. Soc. **1949** 2865/71).
[36] Gutmann, V.; Emeléus, H. J. (J. Chem. Soc. **1950** 1046/50).
[37] Fischer, J.; Rudzitis, E. (J. Am. Chem. Soc. **81** [1959] 6375/7).
[38] Popov, A. J.; Scharabarin, A. V.; Sukhoverkhov, V. F.; Tchumaevsky, N. A. (Z. Anorg. Allgem. Chem. **576** [1989] 242/54).
[39] Tudge, A. P. (Geochim. Cosmochim. Acta **18** [1960] 81/93).
[40] Lange, G.; Dehnicke, K. (Naturwissenschaften **53** [1966] 38).

[41] Sharpe, A. G. (J. Chem. Soc. **1950** 2907/8).
[42] Emeléus, H. J.; Gutmann, V. (J. Chem. Soc. **1949** 2979/82).
[43] Sharpe, A. G.; Woolf, A. A. (J. Chem. Soc. **1951** 798/801).
[44] Fowler, B. R.; Moss, K. C. (J. Fluorine Chem. **14** [1979] 485/94).
[45] Sakurai, T. (J. Nucl. Sci. Technol. **10** [1973] 130/1).
[46] Rakov, E. G.; Melkumyants, M. V.; Sukhoverkhov, V. F. (Zh. Neorg. Khim. **35** [1990] 1123/6; Russ. J. Inorg. Chem. **35** [1990] 632/5).
[47] Clark, H. C.; Sadana, Y. N. (Can. J. Chem. **42** [1964] 702/4).
[48] Hope, E. G.; Jones, P. J.; Levason, W.; Ogden, J. S.; Tajik, M.; Turff, J. W. (J. Chem. Soc. Dalton Trans. **1984** 2445/7).

[49] Clark, H. C.; Sadana, Y. N. (Can. J. Chem. **42** [1964] 30/6).
[50] Cox, B.; Sharp, D. W. A.; Sharpe, A. G. (J. Chem. Soc. **1956** 1242/4).

[51] Sukhoverkhov, V. F.; Melkumyants, M. V. (Tr. Inst. Mosk. Khim. Tekhnol. Inst. im. D. J. Mendeleeva No. 143 [1986] 62/4; C.A. **108** [1988] No. 197206).
[52] Beattie, I. R.; Crocombe, R. A.; Ogden, J. S. (J. Chem. Soc. Dalton Trans. **1977** 1481/9).
[53] Peacock, R. D. (J. Chem. Soc. **1955** 602/3).
[54] Fawcett, J.; Peacock, R. D.; Russell, D. R. (J. Chem. Soc. Dalton Trans. **1987** 567/71).
[55] Gall, J. F.; Miller, H. C.; Verdelli, L. S.; Loomis, F. D. (information presented at Am. Chem. Soc. Meeting, New York 1947; ref. 5 in: Booth, H. S.; Pinkston, J. T.; Fluorine Chem. **1** [1950] 189/200, 197).
[56] Hepworth, M. A.; Robinson, P. L. (J. Inorg. Nucl. Chem. **4** [1957] 24/9).
[57] Sharpe, A. G. (J. Chem. Soc. **1950** 3444/50).
[58] Sharpe, A. G. (J. Chem. Soc. **1953** 197/9).
[59] Hepworth, M. A.; Robinson, P. L.; Westland, G. J. (J. Chem. Soc. **1954** 4269/75).
[60] Preetz, W.; Ruf, D.; Tensfeldt, D. (Z. Naturforsch. **39b** [1984] 1100/9).

[61] Preetz, W.; Petros, Y. (Z. Anorg. Allgem. Chem. **415** [1975] 15/24).
[62] Preetz, W.; Groth, T. (Z. Naturforsch. **41b** [1986] 885/9).
[63] Sinitsyn, M. N.; Svetlov, A. A.; Kokunov, Yu. V.; Fal'kengof, A. T.; Larin, G. M.; Minin, V. V.; Buslaev, Yu. A. (Dokl. Akad. Nauk SSSR **293** [1987] 1144/7; Dokl. Chem. Proc. Acad. Sci. USSR **292/297** [1987] 199/201).
[64] Tensfeldt, D.; Preetz, W. (Z. Naturforsch. **39b** [1984] 1185/92).
[65] Groth, T.; Preetz, W. (Z. Anorg. Allgem. Chem. **548** [1987] 76/88).
[66] Preetz, W.; Tensfeldt, D. (Z. Anorg. Allgem. Chem. **522** [1985] 7/10).
[67] Bartlett, N.; Lohmann, D. H. (J. Chem. Soc. **1962** 5253/61).
[68] Evans, D. F.; Turner, G. K. (J. Chem. Soc. Dalton Trans. **1975** 1238/43).
[69] Cox, B. (J. Chem. Soc. **1954** 3251/2).
[70] Zemskov, S. V.; Levchenko, L. M.; Shipachev, V. A.; Potapova, O. G.; Tkachev, S. V.; Al't, L. Ya. (Zh. Neorg. Khim. **35** [1990] 53/61; Russ. J. Inorg. Chem. **35** [1990] 29/34).

[71] Sharpe, A. G. (J. Chem. Soc. **1949** 2901/2).
[72] Popov, A. J.; Gockler, G. (J. Am. Chem. Soc. **74** [1952] 1357/8).
[73] Chatelet, J.; Luce, M.; Plurien, P.; Rigny, P. (CEA-CONF-3035 [1975] 6 pp.; C.A. **84** [1976] No. 53402).
[74] Sukhoverkhov, V. F.; Takanova, N. D. (Zh. Neorg. Khim. **22** [1977] 2255/60; Russ. J. Inorg. Chem. **22** [1977] 1220/3).
[75] Sukhoverkhov, V. F.; Takanova, N. D.; Uskova, A. A. (Zh. Neorg. Khim. **21** [1976] 2245/9; Russ. J. Inorg. Chem. **21** [1976] 1234/7).
[76] Mit'kin, V. N.; Zemskov, S. V. (Izv. Sibirsk. Otd. Akad. Nauk SSSR Ser. Khim. Nauk **1981** No. 5, pp. 42/6).
[77] Dzevitskii, B. Z.; Sukhoverkhov, V. F. (Izv. Sibirsk. Otd. Akad. Nauk SSSR Ser. Khim. Nauk **1968** No. 2, pp. 54/7).
[78] Sukhoverkhov, V. F.; Dzevitskii, B. Z. (Dokl. Akad. Nauk SSSR **170** [1966] 1099/102; Dokl. Chem. Proc. Acad. Sci. USSR **166/171** [1966] 983/6).
[79] Sukhoverkhov, V. F.; Dzevitskii, B. Z. (Dokl. Akad. Nauk SSSR **177** [1967] 611/4; Dokl. Chem. Proc. Acad. Sci. USSR **172/177** [1967] 1089/91).
[80] Bartlett, N.; Robinson, P. L. (J. Chem. Soc. **1961** 3417).

[81] Mit'kin, V. N.; Nikonorov, Yu. I.; Zemskov, S. V. (Zh. Fiz. Khim. **52** [1978] 486; Russ. J. Phys. Chem. **52** [1978] 276/7).
[82] Mit'kin, V. N.; Zemskov, S. V. (Izv. Sibirsk. Otd. Akad. Nauk SSSR **1980** No. 5, pp. 28/33).

10.10.14.5.5 Reactions with Mixtures of Inorganic Substances

The reactions of mixtures of substances with BrF_3 yield products which are considered to be salts containing fluorinated anions in most cases. The reactions listed in Table 12 are ordered with respect to the anion-forming substances for the individual groups in the periodic system of the elements. In order to obtain the salt, it often suffices to treat equivalent amounts of the acid- and base-forming precursors with BrF_3 and then to remove the excess of the latter under vacuum [1].

Table 12
Reactions of BrF_3 with Mixtures of Inorganic Substances.

reactants	products	reactants	products
B_2O_3-N_2O_4	NO_2BF_4 [2]	B_2O_3-KCl	KBF_4 [2]
-NOCl	$NOBF_4$ [3]	-$KClO_3$	KBF_4 [5]
-BrCN, ICN	$NOBF_4$, NO_2BF_4 [4]		
SiO_2-KCl	K_2SiF_6, $KBrF_4$[a)] [2]	SnF_4-BrCN	$SnF_4 \cdot 1.7\ BrF_3$ [4]
SiF_4-NO	$(NO)_2SiF_6$ [6]	$SnCl_2$-N_2O_4	$(NO_2)_2SnF_6$ [2]
GeO_2-N_2O_4	$(NO_2)_2GeF_6$ [6]	-BrCN	$SnF_4 \cdot 1.7\ BrF_3$ [4]
-NOCl	$(NO_2)_2GeF_6$ [3]	$SnCl_4$-MCl (M = K, Ag)	M_2SnF_6 [9]
-$PdBr_2$	$PdGeF_6$ [7]	-$BaCl_2$	$BaSnF_6$ [9, 10]
SnF_4-N_2O_4	$(NO_2)_2SnF_6$ [8]	$SnBr_4$-$PdBr_2$	$PdSnF_6$ [7]
-NOCl	$(NO)_2SnF_6$ [3, 8]		
P_4O_{10}-BrCN, ICN	$NOPF_6$, NO_2PF_6 [4]	As_2O_3-$K_2(NO)_2SO_3$	$KAsF_6$, $NOAsF_6$ [3]
-Ag	$AgPF_6$ [11]	-$BaCl_2$	$Ba(AsF_6)_2$ [2]
-LiF_6	$LiPF_6$ [12]	MF_5 (M = As, Sb)-Br_2	Br_3MF_6 [13]
-Ag, RbCl, CsCl	MPF_6 [10]	Sb-Ag	$AgSbF_6$ [14]
		Sb_2O_3-N_2O_4	NO_2SbF_6 [2]
PCl_5-KCl, AgCl	MPF_6, $MBrF_4$ [15]	-BrCN, ICN	BrF_2SbF_6 [4]
PBr_5-N_2O_4	NO_2PF_6 [2]	-Li_2CO_3	$LiSbF_6$ [2]
-NOCl	$NOPF_6$ [3]	-$MClO_3$	$MSbF_6$, ClO_2,
-Ag, AgCl	$AgPF_6$ [2]	(M = K, Ag)	BrF_2SbF_6 [5, 16]
-KCl, AgCl	MPF_6 [15]	-$CaCO_3$	$Ca(SbF_6)_2$ [2]
-$BaCl_2$[b)]	$Ba(PF_6)_2$ [2]	SbF_5-N_2O_4	NO_2SbF_6 [2]
As_2O_3-N_2O_4	NO_2AsF_6 [2]	-PF_5, AsF_5	BrF_2SbF_6 [17]
-$NOSO_3F$	$NOAsF_6$ [3]	BiF_5-alkali halide	$MBiF_6$[c)] [18]
-Ag, alkali halide	$MAsF_6$ [2, 10]	BrF_2BiF_6-Ag	$AgBiF_6$ [18]
SO_2-Br_2	SO_2BrF, SO_2F_2[d)] [19]	SO_3-K_2SO_4	KSO_3F [3]
SO_3-NO_2	NO_2SO_3F [3]	-KF	KSO_3F [20]
-Ag	$Ag(SO_3F)_2$ [20]	$(NO)_2S_2O_7$-Ag	$AgSO_3F$ [3, 20]
Br_2-CsF	(no $CsBrF_2$) [21]		
Ti-NOCl	$(NO)_2TiF_6$ [3]	TiO_2-$BaCl_2$	$BaTiF_6$ [10]
TiO_2-KBr	$K_2TiF_6 \cdot 1.1\ BrF_3$, $KBrF_4$ [22]		

Table 12 (continued)

reactants	products	reactants	products
V_2O_5-NOCl	$NOVF_6$ [23]	Nb, Ta-CaF_2, $BaCl_2$	$M(M'F_6)_2$, M = Ca, Ba, M' = Nb, Ta [18]
-BrCN, ICN	$NOVF_6$, NO_2VF_6 [4]		
VF_5-AsF_5	no solid product [17]	Nb_2O_5, Ta_2O_5-N_2O_4	NO_2MF_6 [18]
VCl_3-Ag, alkali halide	MVF_6 [10, 24]	-NOCl	$NOMF_6$ [25]
-$BaCl_2$	$Ba(VF_6)_2$, $Ba(BrF_4)_2$ [24]	Nb_2O_5-Ag, AgCl	$AgNbF_6$ [18]
		TaF_5-SbF_5	$TaF_5 \cdot BrF_3$, $SbF_5 \cdot BrF_3$ [17]
Nb, Ta-alkali halide, Ag	$MM'F_6$ (M = alkali, Ag, M' = Nb, Ta) [14, 18]		
Cr_2O_3-KCl	$KCrOF_4 \cdot 0.5\ BrF_3$ [23]	CrF_4-MCl (M = K, Rb, Cs)	$MCrF_5 \cdot 0.5\ BrF_3$ [e)] [26]
$KMnO_4$-KCl	K_2MnF_6 [23]	ReF_6-$[(CH_3)_3Si]_2NBr$	ReF_5NBr [27]
Ru-KBr, CsCl, AgBr	$MRuF_6$ [f)] [28]	RhF_4-NaF	fluoro complex [30]
-$M(BrO_3)_2$ (M = Ca, Sr, Ba)	$M(RuF_6)_2$ [28]	$RhCl_3$-NaCl, CsCl	M_2RuF_6 [30, 31]
		$PdCl_2$-Ag, KCl	$MPdF_4 \cdot n\ BrF_3$, (n = 0.8, 1.2) [30]
$RuCl_3$-LiCl, NaCl	$MRuF_6$ [29]		
OsO_4-KBr, CsBr, $AgIO_3$	$MOsO_3F$ [32]	Ir_2Br_7-MBr	$MIrF_6$ (M = Li to Cs, Ag) [33]
$OsBr_4$-MBr	$MOsF_6$ (M = Li, Cs, Ag) [33]	-$Ba(BrO_3)_2$	$Ba(IrF_6)_2$ [33]
		$PtCl_4$-NaCl, AgCl	M_2PtF_6 [30]
Ir-KCl	$KIrF_6$ [34]	$PtBr_4$-KBr	$K_2PtF_6 \cdot 1.1\ BrF_3$ [30]
-BaF_2	$Ba(IrF_6)_2$ [34]	-$PdBr_2$	$PdPtF_6$ [7]
		$(BrF_2)_2PtF_6$-KF	$KPtF_6 \cdot 1.1\ BrF_3$ [35]
Au-N_2O_4	NO_2AuF_4 [2]	Au-Ag, AgCl	$AgAuF_4$ [36]
-NOCl	$NOAuF_4$ [2]	-KCl	$KAuF_4 \cdot 0.1\ BrF_3$ [36]

a) In equimolar amounts with excess SiO_2. — b) CaF_2 remains mostly unchanged in the analogous reaction. — c) The products contain BrF_3, the corresponding Ba salt could not be isolated. — d) The SO_2BrF yield is almost quantitative from a saturated solution of SO_2 in Br_2 after adding BrF_3. The main product SO_2F_2 results at high BrF_3 concentration in the Br_2 solution. — e) At molar ratio 1:1, a 1:2 molar ratio yields $M_2CrF_6 \cdot 0.5\ BrF_3$. — f) Reaction of Ru and MBr (M = Na, Tl) yields $RuBrF_8$ and $MBrF_4$, decomposition for M = Tl is extensive.

References:

[1] Stein, L. (Halogen Chem. **1** [1967] 133/224, 162).
[2] Woolf, A. A.; Emeléus, H. J. (J. Chem. Soc. **1950** 1050).
[3] Woolf, A. A. (J. Chem. Soc. **1950** 1053/6).
[4] Woolf, A. A. (J. Chem. Soc. **1954** 252/65).

[5] Woolf, A. A. (Chem. Ind. **1954** 346).
[6] Bouy, P. (Ann. Chim. [Paris] [13] **4** [1959] 853/90).
[7] Bartlett, N.; Rao, P. K. (Proc. Chem. Soc. **1964** 393/4).
[8] Sukhoverkhov, V. F.; Dzevitskii, B. Z. (Dokl. Akad. Nauk SSSR **177** [1967] 611/4; Dokl. Chem. Proc. Acad. Sci. USSR **172/177** [1967] 1089/91).
[9] Woolf, A. A.; Emeléus, H. J. (J. Chem. Soc. **1949** 2865/71).
[10] Cox, B. (J. Chem. Soc. **1956** 876/8).

[11] Sharp, D. W. A.; Sharpe, A. G. (J. Chem. Soc. **1956** 1855/8).
[12] Kemmit, R. D. W.; Russell, D. R.; Sharp, D. W. A. (J. Chem. Soc. **1963** 4408/13).
[13] Glemser, O.; Smalc, A. (Angew. Chem. **81** [1969] 531/2 and unpublished results, ref. 2 in: Smalc, A.; Inst. Josef Stefan I.J.S. Rept. R-612 [1972] 1/7; C.A. **79** [1973] No. 13032).
[14] Muetterties, E. L.; Tullock, C. W. (Preparat. Inorg. React. **2** [1965] 237/99, 262).
[15] Emeléus, H. J.; Woolf, A. A. (J. Chem. Soc. **1950** 164/8).
[16] Woolf, A. A. (J. Chem. Soc. **1954** 4113/6).
[17] Woolf, A. A. (J. Chem. Soc. **1955** 433/43).
[18] Gutmann, V.; Emeléus, H. J. (J. Chem. Soc. **1950** 1046/50).
[19] Jonas, H. (Ger. Appl. R 99870 [1937] ref. 13 in: Jonas, H.; Z. Anorg. Allgem. Chem. **265** [1951] 273/83).
[20] Woolf, A. A. (J. Fluorine Chem. **1** [1971] 127/8).

[21] Naumann, D.; Lehmann, E. (J. Fluorine Chem. **5** [1975] 307/21).
[22] Sharpe, A. G. (J. Chem. Soc. **1950** 2907/8).
[23] Sharpe, A. G.; Woolf, A. A. (J. Chem. Soc. **1951** 798/801).
[24] Emeléus, H. J.; Gutmann, V. (J. Chem. Soc. **1949** 2979/82).
[25] Clark, H. C.; Emeléus, H. J. (J. Chem. Soc. **1958** 190/5).
[26] Clark, H. C.; Sadana, Y. N. (Can. J. Chem. **42** [1964] 50/6).
[27] Fawcett, J.; Peacock, R. D.; Russell, D. R. (J. Chem. Soc. Dalton Trans. **1987** 567/71).
[28] Hepworth, M. A.; Peacock, R. D.; Robinson, P. L. (J. Chem. Soc. **1954** 1197/201).
[29] Boston, J. L.; Sharp, D. W. A. (J. Chem. Soc. **1960** 907/8).
[30] Sharpe, A. G. (J. Chem. Soc. **1950** 3444/50).

[31] Cox, B.; Sharp, D. W. A.; Sharpe, A. G. (J. Chem. Soc. **1956** 1242/4).
[32] Hepworth, M. A.; Robinson, P. L. (J. Inorg. Nucl. Chem. **4** [1957] 24/9).
[33] Hepworth, M. A.; Robinson, P. L.; Westland, G. J. (J. Chem. Soc. **1954** 4269/75).
[34] Zemskov, S. V.; Mit'kin, V. N.; Gornostaev, L. L.; Shipachev, V. A.; Isakova, V. G. (U.S.S.R. 1203021 [1986] 2 pp.; C.A. **104** [1986] No. 112268).
[35] Chernyaev, I. I.; Nikolaev, N. S.; Ippolitov, E. G. (Dokl. Akad. Nauk SSSR **130** [1960] 1041/3; Proc. Acad. Sci. USSR Chem. Sect. **130/135** [1960] 167/9).
[36] Sharpe, A. G. (J. Chem. Soc. **1949** 2901/2).

10.10.14.6 Oxidation of BrF_3 by Fluorine and Fluorides

Oxidation of BrF_3 takes place only with fluorine and its compounds and yields BrF_5. The most common reaction is one with fluorine itself; see the chapter on the formation and preparation of BrF_5, p. 116. Reaction with OF_2 and O_2F_2 yields oxygen and BrF_5. The reaction with OF_2 takes place at temperatures below 470 K [1], while the reaction of cocondensed BrF_3 and O_2F_2 starts during warming at about 130 K with the formation at an unstable intermediate [2]. This was claimed to be BrO_2F_5 and was isolated with 80% yield [3]. Oxidation of BrF_3 by BiF_5 at 450 K leads to BiF_3 and to BrF_5 with about 50% yield at a moderate relative rate [4]. BrF_3 is fluorinated by KrF_2 at room temperature to yield BrF_5 [5]. Oxidation of BrF_3 to BrF_6^+ by $KrF_2 \cdot AuF_5$ was mentioned [6].

$(SeF_3)_2PdF_6$ reacts in warm BrF_3 via: $2\,(SeF_3)_2PdF_6 + 3\,BrF_3 \rightarrow BrF_5 + 4\,SeF_4 + 2\,PdF_3 \cdot BrF_3$ whereby palladium is reduced [7]. By evaporating the solvent from a BrF_3 solution of PtF_5, platinum is reduced and $PtF_4 \cdot 2\,BrF_3$ is formed [8]. The reaction of BrF_3 with excess PtF_6 at ambient temperature with probable formation of BrF_5 was mentioned [9]. A BrF_3 solution of $Ag(SO_3F)_2$ decomposes during evaporation of the solvent with reduction to $AgSO_3F$ and oxidation of BrF_3 [10]. The oxidation of BrF_3 by NpF_6 and PuF_6 with formation of BrF_5 is described in "Transurane" C, 1972, pp. 109, 113. BrF_3 is probably oxidized by CrF_4, CrO_3, and $KRuO_4$ and also by Mn^{IV} ions, when the solvent is evaporated from the BrF_3 solution leaving a residue of MnF_3.

References:

[1] Beal, J.; Pupp, C.; White, W. E. (Inorg. Chem. **8** [1969] 828/30).
[2] Streng, A. G. (J. Am. Chem. Soc. **85** [1963] 1380/5; Chem. Rev. **63** [1963] 607/24, 616).
[3] Grosse, A. V.; Streng, A. G. (U.S. 3341294 [1967] 1 p.; C.A. **67** [1967] No. 110198).
[4] Fischer, J.; Rudzitis, E. (J. Am. Chem. Soc. **81** [1959] 6375/7).
[5] Prusakov, V. N.; Sokolov, V. B. (Zh. Fiz. Khim. **45** [1971] 2950; Russ. J. Phys. Chem. **45** [1971] 1673/4).
[6] Sokolov, V. B.; Prusakov, V. N.; Ryzhkov, A. V.; Drobyshevskii, Yu. V.; Khoroshev, S. S. (Dokl. Akad. Nauk SSSR **229** [1976] 884/7; Dokl. Chem. Proc. Acad. Sci. USSR **226/231** [1976] 525/8).
[7] Bartlett, N.; Quail, J. W. (J. Chem. Soc. **1961** 3728/32).
[8] Bartlett, N.; Lohmann, D. H. (J. Chem. Soc. **1964** 619/26).
[9] Weinstock, B.; Malm, J. G.; Weaver, E. E. (J. Am. Chem. Soc. **83** [1961] 4310/7).
[10] Woolf, A. A. (J. Fluorine Chem. **1** [1971] 127/8).

10.10.14.7 Reactions with Organoelement Compounds

The formation of diaryl bromonium salts from BrF_3 and R_4Sn or R_2Hg is catalyzed by $BF_3 \cdot (C_2H_5)_2O$ and carried out in CH_2Cl_2 below 250 K [1 to 3]. A maximum yield of R_2Br^+ is obtained at reactant ratios given by:

$$R_4Sn + BrF_3 \rightarrow (R_2Br)F + R_2SnF_2 \text{ [2]}$$

$$2\,R_2Hg + BrF_3 \rightarrow (R_2Br)F + 2\,RHgF \text{ [3]}$$

Representative yields of diarylbromonium salts were:

R in R_2Br^+	from R_4Sn [2]	from R_2Hg [3]
C_6H_5	96%	95 to 99%
$4\text{-}CH_3C_6H_4$	68%	80%
$4\text{-}CH_3OC_6H_4$	52%	28%

R_4Sn usually gives higher product yields than R_2Hg. Replacing R_4Sn by R_3SnF results in lower yields; R_2SnF_2 does not react [2]. Oxidation by BrF_3 is extensive in the case of arenes carrying donor substituents, but diminishes when two equivalents of CH_3CN are added to the BrF_3 solution. The diaryl bromonium salts are isolated by crystallization from tetrafluoroborate solutions [2, 3].

The diaryl bromonium salts form in two-step reactions with the intermediate $RBrF_2$ [2, 3]. Asymmetric $RR'Br^+$ salts were obtained in accordance with this mechanism from reactions of BrF_3 with C_6H_6 in equimolar amounts, followed by R_4Sn with $R \neq C_6H_5$ [2] or from reactions at BrF_3 with R_2Hg (mole ratio 1:1) and R'_2Hg addition [3].

The reaction of BrF_3 with bis(9-*m*-carboranyl)mercury in CH_2Cl_2 solution by the described method yields 40% of $(9\text{-}m\text{-}C_2H_2B_{10}H_9)_2Br^+$ [4].

Reactions of BrF_3 with bromodiol or iododiol in $CCl_2F\text{-}CClF_2$ at 253 K yield 99% of the brominane [5] and 80% of the periodinane, respectively [6]:

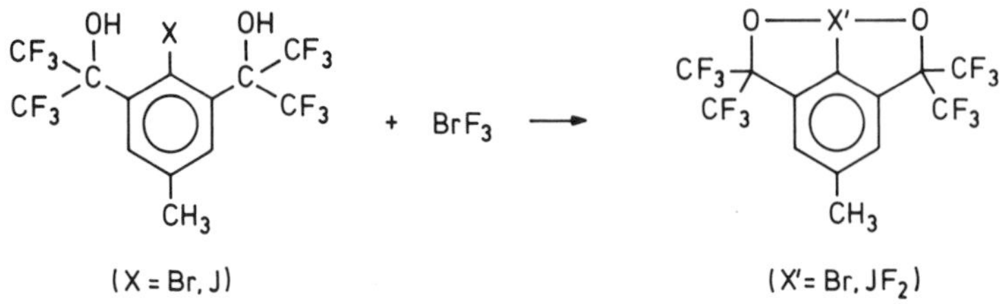

$$+ BrF_3 \longrightarrow$$

(X = Br, J) (X' = Br, JF_2)

Similar reactions of sulfuranes were described [7].

$C_6F_5BrF_2$ forms in a solution of BrF_3 in CCl_3F and added excess $C_6F_5SiF_3$ at 273 K. The product was isolated with 46.7% yield by crystallization at 195 K [8]. The reaction of BrF_3 and $CF_3SeSeCF_3$ yields CF_3SeF_3 and Br_2, when the CCl_3F solution is warmed from 195 to 273 K. The product was separated by vacuum distillation [9].

References:

[1] Nesmeyanov, A. N.; Lisichkina, I. N.; Vanchikov, A. N.; Tolstaya, T. P. (Izv. Akad. Nauk SSSR Ser. Khim. **1977** 1204/5; Bull. Acad. Sci. USSR Div. Chem. Sci. **26** [1977] 1110).

[2] Nesmeyanov, A. N.; Vanchikov, A. N.; Lisichkina, I. N.; Grushin, V. V.; Tolstaya, T. P. (Dokl. Akad. Nauk SSSR **255** [1980] 1386/9; Dokl. Chem. Proc. Acad. Sci. USSR **250/255** [1980] 606/9).

[3] Nesmeyanov, A. N.; Vanchikov, A. N.; Lisichkina, I. N.; Lazarev, V. V.; Tolstaya, T. P. (Dokl. Akad. Nauk SSSR **255** [1980] 1136/40; Dokl. Chem. Proc. Acad. Sci. USSR **250/255** [1980] 594/8).

[4] Grushin, V. V.; Tolstaya, T.P.; Lisichkina, I. N. (Izv. Akad. Nauk SSSR Ser. Khim. **1982** 2412; Bull. Acad. Sci. USSR Div. Chem. Sci. **31** [1982] 2127).

[5] Nguyen, T. T.; Martin, J. C. (J. Am. Chem. Soc. **102** [1980] 7382/3).

[6] Nguyen, T. T.; Amey, R. L.; Martin, J. C. (J. Org. Chem. **47** [1982] 1024/7).

[7] Michalak, R. S.; Martin, J. C. (J. Am. Chem. Soc. **104** [1982] 1683/92).

[8] Frohn, H. J.; Giesen, M. (J. Fluorine Chem. **24** [1984] 9/15).

[9] Lehmann, E. (J. Chem. Res. S **1978** 42).

10.10.14.8 Reactions with Organic Compounds

The reaction of BrF_3 with saturated organic compounds leads to partial substitution by fluorine, primarily; unsaturated compounds react with addition of fluorine and bromine [1]. Solutions of the undiluted starting materials mostly react vigorously, with incandescence, or even explosively [1 to 3]. The reactions are difficult to control and frequently numerous coproducts form [2]. However, some compounds, e.g. highly fluorinated, unsaturated compounds, are quite unreactive in which case BrF_3 is replaced by a 1:1 mixture of BrF_3 and Br_2. The increased reactivity of the mixture may result from the intermediate formation of BrF [4].

Unreactive solvents are required to better control the liquid-phase reactions [2]. The common feature of the solvents used is their low melting point which allows reactions at temperatures below 273 K. The most used solvents are polyhalogenated alkanes. The methane derivatives CH_2Cl_2 [5 to 7], CCl_4 [7, 8], and CCl_3F [9 to 11] need to be cooled when used as solvents in reactions of BrF_3, since CCl_4 and BrF_3 start to react at ambient temperature [12], the CH_2Cl_2 reactivity is similar to that of CCl_4 [7], and the reaction of CCl_3F with BrF_3 at 250 to 270 K is slow, but becomes rather violent at ambient temperature [10]. Ethane derivatives used as solvents are CCl_2FCCl_2F (Freon-112) [13, 14] and CCl_2FCClF_2 (Freon-113) [13, 15, 16]. The latter dissolves slightly more than 0.5 mol of BrF_3/L and does not react with BrF_3 in three weeks at room temperature [17]. The acid anhydrides $(CH_3CO)_2O$ and $(CF_3CO)_2O$ were used at 240 K [8], and decafluoro-*p*-dimethylcyclohexane at about 273 K [4]. Other inorganic solvents used include bromine [18 to 21], HF [22 to 24], SO_2ClF [25], and IF_5; the latter can only be used above its melting point at 283 K [23].

Controlled reactions of BrF_3 with saturated hydrocarbons lead to partial substitution primarily by fluorine, but also by bromine, as shown in reactions with kerosene fractions [6]. The selective substitution of H atoms at tertiary carbon atoms with BrF_3 required a reactant ratio of 3:2. Better yields resulted with excess BrF_3. Formed HF occasionally accelerated the reaction [26]. Selective fluorination of C-H bonds in partly fluorinated ethers is described in [27]. H and Cl in partly chlorinated heptanes were substituted [3]. A cationoid intermediate was indicated by 1,2-migration of the methyl group in 2-methyl-2-chlorobutane [28]. Substitution in the case of CX_4 (X = Cl, Br, I) proceeds stepwise [12]; see "Kohlenstoff" D 2, 1974, pp. 243, 272, 282 for details. Chlorine is selectively substituted by fluorine in the presence of C-H bonds and ether groups [27, 29]. The reaction proceeds stepwise when increasing amounts of BrF_3 are added [27]. The halogen exchange between BrF_3 and primary monochloroalkanes is rapid. To accelerate the reaction with the relatively inert polychloroalkanes, Lewis acids $SnCl_4$ or $SbCl_5$ are required. By-products result from hydride transfer and occasionally from H substitution [14].

The mechanism of fluorination of chloro- and bromoalkanes by BrF_3 was investigated [30]. Polyhalogenated hydrocarbons containing bromine substituents react stepwise [31, 32]. The reaction is selective in the presence of C-H bonds [13, 15, 18, 31], C-Cl bonds, ester groups [13, 15, 18], CN-, C(O)R-, CHO-, C(O)F-, NO_2-, and ether groups [18], for example:

$CH_2Cl\text{-}CHBr\text{-}CH_2Br + 1/3\ BrF_3 \rightarrow CH_2Cl\text{-}CHF\text{-}CH_2Br\ (60\%) + CH_2Cl\text{-}CHBr\text{-}CH_2F\ (20\%)$ [13]

However, the NO_2 group was preferentially substitued in the presence of a bromine substituent under more drastic conditions [33]. Iodine is substituted at lower temperatures than bromine [34, 35]; the reaction is selective in the presence of C-Cl bonds [35].

Only acetone was fluorinated by BrF_3 to yield the expected $CH_3CF_2CH_3$ [23], while other ketones with longer carbon chains reacted with cleavage of C-C bonds [23, 24]. Nitriles, R-CN, were fluorinated to R-CF_3. The fate of the nitrogen atom is not known; C-Cl and C-H bonds in the organic group remained unchanged in some cases [23]. Nitroalkanes yielded fluorinated hydrocarbons, NO_2, and other products in a slightly exothermic reaction [7]. Methanol reacts violently with BrF_3 [36].

Alkenes and BrF_3 react by bromofluorination of the double bond [16, 20, 27, 37, 38]. The ratio of the Markovnikov and anti-Markovnikov product [16, 20, 37] varies with the electron-donating abilities of the carbon substituents. Vicinal or geminal difluorides are occasionally obtained, for example [16]:

$$\begin{array}{llll} CH_2{=}CHR + 1/2\ BrF_3 \rightarrow & CH_2Br\text{-}CHFR & + CH_2F\text{-}CHBrR & + CH_2F\text{-}CHFR \\ (R = CH_2Cl) & 25\% & 11\% & 30\% \end{array}$$

Under appropriate conditions, C-H [16, 27] and C-Cl bonds [37] as well as ether groups [27] are largely unchanged. Highly fluorinated perhalogen alkenes occasionally react with BrF_3 under drastic conditions only [4, 19, 37].

Tetrachloroquinone and BrF_3 yield products that result from the addition of BrF and from the addition at BrF and 2 F [19]. Conjugated dienes react in the same way as alkenes, the addition of BrF and 2 F to the double bonds in cyclo-C_5Br_6 at 288 K is accompanied by bromine substitution [21], while cyclo-C_6F_8 at 273 K yields an equimolar mixture of cyclohexane by addition of BrF and 2 F, and cyclohexene by addition of 2 F atoms [4].

Reactions of BrF_3 with benzene derivatives in solution at about 200 K yield diarylbromonium salts by nucleophilic substitution of fluorine in BrF_3 [5, 8], while the reaction of frozen BrF_3 with toluene at 190 K is violent in the absence of a solvent [39]. Formation of the diarylbromonium salts is a two-step reaction and allows the preparation of various mixed products, for example:

$$BrF_3 + C_6H_6\ (+\ BF_3 \cdot (C_2H_5)_2O,\ CH_3CN) \rightarrow [C_6H_5BrF_2] + HF$$

$$[C_6H_5BrF_2] + C_6H_5F \rightarrow [C_6H_5(4\text{-}FC_6H_4)Br]F + HF$$

The yield is 49% after addition of alkali fluoroborate solution and crystallization [5]. The syntheses of mixed bromonium salts $(RR'Br)BF_4$, with R = phenyl and R' = 9-*m*-carboranyl were carried out by the same method with R'_2Hg [40, 41].

Reactions of benzene derivatives in solution between 273 K and ambient temperature proceed with attack on the aromatic system and bromofluorination. Deactivated (electron-poor) compounds require the more reactive BrF_3-Br_2 mixture or drastic conditions [4]. Bromination of deactivated arenes by BrF_3-Br_2 mixtures was also observed [42]. Monofluorination resulted when BrF_3 was added by a flow of N_2 to excess nitrobenzene containing BF_3 at ambient temperature [45].

Electron-rich compounds, like C_6Br_6, yield mixtures of saturated cyclohexanes [4, 21]. Reactions of other arenes lead to mixtures of cyclohexa-1,4-dienes and cyclohexenes by primary addition of two fluorine atoms and secondary BrF addition [4, 25]. Cyclohexanes were also identified in the product mixtures [19, 25]. Oxidation of pentafluorophenol by BrF_3 yields an isomeric mixture of cyclohexadienones [11].

Bulk reactions of perhalogenated [3, 43] or partly halogenated benzenes [3] require heating and yield mixtures of addition products. However, these reactions frequently lead to explosions in the later heating stages [19].

Perhalogenated naphthalenes react under mild conditions in the same way as benzenes and yield 1,4-dihydronaphthalene or tetralin derivatives [25, 44]. Formation of more highly saturated products by attack on the electron system of the remaining aromatic ring was also observed [44], especially in bulk reactions with heating [3]. A radical mechanism in reactions of BrF_3 and arenes was deduced from the observation of ESR signals of the radical cation of octafluoronaphthalene after treating its SO_2ClF solution with BrF_3 at 223 K [25].

The heteroarene pyridine forms a colorless solid with BrF_3 in CCl_4 [7] or in CCl_3F at 195 K [9]. Thiophene addition to a CH_2Cl_2 solution of BrF_3 and benzene at 200 K yields the phenyl-2-thienyl brominium salt [5].

References:

[1] Gubkina, N. I.; Sokolov, S. V.; Krylov, E. I. (Usp. Khim. **35** [1966] 2219/42; Russ. Chem. Rev. **35** [1966] 930/42, 937).
[2] Halbedel, H. S. (Kirk-Othmer Encycl. Chem. Technol. 2nd Ed. **9** [1966] 585/98, 590).
[3] McBee, E. T.; Lindgren, V. V.; Ligett, W. B.; Purdue Research Foundation (U.S. 2488216 [1949] 3 pp.; C.A. **1950** 1629; U.S. 2489969 [1949] 4 pp.; C.A. **1950** 2019; U.S. 2471831 [1949] 4 pp.; C.A. **1949** 3523 see also Ind. Eng. Chem. **39** [1947] 378).
[4] Bastock, T. W.; Harley, M. E.; Pedler, A. E.; Tatlow, J. C. (J. Fluorine Chem. **6** [1975] 331/55).
[5] Nesmeyanov, A. N.; Vanchikov, A. N.; Lisichkina, I. N.; Krushcheva, N. S.; Tolstaya, T. P. (Dokl. Akad. Nauk SSSR **254** [1980] 652/6; Dokl. Chem. Proc. Acad. Sci. USSR **250/255** [1980] 445/8).
[6] Shenk, W. J., Jr. (private communication, ref. 30 in: Booth, H. S.; Pinkston, J. T.; Fluorine Chem. **1** [1950] 189/200, 199).
[7] Shenk, W. J., Jr.; Pellon, G. R. (personal communication [1947], ref. 93 in [2, pp. 590/2]).
[8] Nesmeyanov, A. N.; Lisichkina, I. N.; Vanchikov, A. N.; Tolstaya, T. P. (Izv. Akad. Nauk SSSR Ser. Khim. **1976** 228/9; Bull. Acad. Sci. USSR Div. Chem. Sci. **25** [1976] 224).
[9] Naumann, D.; Lehmann, E. (J. Fluorine Chem. **5** [1975] 307/21).
[10] Lehmann, E.; Naumann, D.; Schmeisser, M. (Z. Anorg. Allgem. Chem. **388** [1972] 1/3).

[11] Soelch, R. R.; Mauer, G. W.; Lemal, D. M. (J. Org. Chem. **50** [1985] 5845/52).
[12] Banks, A. A.; Emeléus, H. J.; Haszeldine, R. N.; Kerrigan, V. (J. Chem. Soc. **1948** 2188/90).
[13] Chuvatkin, N. N.; Kartashov, A. V.; Morozova, T. V.; Boguslavskaya, L. S. (Zh. Org. Khim. **23** [1987] 269/74; J. Org. Chem. [USSR] **23** [1987] 237/42).
[14] Kartashov, A. V.; Chuvatkin, N. N.; Boguslavskaya, L. S. (Zh. Org. Khim. **24** [1988] 2522/5; J. Org. Chem. [USSR] **24** [1988] 2276/8).
[15] Boguslavskaya, L. S.; Chuvatkin, N. N.; Morozova, T. V.; Panteleeva, I. Yu.; Kartashov, A. V.; Sineokov, A. P. (Zh. Org. Khim. **23** [1987] 1173/7; J. Org. Chem. [USSR] **23** [1987] 1060/4).
[16] Boguslavskaya, L. S.; Chuvatkin, N. N.; Kartashov, A. V.; Ternovskoi, L. A. (Zh. Org. Khim. **23** [1987] 262/9; J. Org. Chem. [USSR] **23** [1987] 230/6).
[17] Surles, T.; Quarterman, L. A.; Hyman, H. H. (J. Fluorine Chem. **3** [1973] 453/6).
[18] Davis, R. A.; Larsen, E. R.; Dow Chemical Co. (Brit. 1059234 [1967] 6 pp.; C.A. **66** [1967] No. 75657).
[19] Florin, R. E.; Pummer, W. J.; Wall, L. A. (J. Res. Natl. Bur. Std. **62** [1959] 107/12).
[20] Lo, E. S.; Readio, J. D.; Iserson, H. (J. Org. Chem. **35** [1970] 2051/3).

[21] Pews, R. G.; Davis, R. A.; Dow Chemical Co. (U.S. 3651155 [1972] 3 pp.; C.A. **76** [1972] No. 140046).
[22] Gall, J. F.; Inman, C. E.; Pennsalt Chemicals Corp. (U.S. 2918434 [1959] 5 pp.; C.A. **1960** 6548).
[23] Stevens, T. E. (J. Org. Chem. **26** [1961] 1627/30).
[24] Stevens, T. E.; Rohm and Haas Co. (U.S. 3068299 [1962] 2 pp.; C.A. **58** [1963] 10076).
[25] Bardin, V. V.; Furin, G. G.; Yakobson, G. G. (J. Fluorine Chem. **23** [1983] 67/86).
[26] Boguslavskaya, L. S.; Kartashov, A. V.; Chuvatkin, N. N. (Zh. Org. Khim. **25** [1989] 2029/30; J. Org. Chem. [USSR] **25** [1989] 1835/6).

[27] Yuminov, V. S.; Pushkina, L. N.; Sokolov, S. V. (Zh. Obshch. Khim. **37** [1967] 375/80; J. Gen. Chem. [USSR] **37** [1967] 350/4).
[28] Kartashov, A. V.; Chuvatkin, N. N.; Boguslavskaya, L. S. (Zh. Org. Khim. **25** [1989] 2452/3; J. Org. Chem. [USSR] **25** [1989] 2209).
[29] Terrell, R. C.; BOC, Inc. (U.S. 4762856 [1988] 4 pp.; C.A. **109** [1988] No. 189832).
[30] Kartashov, A. V.; Chuvatkin, N. N.; Kurskii, Yu. A.; Boguslavskaya, L. S. (Zh. Org. Khim. **24** [1988] 2525/31; J. Org. Chem. [USSR] **24** [1988] 2279/84).

[31] Davis, R. A.; Larsen, E. R. (J. Org. Chem. **32** [1967] 3478/81).
[32] Haszeldine, R. N. (Nature **167** [1951] 139/40).
[33] Haszeldine, R. N. (J. Chem. Soc. **1953** 2075/81).
[34] Haszeldine, R. N. (J. Chem. Soc. **1953** 3559/64).
[35] Haszeldine, R. N. (J. Chem. Soc. **1953** 1592/600).
[36] Rosen, F. D. (NAA-SR-213 [1957] 34 pp.; N.S.A. **11** [1957] No. 9637).
[37] Chambers, R. D.; Musgrave, W. K. R.; Savory, J. (J. Chem. Soc. **1961** 3779/86).
[38] Hauptschein, M.; Braid, M. (unpublished work, ref. 10 in: Hauptschein, M.; Braid, M.; J. Am. Chem. Soc. **83** [1961] 2383/6).
[39] Simons, J. H. (Inorg. Synth. **3** [1950] 184/6).
[40] Grushin, V. V.; Demkina, I. I.; Tolstaya, T. P.; Galakhov, M. V.; Bakhmutov, V. I. (Metalloorg. Khim. **2** [1989] 727/36).

[41] Grushin, V. V.; Tolstaya, T. P.; Lisichkina, I. N. (Izv. Akad. Nauk SSSR Ser. Khim. **1982** 2412; Bull. Acad. Sci. USSR Div. Chem. Sci. **31** [1982] 2127).
[42] Lerman, O.; Rozen, S.; ICL Industries Ltd. (Eur. Appl. 344936 [1989] 7 pp. from C.A. **113** [1990] No. 5891).
[43] Ligett, W. B.; McBee, E. T.; Lindgren, V. V.; Purdue Research Foundation (U.S. 2432997 [1947] 1 p.; C.A. **1948** 2618; U.S. 2480080 [1949] 2 pp.; C.A. **1950** 2020; U.S. 2461554 [1949] 2 pp.; C.A. **1949** 3845; U.S. 2498891 [1950] 2 pp.; C.A. **1950** 4500).
[44] Bastock, T. W.; Pedler, A. E.; Tatlow, J. C. (J. Fluorine Chem. **8** [1976] 11/22).
[45] Pavlath, A. E.; Stauffer Chemical Company (U.S. 2993937 [1961] 4 pp.; C.A. **56** [1962] 409).

10.10.15 The Solubility of Metal Fluorides in BrF_3

Solubilities of metal fluorides in BrF_3 are compiled in Table 13. The purity of BrF_3 is important in these investigations, since the presence of HF may increase the solubility, as noted for RbF [1] and CsF [2]. A qualitative estimate of the solubility of metal and nonmetal halides is given in [3, 4].

Table 13
Solubility of Metal Fluorides in BrF_3 [5].

metal fluoride		LiF	NaF	KF	RbF[b)]	CsF
solubility in wt%	at 298 K	0.125(3)[a)]	2.08(20)	4.73(1)	9.20	18.8(1.0)[c)]
	at 343 K	0.081(7)	2.55(2)	5.38(3)	–	19.5(2)

metal fluoride		CaF_2	BaF_2	AlF_3	BiF_3[d)]	LaF_3
solubility in wt%	at 298 K	0.017(9)	3.53(20)	0.0195(2)	~0.7	<0.02
	at 343 K	<0.001	5.16(7)	0.0038(3)	–	<0.02

Table 13 (continued)

metal fluoride		ZrF_4	NbF_5	NiF_2	CuF_2	AgF	ThF_4
solubility in wt%	at 298 K	<0.0005	15.70(1)	<0.002	–	3.22(30)	<0.01
	at 343 K	<0.001	–	<0.001	<0.002	4.08(5)	–

[a] Confirmed in [6]. – [b] At 293 K, from [1]. – [c] This value may possibly result from the presence of HF, a solubility of 16.46% at 293 K was found in [2]. – [d] At ambient temperature, from [7].

References:

[1] Sukhoverkhov, V. F.; Takanova, N. D. (Zh. Neorg. Khim. **22** [1977] 2255/60; Russ. J. Inorg. Chem. **22** [1977] 1220/3).

[2] Sukhoverkhov, V. F.; Takanova, N. D.; Uskova, A. A. (Zh. Neorg. Khim. **21** [1976] 2245/9; Russ. J. Inorg. Chem. **21** [1976] 1234/7).

[3] Audrieth, L. F.; Kleinberg, J. (Non-Aqueous Solvents, Wiley, New York 1953, pp. 258/9).

[4] Brasted, R. C. (in: Sneed, M. C.; Maynard, J. L.; Brasted, R. C.; Comprehensive Inorganic Chemistry, Vol. 3, Van Nostrand, New York 1954, pp. 205/6).

[5] Sheft, I.; Hyman, H. H.; Katz, J. J. (J. Am. Chem. Soc. **75** [1953] 5221/3).

[6] Richards, G. W.; Woolf, A. A. (J. Fluorine Chem. **1** [1971] 129/39).

[7] Gutmann, V.; Emeléus, H. J. (J. Chem. Soc. **1950** 1046/50).

10.10.16 Systems of BrF_3

10.10.16.1 The BrF_3-Br_2 System

The solid-liquid and liquid-liquid equilibrium curves of the BrF_3-Br_2 system are shown in **Fig. 3.** The pure components are the only solid phases. There are no indications for the presence of BrF in the solid phase. The mixture of the liquids has an immiscibility gap which intersects the BrF_3 solubility curve. The consolute temperature is 328.7 ± 0.2 K at a consolute composition of about 61 mol% Br_2 [1]. An earlier paper had already stated that the solubility of bromine in BrF_3 is slightly larger than the solubility of BrF_3 in bromine,

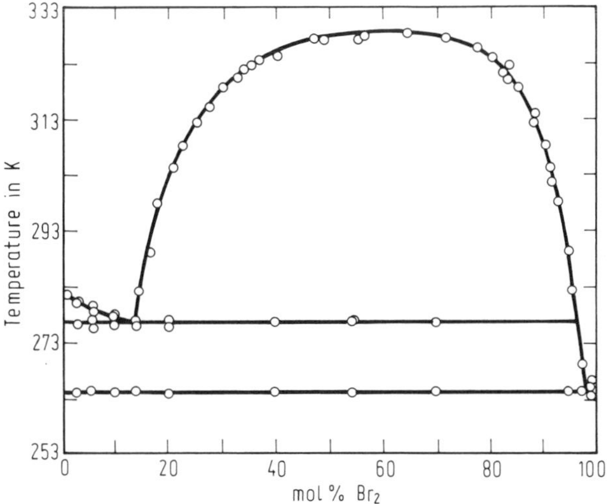

Fig. 3. Liquid-liquid and solid-liquid equilibria of the system BrF_3-Br_2 (from [1]).

but both solubilities (near the melting temperatures) are quite small. Both solutions tend to supercool; melting points were found to be 273 to 275 K for the BrF_3 layer and 263 to 265 K for the bromine layer [2].

Isotherms of the vapor-liquid equilibria in mixtures of BrF_3 and bromine are given in **Fig. 4.** The observed azeotropes at pressures greater than the sum of the pressures of the pure components suggest that at least one other component, probably BrF, forms in the vapor. The shift in the maximum vapor pressure point in going from one isotherm to the next eliminates any contention that such a maximum point represents the presence of another pure compound in the system and supports the theory that the vapor is at least a mixture of Br_2, BrF_3, and BrF. However, the equilibrium constants of BrF formation could not be calculated from the measured data [3]. A temperature-dependent equilibrium constant $\log(K_p/atm) = 7.83262 - 2597.06/T$ for the reaction $BrF_3 + Br_2 \rightleftharpoons 3\,BrF$ was obtained from manometric measurements between 328 and 380 K [4]. This equilibrium constant must be increased by 10 to 30% in order to account for the presence of 0.5 to 2.0 Torr of BrF_5 formed by disproportionation of BrF_3 [5]. The formation of BrF is accelerated on aluminium surfaces at ambient temperature, but not by Monel [6].

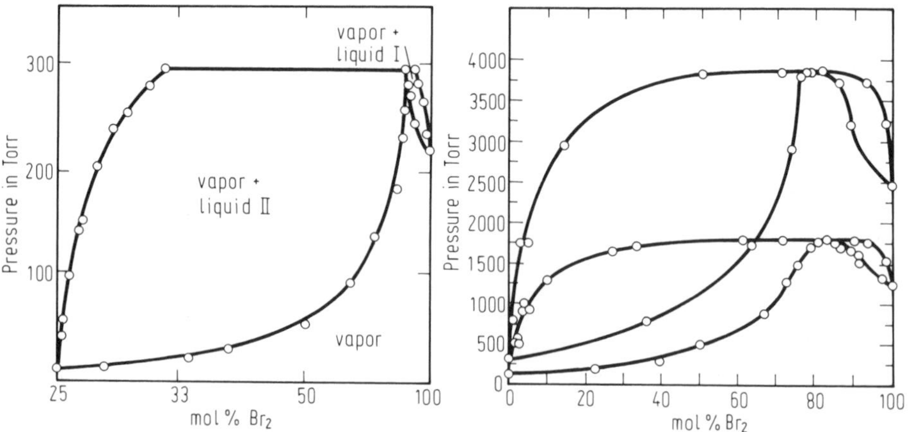

Fig. 4. Vapor-liquid phase diagram of the system BrF_3-Br_2. Left: at 298 K (from [7]); right: at 373.16±0.01 K (upper curves) and 348.16±0.01 K (from [3]). Liquid I: lower phase rich in Br_2, liquid II: upper phase rich in BrF_3.

Prolonged shaking of liquid BrF_3-Br_2 mixtures at ambient temperature leads to the equilibrium concentration of BrF [8]. UV spectra indicated that the BrF concentration increased with temperature on heating to 340 K [9]. The higher vapor pressure observed after adding Br_2 to BrF_3 is presumably due to BrF formation [2, 10]. Attempts to shift the equilibrium $BrF_3 + Br_2 \rightleftharpoons 3\,BrF$ to the right side by adding catalytic amounts of CsF or pyridine to the CCl_3F solution at 233 to 195 K failed [11]. The conductivity of solutions of Br_2 in BrF_3 is larger than that of pure BrF_3 which may result from ionic species formed in addition to BrF_2^+ and BrF_4^- [10]. However, these additional ionic species are minor constituents at 298 K [12].

References:

[1] Fischer, J.; Steunenberg, R. K.; Vogel, R. C. (J. Am. Chem. Soc. **76** [1954] 1497/8).

[2] Ruff, O.; Braida, A. (Z. Anorg. Allgem. Chem. **214** [1933] 81/90).

[3] Fischer, J.; Bingle, J.; Vogel, R. C. (J. Am. Chem. Soc. **78** [1956] 902/4).
[4] Steunenberg, R. K.; Vogel, R. C.; Fischer, J. (J. Am. Chem. Soc. **79** [1957] 1320/3).
[5] Stein, L. (J. Am. Chem. Soc. **81** [1959] 1273/6).
[6] Irsa, A. P.; Friedman, L. (J. Inorg. Nucl. Chem. **6** [1958] 77/90).
[7] Stein, L. (J. Am. Chem. Soc. **81** [1959] 1269/73).
[8] Wessel, J.; Kleemann, G.; Seppelt, K. (Chem. Ber. **116** [1983] 2399/407).
[9] White, C. F.; Goodeve, C. F. (Trans. Faraday Soc. **30** [1934] 1049/51).
[10] Quarterman, L. A.; Hyman, H. H.; Katz, J. J. (J. Phys. Chem. **61** [1957] 912/7).
[11] Naumann, D.; Lehmann, E. (J. Fluorine Chem. **5** [1975] 307/21).
[12] Richards, G. W.; Woolf, A. A. (J. Chem. Soc. A **1969** 1072/6).

10.10.16.2 The BrF_3-HF System

BrF_3 and HF are completely miscible; a description of the system is given in "Fluorine" Suppl. Vol. 3, 1982, p. 192.

10.11 Trifluorobromine(1+), BrF_3^+

CAS Registry Number: –

The BrF_3^+ cation was observed in 50 eV mass spectra of BrF_3 (see also p. 75) and BrF_5 (see also p. 128) at m/z 136. The appearance potentials are 12.9 ± 0.3 and 15.5 ± 0.2 eV, respectively [1]. The ion was also found in the mass spectra of partly hydrolyzed BrF_3 and BrF_5 [2]. The ion was identified in the mass spectrum of BrF_7 [3]; see p. 144.

References:

[1] Irsa, A. P.; Friedman, L. (J. Inorg. Nucl. Chem. **6** [1958] 77/90).
[2] Sloth, E. N.; Stein, L.; Williams, C. W. (J. Phys. Chem. **73** [1969] 278/80).
[3] Fogle, C. E.; Rewick, R. T.; United Aircraft Corp. (U.S. 3615206 [1971] 4 pp.; C.A. **76** [1972] No. 5447).

10.12 Heptafluorodibromate(1–) and Decafluorotribromate(1–), $Br_2F_7^-$ and $Br_3F_{10}^-$

CAS Registry Numbers: –

Formation. The salts MBr_2F_7, with M = Rb [1], Cs [2], form in the BrF_3-MF-HF systems at 293 K and can be isolated by crystallization. Both salts were prepared from 1:2 ratios of MF and BrF_3 in the presence of HF at 323 to 343 K [3]. Heating equimolar amounts of $MBrF_4$ and BrF_3 or pumping on molten MBr_3F_{10} yields MBr_2F_7 as a single-phase product when M = Cs, but not if M = Rb [4].

The reaction of $MBrF_4$ (M = Rb, Cs) and excess BrF_3 produces MBr_3F_{10} upon heating. The products crystallize during cooling [4].

$KBrF_4$ and BrF_3 did not form products richer in BrF_3 [4].

Vibrational Spectra. The vibrational spectra of MBr_2F_7 for M = Rb and Cs were found to be almost identical in older publications [1, 2] and in the more recent one [4]. However, the IR band positions reported in the earlier papers are similar to those given in the later publication in part at most; the earlier spectra were not discussed in [4].

The recent and more detailed investigation assigned ionic structures to MBr_2F_7 and MBr_3F_{10} (M = Rb, Cs) because their Raman spectra were similar to those of the ionic $CsBrF_4$. The spectra become more complex with increasing BrF_3 content of the compounds. The similarity of the spectra suggests anion structures involving two or three BrF_4 groups joined by common F atoms. A trans-bridged structure is favored for $Br_3F_{10}^-$. The very strong Raman bands of $CsBr_2F_7$ at 608 and 503 cm^{-1} are assigned to the in-phase and out-of-phase stretching modes of four planar F atoms, and a very strong IR band at 415 cm^{-1} to the Br-F-Br bridge. Other unassigned Raman bands of MBr_2F_7 and MBr_3F_{10} in the range 650 to 70 cm^{-1}, IR bands of MBr_3F_{10} from 615 to 335 cm^{-1} and of $CsBr_2F_7$ from 850 to 340 cm^{-1} are listed in [4].

The earlier papers list unassigned IR bands of $CsBr_2F_7$ from 840 to 285 cm^{-1} [2] and display the IR spectrum of $RbBr_2F_7$ which agrees with that of the Cs salt [1].

Reactions. Solid $RbBr_2F_7$ is hygroscopic, but can be stored at ambient temperature in the absence of moisture [1]. Reactions of MBr_2F_7 (M = Rb, Cs) with organic solvents, including perfluorinated lubricants, and with H_2O are violent [1, 2]; polyethylene [4] and even fluoroplastics are frequently attacked by $RbBr_2F_7$ [1].

References:

[1] Sukhoverkhov, V. F.; Takanova, N. D. (Zh. Neorg. Khim. **22** [1977] 2255/60; Russ. J. Inorg. Chem. **22** [1977] 1220/3).

[2] Sukhoverkhov, V. F.; Takanova, N. D.; Uskova, A. A. (Zh. Neorg. Khim. **21** [1976] 2245/9; Russ. J. Inorg. Chem. **21** [1976] 1234/7).

[3] Sukhoverkhov, V. F.; Takanova, N. D. (U.S.S.R. 559897 [1977] 2 pp.; C.A. **87** [1977] No. 119976).

[4] Stein, L. (J. Fluorine Chem. **27** [1985] 249/56).

10.13 Bromine Tetrafluoride, BrF_4

CAS Registry Number: *[21728-87-0]*

The reaction $BrF_3 + F \rightarrow BrF_4$ was proposed to form the BrF_4 radical in the gas phase from Br_2 and excess F_2. The reaction rate between 291 and 310 K is at least 10^{-11} cm^3/s at total pressures up to ~200 Torr. Disproportionation of the radical to BrF_3 and BrF_5 is believed to be likely [1]. An upper limit of 15.6 eV for the ionization potential of BrF_4 was deduced from mass spectra of BrF_5 [2].

References:

[1] Arutyunov, V. S.; Buben, S. N.; Trofimova, E. M.; Chaikin, A. M. (Kinet. Katal. **21** [1980] 337/42; Kinet. Catal. [USSR] **21** [1980] 258/62).

[2] Irsa, A. P.; Friedman, L. (J. Inorg. Nucl. Chem. **6** [1958] 77/90).

10.14 Tetrafluorobromine(1+), BrF_4^+

CAS Registry Number: *[41646-52-0]*

Formation. The signal of BrF_4^+ at m/z 155 is the most intense one in the mass spectrum of BrF_5 (see also p. 128) at 50 eV and has an appearance potential of 14.0 ± 0.3 eV [1]. The signal was also observed in the mass spectrum of partly hydrolyzed BrF_5 [2] and in the mass spectrum of BrF_7 [3]; see p. 144.

The cation BrF_4^+ results from reactions of BrF_5 with Lewis acids acting as fluoride acceptors. The solid $Sb_2F_{11}^-$ salt [4 to 8] was isolated from a solution of BrF_5 and SbF_5 [9]. The solid was also obtained by decomposing $KrFSb_2F_{11}$ in BrF_5 and evaporating the volatiles at ambient temperature [25]. Salts with the anions $Sb_3F_{16}^-$ and $Sb_7F_{36}^-$ were also prepared [6, 8]. However, BrF_4SbF_6 could not be isolated [7] due to its low stability; instead, $BrF_5 \cdot BrF_4Sb_2F_{11}$ was obtained [6]. The reaction of BrF_5 and SnF_4 produces solid $(BrF_4)_2SnF_6$ [10, 11]. The Lewis acids BF_3 [5, 12], PF_5 [12], AsF_5 [5, 7], BiF_5, TiF_4 [12], and NbF_5 [5] do not yield stable BrF_4^+ salts at ambient or higher temperatures, but the formation of instable salts with BF_3 [13], AsF_5 [7], and BiF_5 [5] was discussed. The reaction of BrO_2F and SbF_5 yields BrF_4SbF_6 as a by-product [14]. The reported formation of BrF_4SO_3F from BrF_5 and SO_3 [4] actually leads to the BrF_2^+ salt [15].

Structural Investigations. The ionic structure of $BrF_4Sb_2F_{11}$ with discrete BrF_4^+ ions is known from a single-crystal X-ray investigation using photographic techniques. The structure of BrF_4^+ can be described as a distorted trigonal bipyramid with two axial F atoms, the equatorial positions being occupied by two F atoms and the free electron pair being on the Br atom. The Br-F distances are 1.80 ± 0.12 and 1.91 ± 0.13 Å (Br-F_{ax}) and 1.76 ± 0.12 and 1.78 ± 0.09 Å (Br-F_{eq}). The average Br-F distance of 1.81 Å in BrF_4^+ is smaller than the distance of 1.89 Å in BrF_4^-. Angles in BrF_4^+ are $F_{ax}BrF_{ax} = 173.5° \pm 6.1°$ and $F_{eq}BrF_{eq} = 95.5° \pm 6.1°$. The BrF_4^+ ions are coupled to the $Sb_2F_{11}^-$ ions by two bridging F atoms at distances of 2.49 and 2.24 Å. Their inclusion into the Br coordination sphere results in a puckered pentagonal-bipyramidal structure of the cation [16]. Fluorine bridges are also known for $(BrF_4)_2SnF_6$ from Mössbauer investigations [10].

^{19}F NMR and NQR Spectra. The ^{19}F chemical shift of BrF_4^+ (positive low field shifts) in molten $BrF_4Sb_2F_{11}$ is $\delta = 244$ ppm with respect to CF_3COOH [12] (about $\delta = 167$ ppm with respect to CCl_3F [7]). Spectra of SbF_5 solutions at 299 K gave a chemical shift of $\delta = 180.4$ ppm [9] and those of solutions in HF or HF-AsF_5 $\delta = 197$ ppm with respect to CCl_3F in the temperature range 193 to 293 K. The differences of the chemical shifts are probably caused by differences in the composition of the solutions [7]. The absence of splittings by Br-F spin-spin coupling is due to rapid relaxation caused by the interaction of the quadrupole moment of Br with the asymmetric electric field gradient [17].

NQR resonances of BrF_4^+ at 77 K were observed for $BrF_5 \cdot BrF_4Sb_2F_{11}$ at 124.320 MHz, for $BrF_4Sb_2F_{11}$ at 128.153 MHz, and for $BrF_4Sb_3F_{16}$ at 127.039 MHz [8].

Vibrational Spectra. The vibrational spectra of BrF_4^+ in $BrF_4Sb_2F_{11}$ were assigned consistent with C_{2v} symmetry giving nine fundamentals: $4\,A_1 + A_2 + 2\,B_1 + 2\,B_2$ [5, 7]. The wave numbers are listed in Table 14. The identification of the cation bands is difficult due to interfering anion bands; thus the assignment must be considered tentative [7]. For example, the 385 cm^{-1} band was assigned to ν_3 in [7] and [9] and to ν_7 in [5]. Bands assigned to ν_8 in [7] were not identified in [5, 9].

The symmetric stretching vibrations of the BrF_2 unit were identified by the polarization of the bands. Bands of an earlier Raman spectrum of solid $BrF_4Sb_2F_{11}$ [5] agree with the later results [7, 9].

Mean Amplitudes of Vibration. Vibrational data of [7] were used to calculate mean amplitudes of vibration for BrF_4^+ from 0 to 500 K. Values at 298 K in Å are:

u(Br-F_{ax})	u(Br-F_{eq})	u($F_{ax} \cdots F_{ax}$)	u($F_{eq} \cdots F_{eq}$)	u($F_{ax} \cdots F_{eq}$)
0.0426	0.0399	0.058	0.065	0.071

The larger Br-F_{ax} value was attributed to the more polar character of these bonds [18].

Table 14
Fundamental Vibrations of BrF_4^+ in $BrF_4Sb_2F_{11}$.

vibration (approximate description)	wave numbers in cm^{-1}			
ν_1 (A_1, $\nu_s(BrF_{2eq})$)	725 (10)	723 (10)	680 sh	730 sh
ν_2 (A_1, $\nu_s(BrF_{2ax})$)	601 (6.1)	606 (4.8)	575 (5)	606 mw
ν_3 (A_1, $\delta_{sciss,s}(BrF_{2(eq+ax)})$)	382 (0.5)	385 (0.5)	–	–
ν_4 (A_1, $\delta_{sciss,as}(BrF_{2(eq+ax)})$)	214 (0.6)	219 (0.2)	–	–
ν_6 (B_1, $\nu_{as}(BrF_{2ax})$)	710 (1.0), 705 (1.1), 700 (2.1)	704 (2.4)	–	690 vs
ν_7 (B_1, $\omega(BrF_{2eq})$)	418 (0.4)	–	426 (2)	419 m
ν_8 (B_2, $\nu_{as}(BrF_{2eq})$)	–	736 sh	–	730 sh
ν_9 (B_2, $\delta_{sciss}(BrF_{2ax})$)	371 (0.5)	369 (0.5)	363 (5)	369 m
spectrum	Raman	Raman	Raman	IR
sample	solid, 183 K	solid, 298 K	in HF	solid
Ref.	[9]	[7]	[7]	[7]

Chemical Behavior. The pure BrF_4^+ salts with the anions $Sb_2F_{11}^-$ [6, 7], $Sb_3F_{11}^-$, $Sb_7F_{36}^-$ [6], and SnF_6^{2-} [10] are stable, crystalline, white solids at ambient temperature; $BrF_5 \cdot BrF_4Sb_2F_{11}$ is liquid [8]. The conductivity of BrF_5 increases after dissolving the BrF_4^+ salts and confirms the ionic nature of the compounds [5].

Contact of Rn with $BrF_4Sb_2F_{11}$ results in oxidation of the former [19]. Reaction of $BrF_4Sb_2F_{11}$ with moisture leads to Br_2^+ salts [7]. The exothermal hydrolysis in a dilute ammonia solution produces F^-, BrO^-, and BrO_3^- and can be used for chemical analysis [20]. The equilibrium $BrF_4^+ + 2\,HF \rightleftharpoons BrF_5 + H_2F^+$ lies far to the right, but the strongest Raman bands of BrF_4^+ can still be identified [21]. Neutralization reactions of the Lewis acid BrF_4^+ with the Lewis bases F^- and BrF_6^- yield BrF_5 [12]. Reacting $BrF_4Sb_2F_{11}$ with KrF_2 in a BrF_5 solution frees BrF_5 at 233 K [22]; oxidation to $BrF_6SbF_6 \cdot x\ SbF_5$ ($x<1$) was found after warming to 298 K [23].

The reaction of $BrF_4Sb_2F_{11}$ with CH_3CN is violent [12]. BrF_4BF_4 and $BrF_4Sb_2F_{11}$ fluorinate benzene, naphthalene, and pyridine derivatives in SO_2ClF solution at 240 to 190 K and yield product mixtures containing compounds with cyclohexadiene and cyclohexane systems as well as products that formally result by substituting H in reactants and products by Br [13].

The removal of Rn from dry air by oxidation with powdered crystalline $BrF_4Sb_2F_{11}$ was proposed for the purification of air in uranium mining areas [19, 24].

References:

[1] Irsa, A. P.; Friedman, L. (J. Inorg. Nucl. Chem. **6** [1958] 77/90).
[2] Sloth, E. N.; Stein, L.; Williams, C. W. (J. Phys. Chem. **73** [1969] 278/80).
[3] Fogle, C. E.; Rewick, R. T.; United Aircraft Corp. (U.S. 3615206 [1971] 4 pp.; C.A. **76** [1972] No. 5447).

[4] Schmeisser, M.; Pammer, E. (Angew. Chem. **69** [1957] 781).
[5] Surles, T.; Perkins, A.; Quarterman, L. A.; Hyman, H. H.; Popov, A. I. (J. Inorg. Nucl. Chem. **34** [1972] 3561/4).
[6] Sukhoverkhov, V. F.; Shpanko, V. I.; Takanova, N. D. (Zh. Neorg. Khim. **22** [1977] 2534/8; Russ. J. Inorg. Chem. **22** [1977] 1371/4).
[7] Christe, K. O.; Sawodny, W. (Inorg. Chem. **12** [1973] 2879/86).
[8] Kuz'min, A. I.; Shpanko, V. I.; Zvidadze, G. N.; Sukhoverkhov, V. F.; Dzevitskii, B. E. (Zh. Neorg. Khim. **24** [1979] 2127/33; Russ. J. Inorg. Chem. **24** [1979] 1178/82).
[9] Gillespie, R. J.; Schrobilgen, G. J. (Inorg. Chem. **13** [1974] 1230/5).
[10] Sukhoverkhov, V. F.; Dzevitskii, B. E. (Dokl. Akad. Nauk SSSR **170** [1966] 1099/102; Dokl. Chem. Proc. Acad. Sci. USSR **166/171** [1966] 983/6).

[11] Sukhoverkhov, V. F.; Dzevitskii, B. Z. (Dokl. Akad. Nauk SSSR **177** [1967] 611/4; Dokl. Chem. Proc. Acad. Sci. USSR **172/177** [1967] 1089/91).
[12] Meinert, H.; Gross, U.; Grimmer, A.-R. (Z. Chem. [Leipzig] **10** [1970] 226/7).
[13] Bardin, V. V.; Furin, G. G.; Yakobson, G. G. (J. Fluorine Chem. **23** [1983] 67/86).
[14] Jacob, E. (Angew. Chem. **88** [1976] 189/90).
[15] Gross, U.; Meinert, H.; Grimmer, A.-R. (Z. Chem. [Leipzig] **10** [1970] 441/3).
[16] Lind, M.; Christe, K. O. (Inorg. Chem. **11** [1972] 608/12).
[17] Shamir, J. (Struct. Bonding [Berlin] **37** [1979] 162/210, 195).
[18] Baran, E. J. (J. Fluorine Chem. **17** [1981] 543/8).
[19] Stein, L. (Science **175** [1972] 1463/5, Noble Gases Symp., Las Vegas 1973, pp. 376/85, CONF-730915).
[20] Sukhoverkhov, V. F.; Takanova, N. D. (Zh. Anal. Khim. **33** [1978] 1365/9; J. Anal. Chem. [USSR] **33** [1978] 1070/3).

[21] Surles, T.; Quarterman, L. A.; Hyman, H. H. (J. Fluorine Chem. **3** [1973] 293/306).
[22] Frlec, B.; Holloway, J. H. (Inorg. Chem. **15** [1976] 1263/70).
[23] Christe, K. O.; Wilson, R. D. (Inorg. Chem. **14** [1975] 694/6).
[24] Stein, L. (Can. 952290 [1974] 11pp.; C.A. **81** [1974] No. 175632 and U.S. 3784674 [1974] 3pp.).
[25] McKee, D. E. (LBL-1814 [1973] 1/93, 65; C.A. **80** [1974] No. 66301).

10.15 Tetrafluorobromate(1−), BrF_4^-

CAS Registry Number: *[19 702-38-6]*

Formation. The formation of BrF_4^- by dissociation of BrF_3 via $2\,BrF_3 \rightleftharpoons BrF_4^- + BrF_2^+$ is described on p. 66. The concentration of BrF_4^- increases upon adding a fluoride donor (Lewis base) [1]. A small amount of BrF_4^- forms in the equilibrium established in a mixture of BrF_3 and the weaker fluoride acceptor ClF_3 [2]; see p. 82.

BrF_3 reacts with alkali fluorides to yield $MBrF_4$ with M=K [3, 4], Rb [3], and Cs [4, 5]. Similar reactions with FNO and FNO_2 lead to $NOBrF_4$ [4, 6, 7] (see p. 168) and NO_2BrF_4 [4, 8] (see p. 169). The reactions with the fluorides can be described as solvolytic in nature [1]. The reaction with KF with formation of $KBrF_4$ also takes place when the amphoteric BrF_3 is replaced by the acidic compounds BrF_2SO_3F [9] or $(BrF_2)_2PtF_6$ [10].

When nonfluoride compounds are fluorinated by excess BrF_3, they also yield the BrF_4^- salts [1]. $NaBrF_4$ is obtained from Na_2CO_3 or Na_3PO_4 [11]. $KBrF_4$ is formed with KNO_3 [12], $K_2(NO)_2SO_3$ [13], $K_2S_2O_3$ or $K_2S_2O_7$ [11], KCl [12, 14], $KClO_3$ [11, 15, 16], KIO_3 [11], or with K_2MF_8 (M=Mo, W) by solvolysis [3]. Solvolysis of Rb_2MF_8 with M=Mo, W in BrF_3 gives $RbBrF_4$ [3]. Fluorination of Cs_2CrO_4 and Cs_2CrOCl_4 with BrF_3 yields $CsBrF_4$ [17]

as does the solvolysis of Cs_2MF_8 with M = Mo, W [3]. $Ba(BrF_4)_2$ is obtained from BrF_3 and $BaCl_2$ [14]. Reactions with AgCl [14] and Ag_3AsO_4 [11] yield $AgBrF_4$. $NOBrF_4$ is formed from NO and BrF_3 [6, 7].

The reaction of alkali bromides and F_2 yields the BrF_4^- salts of K^+, Rb^+ [18, 19], and Cs^+ [18]. Fluorination of alkali bromides by XeF_2 [20] readily leads to $MBrF_4$, where M = K, Rb, and Cs, while for M = Na the product is $NaBrF_4 \cdot 2\,NaF$ [21].

Formation of $KBrF_4$ from Br(V) compounds occurs during the decomposition of $KBrF_6$ in CH_3CN solution [22]. The formation of $NaBrF_4$ and $KBrF_4$ by reactions of BrF_5 with sodium or potassium salts is described on p. 131; examples are the reduction of BrF_5 by KNO_2 [23] and the reaction of $KBrO_3$ and BrF_5 [24, 25] with an excess of the latter [26]. This reaction does not always proceed with formation of $KBrF_4$ [22]. An equimolar mixture of $CsBrF_4$ and $CsBrF_6$ is obtained from CsN_3 in excess BrF_5 via intermediate formation of CsF and partial reduction of BrF_5 [23]. The slow decomposition of solid $(CH_3)_4NBrF_6$ at ambient temperature yields $(CH_3)_4NBrF_4$, probably by partial oxidation of the cation [27]. $CsBrF_6$ is reduced to $CsBrF_4$ with liberation of O_2 when BrF_3 is reacted with oxides, such as Al_2O_3 or SiO_2, at 420 to 670 K [28]. Thermal decomposition of BrO_2F in the presence of KF yields $KBrF_4$ and $KBrO_2F_2$ at room temperature [29].

Metathetical reactions yielded NF_4BrF_4 from NF_4SbF_6 and $CsBrF_4$ [30] and $(CH_3)_4NBrF_4$ from $(CH_3)_4NF$ and $CsBrF_4$ [27]. Electrochemical oxidation of HBr in anhydrous HF containing NaF can lead to BrF_4^- via Br_2 [31].

Uses. The melts of $MBrF_4$ with M = K, Rb, Cs are good fluorination agents for inorganic oxides [32, 33]. The "basic" salts readily react with amphoteric oxides, for example ZrO_2, and less so with basic ones, like CaO [32]. Oxygen can be quantitatively determined by high-temperature fluorination of BeO, MgO, Y_2O_3, ZrO_2, and MoO_3 with $KBrF_4$ [34]. The oxygen isotopes were analyzed by fluorinating $CaSO_4$ with $CsBrF_4$ [35]. Small quantities of O_2 and N_2 in metal fluorides, Mg-Y alloys, and Li metal can also be determined by this method [34, 36].

The oxidants $NOBrF_4$ and $KBrF_4$ give rise to a vigorous, hypergolic reaction when mixed with rocket fuels, like amines and boranes [37, 38]. Oxidative removal of Rn from dry air was achieved in contact with solid $KBrF_4$ [39]; additional details are described in [40].

Structure. The square-planar structure of the BrF_4^- ion in isomorphous $KBrF_4$ [41] and $RbBrF_4$ [42] was confirmed by neutron and X-ray diffraction studies, respectively. The distances and angles of the anion are identical in both salts within the standard deviations. A Br-F distance of 1.890(4) Å and FBrF angles of 89.9(2)° and 90.1(2)° were found for $RbBrF_4$ by a single-crystal structure determination at 213 K [42]. $NaBrF_4$, $CsBrF_4$ [21], and $AgBrF_4$ [43] are isomorphous with $KBrF_4$ and $RbBrF_4$ because of the similar X-ray powder patterns.

The results of early X-ray investigations of $KBrF_4$ powder and single crystals [44] are compatible with both a square-planar [45] and tetrahedral [44, 46] arrangement of the F atoms around the Br atom.

^{19}F NMR Spectrum. The ^{19}F NMR spectra of CH_3CN solutions of $CsBrF_4$ and $(CH_3)_4NBrF_4$ at 233 K contain sharp signals with a high field shift of $\delta = 37$ ppm relative to CCl_3F. Significant line broadening was observed with increasing temperature [27].

Vibrational Spectra. Vibrational spectra of BrF_4^- in solids and in solution with the exception of NF_4BrF_4 are similar and were assigned to the point group D_{4h} [4, 21, 27, 43, 47] consistent with the ion's known square-planar symmetry. The nine normal modes of vibration are classified as $A_{1g}(Ra) + A_{2u}(IR) + B_{1g}(Ra) + B_{2g}(Ra) + B_{2u}(inactive) + 2\,E_u(IR)$; bands are listed in Table 15. Additional Raman data are given in [21, 48]. The ionic structure of BrF_4^- salts with the cations NO^+, NO_2^+ [4], and $N(CH_3)_4^+$ [27] is confirmed by the typical cation bands.

Table 15
Fundamental Vibrations of BrF_4^- in Solids (in cm^{-1}).

BrF_4^- salt	$\nu_1(A_{1g})$ ν_s in phase	$\nu_2(A_{2u})$ δ_s out of plane	$\nu_3(B_{1g})$ δ_s in plane	$\nu_4(B_{2g})$ ν_s out of phase	$\nu_6(E_u)$ ν_{as}	$\nu_7(E_u)$ δ_{as} in plane	Ref.
$NaBrF_4$ [a)]	530	–	260 [b)]	461	–	–	[21]
$KBrF_4$	531 (10)	325 s	242 (0.9)	454 (7.9)	410 to 480 vs, br	188 [c)]	[4]
$RbBrF_4$	527	–	238 [b)]	450	–	–	[21]
$CsBrF_4$	523 (10)	317 s	246 (0.7)	449 (7.9)	478 vs [d)]	194 [c)]	[4]
$AgBrF_4$	524 (10)	353 s	244 (1.3)	443 (7.1)	471 vs	183 (14)	[43]
$NOBrF_4$	527 (10)	–	235 (3)	461 (4)	542 s [e)], 488 vs	183 (0.4) [f)]	[4]
$N(CH_3)_4BrF_4$	520 (10)	315 m	240 (1)	448 (6)	480 vs, br	–	[27]
BrF_4^- [g)]	528	302 [h)]	249	455	570 [h)]	–	[49]

[a)] The Raman bands of $NaBrF_4 \cdot 2\,NaF$ are similar [21]. – [b)] An additional band is attributed to the influence of the crystal lattice. – [c)] Estimated from combination bands. – [d)] In CH_3CN solution. – [e)] Sample prepared and measured at low temperature. – [f)] The band should be active only in the IR spectrum, but was identified in the Raman spectrum [1]. – [g)] In BrF_3 solution. – [h)] IR data from [47].

The vibrational bands of BrF_4^- in solid NF_4BrF_4 agree with the selection rules for C_{4v} symmetry and indicate that BrF_4^- is distorted by the cation. Selection rules are: $2\,A_1$(IR, Ra) + $2\,B_1$(Ra) + B_2(Ra) + 2 E(IR, Ra). Bands [30] are listed in Table 16.

Table 16
Fundamental Vibrations of BrF_4^- in Solid NF_4BrF_4 (in cm^{-1}).

assignment	description	IR	Raman
$\nu_1(A_1)$	ν_s	530 sh	535 (10)
$\nu_2(A_1)$	δ_s out of plane	325 vw	320 (0.2)
$\nu_3(B_1)$	ν_s	–	466 (7.2)
$\nu_5(E)$	δ_s in plane	–	258 (1.0)
$\nu_6(E)$	ν_{as}	500 sh, 452 vs, 430 sh	505 sh
$\nu_7(E)$	δ_{as} in plane	–	202 (0.5)

A very weak band at 363 cm^{-1} was not assigned.

Vibrational Amplitudes. Mean amplitudes of vibrations were calculated from revised fundamental frequencies of BrF_4^- [4, 50] by the method of characteristic vibrations. Values in Å are $u_{Br-F} = 0.0526$, $u_{F \cdots F(short)} = 0.115$, and $u_{F \cdots F(long)} = 0.067$ at 298.16 K. The large value for the Br–F bond possibly results from the bonding properties of the ion [51].

Force Constants. A calculation of the internal force constants of BrF_4^- in D_{4h} symmetry is based on revised data for the fundamental vibrations and gives the following values in mdyn/Å: $f_r = 2.227 \pm 0.047$, $f_{rr} = 0.20$, $f_{rr'} = 0.433 \pm 0.047$ [50]. Earlier results are given in [4]. An earlier calculation of the kinetic constants [52] apparently was also based on unrevised fundamental frequencies.

Bonding. The square-planar structure of BrF_4^- was correctly predicted by the VSEPR model [53] and an electrostatic model [54], whereas this configuration was found to be one of several, approximately equienergetic configurations using the LCAO-MO method [55].

The electronic structure of BrF_4^- and atomic charges were calculated by the NDDO-2(α, β) method [56]. The atomic charges and bond order were also calculated by a modified HMO theory [57].

Hybridization at Br is either sp^3d^2 [53] or d^2p^2, whereby four identical σ bonds form from the $d_{x^2-y^2}$, d_{z^2}, p_x, and p_y orbitals of Br [58]. A semiempirical MO model indicates two semi-ionic 3-center-4-electron p-σ bonds in BrF_4^- [51].

Chemical Behavior. Thermal decomposition of BrF_4^- salts usually yields BrF_3 and a fluoride derived from the cation. The temperature of the maximum rate of decomposition of $MBrF_4$ is 590 K for M = Na and increases in the series Na < K < Rb < Cs [21]. The thermal decomposition of sodium and barium tetrafluorobromates was investigated at 400 to ~580 K [59]. The thermal stability of NO_2BrF_4 is low [8], while $(CH_3)_4NBrF_4$ decomposes at about 530 K [27]. Thermal decomposition of NF_4BrF_4 at 298 K yields BrF_5 and NF_3 by oxidative fluorination [30].

The conductivity of BrF_3 solutions increases with increasing concentration of BrF_4^- salts which proves dissociation into ions; see for example [7, 60]. Electrochemical reduction of BrF_4^- in HF at 248 K yields HBr via Br_2 [31]. The electrolysis of $KBrF_4$ in BrF_3 solution was described in a report [61].

The chemical reactions of BrF_4^- are largely those of the dissociation products and therefore reflect the stability of the complex with respect to dissociation [62]. Solid $KBrF_4$ oxidizes Rn [39]. Molten $MBrF_4$ (M = K, Rb, Cs) reacts with Ni [21]; $KBrF_4$ attacks Pt when heated to 550 K [14].

The BrF_4^- salts are sensitive to moisture [39, 63]; hydrolysis yields HF, Br_2 [6, 14], $HBrO_3$, and O_2 [14]. The equilibrium constant of the reaction $BrF_4^- + HF \rightleftharpoons BrF_3 + HF_2^-$ was found to be in the range of 0.12 to 0.45 for solutions of $KBrF_4$ and $CsBrF_4$, probably at room temperature [64]. MBr_2F_7 and MBr_3F_{10} with M = Rb, Cs were obtained from $MBrF_4$ and BrF_3 [5, 65]; see p. 105.

The most extensively investigated reactions of the "basic" BrF_4^- salts are neutralization reactions with "acidic" salts of BrF_2^+ or their "anhydrides". General remarks on the reactions are given in connection with BrF_2^+ on p. 76. The reaction $MBrF_4 + BrF_2EF_n \rightarrow 2\,BrF_3 + MEF_n$ with M = alkali, Ag, 1/2 Ba, NO, or NO_2 was observed in solutions of BrF_3 with the acids or acid-forming compounds SiF_4 [4, 6, 7], GeF_4, SnF_4 [6, 7] and $(BrF_2)_2SnF_6$ [7, 66], BrF_2SbF_6 [66], BrF_2BiF_6 [67], TiF_4 [6, 7, 68], BrF_2MF_6 (M = Nb, Ta) [67], solutions of Mn(IV) [7, 69], $(BrF_2)_2PtF_6$ [70], BrF_2AuF_4 [71], and UF_4 [60, 72]. Similar reactions of $CsBrF_4$ and transition metal fluorides in the absence of a solvent yield ternary fluoro complexes. The reactions start at the melting point of $CsBrF_4$ at 480 K and lead to Cs_3LnF_6 at 570 K for the lanthanide trifluorides, except for CeF_3 and TbF_3 where Cs_3LnF_7 is formed. The reaction of $CsBrF_4$ and MF_2 (M = Cu, Ag) at 870 K yields $CsMF_3$ or Cs_2MF_4, depending on whether a molar reactant ratio of 1:1 or 2:1 is used [73]. The reactions of $KBrF_4$ and

CuF_2 at 750 to 790 K quantitatively produce $KCuF_3$ and K_2CuF_4, depending on the reactant ratios 1:1 and 2:1 used [74].

Molten alkali tetrafluorobromates fluorinate ZrO_2, HfO_2, and rare earth oxides explosively at 410 to 600 K. Oxides of rare earth elements yield trifluorides in most cases; tetrafluorides are isolated for Ce and Tb and also for Zr and Hf [33, 75]. The oxygen content of the products does not exceed 10^{-2} mass%. The reaction temperature is practically independent of the alkali cation in $MBrF_4$ and the reactant ratio [75]. Metathetical reactions of tetrafluorobromates are described in the section on preparation and formation.

The substitution of H by F in arenes, such as benzene, toluene, and chlorobenzene, with alkali tetrafluorobromates proceeds with low yields at ambient temperature [76]. Reactions of organic liquids with BrF_4^- salts in general tend to be less violent than with BrF_3 [14].

References:

[1] Shamir, J. (Israel J. Chem. **17** [1978] 37/47).
[2] Surles, T.; Hyman, H. H.; Quarterman, L. A.; Popov, A. I. (Inorg. Chem. **10** [1971] 913/6).
[3] Cox, B.; Sharp, D. W. A.; Sharpe, A. G. (J. Chem. Soc. **1956** 1242/4).
[4] Christe, K. O.; Schack, C. J. (Inorg. Chem. **9** [1970] 1852/8).
[5] Sukhoverkhov, V. F.; Takanova, N. D. (U.S.S.R. 559897 [1977] 2 pp.; C.A. **87** [1977] No. 119976).
[6] Chrétien, A.; Bouy, P. (Compt. Rend. **246** [1958] 2493/5).
[7] Bouy, P. (Ann. Chim. [Paris] [13] **4** [1959] 853/90).
[8] Aynsley, E. E.; Hetherington, G.; Robinson, P. L. (J. Chem. Soc. **1954** 1119/24).
[9] Gross, U.; Meinert, H.; Grimmer, A.-R. (Z. Chem. [Leipzig] **10** [1970] 441/3).
[10] Chernyaev, I. I.; Nikolaev, N. S.; Ippolitiv, E. G. (Dokl. Akad. Nauk SSSR **130** [1960] 1041/3; Proc. Acad. Sci. USSR Chem. Sect. **130/135** [1960] 167/9).

[11] Emeléus, H. J.; Woolf, A. A. (J. Chem. Soc. **1950** 164/8).
[12] Woolf, A. A.; Emeléus, H. J. (J. Chem. Soc. **1950** 1050/2).
[13] Woolf, A. A. (J. Chem. Soc. **1950** 1053/6).
[14] Sharpe, A. G.; Emeléus, H. J. (J. Chem. Soc. **1948** 2135/8).
[15] Woolf, A. A. (Chem. Ind. [London] **1954** 346).
[16] Woolf, A. A. (J. Chem. Soc. **1954** 4113/6).
[17] Hope, E. G.; Jones, P. J.; Levason, W.; Ogden, J. S.; Tajik, M.; Turff, J. W. (J. Chem. Soc. Dalton Trans. **1984** 2445/7).
[18] Asprey, L. B.; Margrave, J. L.; Silverthorne, M. E. (J. Am. Chem. Soc. **83** [1961] 2955/6).
[19] Bode, H.; Klesper, E. (Z. Anorg. Allgem. Chem. **313** [1961] 161/9).
[20] Legasov, V. A.; Prusakov, V. I.; Chaibanov, B. V. (preprint of paper, ref. 11 in [21]).

[21] Popov, A. I.; Kiselev, Yu. M.; Sukhoverkhov, V. F.; Chumaevskii, N. A.; Krasnyanskaya, O. A.; Sadikova, A. T. (Zh. Neorg. Khim. **32** [1987] 1007/12; Russ. J. Inorg. Chem. **32** [1987] 619/22).
[22] Gillespie, R. J.; Spekkens, P. (J. Chem. Soc. Dalton Trans. **1976** 2391/6).
[23] Christe, K. O.; Wilson, W. W.; Schack, C. J. (J. Fluorine Chem. **43** [1989] 125/9).
[24] Schmeisser, M.; Pammer, E. (Angew. Chem. **69** [1957] 781).
[25] Christe, K. O.; Wilson, R. D.; Curtis, E. C.; Kuhlmann, W.; Sawodny, W. (Inorg. Chem. **17** [1978] 533/8).
[26] Tantot, G.; Bougon, R. (Compt. Rend. C **281** [1975] 271/3).
[27] Wilson, W. W.; Christe, K. O. (Inorg. Chem. **28** [1989] 4172/5).

[28] Sukhoverkhov, V. F.; Ustinov, V. I.; Grinenko, V. A. (Zh. Anal. Khim. **26** [1971] 2166/71; J. Anal. Chem. [USSR] **26** [1971] 1934/8).
[29] Gillespie, R. J.; Spekkens, P. H. (J. Chem. Soc. Dalton Trans. **1977** 1539/46).
[30] Christe, K. O.; Wilson, W. W. (Inorg. Chem. **25** [1986] 1904/6).

[31] Thiébault, A.; Herlem, M. (Compt. Rend. C **278** [1974] 443/4).
[32] Sheft, I.; Martin, A. F.; Katz, J. J. (J. Am. Chem. Soc. **78** [1956] 1557/9).
[33] Brekhovskikh, M. N.; Popov, A. I.; Fedorov, V. A.; Kiselev, Yu. M. (Mater. Res. Bull. **23** [1988] 1417/21).
[34] Goldberg, G.; Meyer, A. S., Jr.; White, J. C. (Anal. Chem. **32** [1960] 314/7).
[35] Ustinov, V. I.; Sukhoverkhov, V. F.; Podzolko, L. G. (Zh. Fiz. Khim. **52** [1978] 610/4; Russ. J. Phys. Chem. **52** [1978] 344/7).
[36] Goldberg, G. (Anal. Chem. **34** [1962] 1343/4).
[37] Olah, G. A.; Kuhn, S. J.; Dow Chemical Co. (U.S. 3103782 [1963] 2 pp.; C.A. **60** [1963] 2718).
[38] Iwanciow, B. L.; Lawrence, W. J.; United Aircraft Corp. (U.S. 3797238 [1974] 5 pp.; C.A. **81** [1974] No. 15274).
[39] Stein, L. (Science **175** [1972] 1463/5, Noble Gases Symp., Las Vegas 1973, pp. 376/85, CONF-730915).
[40] Stein, L. (Can. 952290 [1974] 11 pp.; C.A. **81** [1974] No. 175632; U.S. 3784674 [1974] 3 pp.).

[41] Edwards, A. J.; Jones, G. R. (J. Chem. Soc. A **1969** 1936/8).
[42] Mahjoub, A. R.; Hoser, A.; Fuchs, J.; Seppelt, K. (Angew. Chem. **101** [1989] 1528/9).
[43] Sukhoverkhov, V. F.; Moltasova, J. (Zh. Neorg. Khim. **25** [1980] 3041/5; Russ. J. Inorg. Chem. **25** [1980] 1671/3).
[44] Siegel, W. (Acta Crystallogr. **9** [1956] 493/5).
[45] Sly, W. G.; Marsh, R. E. (Acta Crystallogr. **10** [1957] 378/9).
[46] Siegel, W. (Acta Crystallogr. **10** [1957] 380).
[47] Surles, T.; Hyman, H. H.; Quarterman, L. A.; Popov, A. I. (Inorg. Chem. **9** [1970] 2726/30).
[48] Shamir, J.; Yaroslavsky, I. (Israel J. Chem. **7** [1969] 495/7).
[49] Surles, T.; Quarterman, L. A.; Hyman, H. H. (J. Fluorine Chem. **3** [1973] 453/6).
[50] Christe, K. O.; Naumann, D. (Inorg. Chem. **12** [1973] 59/62).

[51] Baran, E. J. (J. Mol. Struct. **21** [1974] 461/3).
[52] Sanyal, N. K.; Dixit, L.; Pandey, A. N.; Singh, H. S.; Singh, B. P. (Indian J. Pure Appl. Phys. **10** [1972] 493/4).
[53] Gillespie, R. J.; Nyholm, R. S. (Quart. Rev. Chem. Soc. **11** [1957] 339/80, 373) and Gillespie, R. J. (J. Chem. Educ. **40** [1963] 295/301).
[54] Searcy, A. W. (J. Chem. Phys. **31** [1959] 1/4).
[55] Havinga, E. E.; Wiebenga, E. H. (Recl. Trav. Chim. **78** [1959] 724/38).
[56] Smolyar, A. E.; Charkin, O. P.; Klimenko, N. M. (Zh. Strukt. Khim. **15** [1974] 993/1003; J. Struct. Chem. [USSR] **15** [1974] 885/93).
[57] Wiebenga, E. H.; Kracht, D. (Inorg. Chem. **8** [1969] 738/46).
[58] Gabuda, S. P.; Zemskov, S. V.; Mit'kin, V. N.; Obmoin, B. I. (Zh. Strukt. Khim. **18** [1977] 515/24; J. Struct. Chem. [USSR] **18** [1977] 413/20).
[59] Kiselev, N. I.; Lapshin, O. N.; Sadikova, A. T.; Sukhoverkhov, V. F.; Churbanov, M. F. (Vysokochist. Veshchestva **1987** No. 3, pp. 178/82 from C.A. **108** [1988] No. 44536).
[60] Martin, D. (Rev. Chim. Miner. **4** [1967] 367/97).

[61] Didchenko, R.; Toeniskoetter, R. H. (AD-433910 [1964] 17 pp. from C.A. **62** [1965] 8673).

[62] Downs, A. J.; Adams, C. J. (in: Bailar, J. C., Jr.; Emeléus, H. J.; Nyholm, R., Trotman-Dickenson, A. F.; Comprehensive Inorganic Chemistry, Vol. 2, Pergamon, Oxford 1973, pp. 1476/1563, 1555).
[63] Sukhoverkhov, V. F.; Takanova, N. D. (Zh. Anal. Khim. **33** [1978] 1365/9; J. Anal. Chem. [USSR] **33** [1978] 1070/3).
[64] Surles, T.; Quarterman, L. A.; Hyman, H. H. (J. Fluorine Chem. **3** [1973] 293/306).
[65] Stein, L. (J. Fluorine Chem. **27** [1985] 249/56).
[66] Woolf, A. A.; Emeléus, H. J. (J. Chem. Soc. **1949** 2865/71).
[67] Gutmann, V.; Emeléus, H. J. (J. Chem. Soc. **1950** 1046/50).
[68] Sharpe, A. G. (J. Chem. Soc. **1950** 2907/8).
[69] Sharpe, A. G.; Woolf, A. A. (J. Chem. Soc. **1951** 798/801).
[70] Sharpe, A. G. (J. Chem. Soc. **1950** 3444/50).

[71] Sharpe, A. G. (J. Chem. Soc. **1949** 2901/2).
[72] Chrétien, A.; Martin, D. (Compt. Rend. C **263** [1966] 235/8).
[73] Kiselev, Yu. M.; Popov, A. I.; Goryachenkov, S. A.; Brekhovskikh, M. N.; Fadeeva, N. E. (Vysokochist. Veshchestva **1988** No. 6, pp. 105/12; High Purity Subst. [USSR] **2** [1988] 1032/8).
[74] Kiselev, Yu. M.; Popov, A. I.; Martynenko, L. I. (U.S.S.R. 1281519 [1987] 2 pp.; C.A. **106** [1987] No. 140540).
[75] Brekhovskikh, M. N.; Popov, A. I.; Kiselev, Yu. M.; Fedorov, V. A. (Vysokochist. Veshchestva **1988** No. 5, pp. 107/8).
[76] Pavlath, A. E.; Stauffer Chemical Co. (U.S. 2993937 [1961] 4 pp.; C.A. **56** [1962] 409).

10.16 Bromine Pentafluoride, BrF_5

CAS Registry Numbers: BrF_5 *[7789-30-2]*, $^{79}BrF_5$ *[33744-63-7]*, $^{81}BrF_5$ *[33744-64-8]*

Bromine pentafluoride is a toxic, colorless liquid which is formed by the reaction of Br_2 or BrF_3 with F_2 at 470 to 570 K. It is stable up to 823 K. BrF_5 hydrolyzes violently and fluorinates many other compounds in exothermic reactions. The BrF_5 molecule forms a square pyramid with the bromine atom below, four equal equatorial fluorine atoms at the base, and one axial fluorine atom at the top of the pyramid. The first work on BrF_5 was done in 1931 and is described in "Brom" 1931, pp. 338/9. Some of its characteristic data are given below.

molecular weight	174.9
melting point	212.7 K
boiling point	313.9 K
vapor pressure	406 Torr (298 K)
density (l)	2.4616 g/cm^3 (298.15 K)
surface tension	22.4 dyn/cm (300.0 K)
viscosity	0.62 cP (297.4 K)
dipole moment (g)	1.51 D
dielectric constant	7.91 (297.7 K)
electric conductivity	9.1×10^{-8} S/cm (298 K)
magnetic susceptibility	-4.51×10^{-5} cm^3/mol (298 K)
refractive index	1.3529 (298 K, n_D)
standard enthalpy of formation	-428.72 kJ/mol (298.15 K, 0.1 MPa)

10.16.1 Preparation. Formation

BrF_5 was first synthesized in 1931 from Br_2 and F_2 or BrF_3 and F_2; see "Brom" 1931, p. 338. Based on this work, Br_2 is reacted with a fivefold excess of F_2 at 473 K in an iron vessel. Subsequent distillation of the raw product (95% BrF_5, 5% BrF_3) in an F_2 stream yields 87% BrF_5 [1]. High pressures (40 to 50 atm) and high temperatures (473 to 523 K) have proven advantageously [2, 3]. Glow discharge of Br_2 and F_2 (current: 20 to 30 mA, potential: 3.5 to 3.0 kV) in an U-shaped quartz tube at 195 K results in quantitative formation of BrF_5 [4]. The reaction system Br_2-F_2 is described in "Fluorine" Suppl. Vol. 2, 1980, pp. 143/4; for more recent results see "Bromine" Suppl. Vol. A, 1985, pp. 373/4.

The reaction of BrF_3 with F_2 forming BrF_5 was studied between 341 and 394 K and intitial pressures of 30 to 300 Torr in a nickel reactor. An activation energy of 68.6 kJ/mol and Arrhenius factor of 2.6×10^{11} $cm^3 \cdot mol^{-1} \cdot s^{-1}$ were obtained for the homogeneous, bimolecular reaction. The derived probability factor of 10^{-3} indicates that only part of the bimolecular collisions have sufficient energy to cause BrF_5 formation under the experimental conditions [5].

A simple laboratory method for preparing BrF_5 with about 50% yield is the exothermic reaction of solid KBr with F_2 at room temperature [6].

An electric discharge through a mixture of NF_3 and Br_2 at 7 K gives BrF_5 and BrF_3 as the main products [4]. BrF_5 and BrF are formed in purified BrF_3 after long storage in nickel vessels due to the spontaneous but slow disproportionation of BrF_3 [7]. The disproportionation of BrF_3 is catalyzed by small amounts of CsF [9]. The reaction of BrF_3 with Br_2 yields BrF_5 according to $3\,BrF_3 + Br_2 \rightarrow BrF_5 + 4\,BrF$ [7]. β Irradiation of liquid BrF_3 also gives BrF_5 [8]. BrF_3 is fluorinated at 37 K with BiF_5 [10]. The fluorination at room temperature with PtF_6 probably leads to BrF_5 also [11]. Frozen BrF_3 reacts on warming with PuF_6 to give BrF_5 [12]. Bromylfluoride, BrO_2F, is fluorinated with krypton difluoride to $BrOF_3$ and subsequently to BrF_5 [13]. Other methods for the fluorination of BrF_3 are described on pp. 96/7.

Purification. BrF_5 containing BrF_3 as impurity is purified by addition of KF which gives solid $KBrF_4$ by reaction with BrF_3 [14, 15]. Bromine-colored BrF_5 is purified as follows: Transfer to a stainless steel cylinder, which is previously passivated with BrF_5 and loaded with KF, and shaking the closed cylinder at room temperature for 2 days or longer. The only liquid obtained is clear BrF_5 [15].

References:

[1] Kwasnik, W. (in: Brauer, G.; Handbuch der Präparativen Anorganischen Chemie, Vol. 1, Enke, Stuttgart 1975, pp. 170/1).
[2] Iwasaki, M.; Yawata, T.; Suzuki, K.; Tsujimura, S.; Oshima, K. (Nippon Kagaku Zasshi **83** [1962] 36/9 from C.A. **58** [1963] 7600).
[3] Meinert, H.; Gross, U. (Z. Chem. [Leipzig] **9** [1969] 455/6).
[4] Nikitin, I. V.; Rosolovski, V. Y. (Zh. Neorg. Khim. **20** [1975] 263/4; Russ. J. Inorg. Chem. **20** [1975] 143/4).
[5] Kluksdahl, H. E.; Cady, G. H. (J. Am. Chem. Soc. **81** [1959] 5285/6).
[6] Hyde, G. A.; Boudakian, M. M. (Inorg. Chem. **7** [1968] 2648/9).
[7] Stein, L. (J. Am. Chem. Soc. **81** [1959] 1273/6).
[8] Yosmin, S. J. (J. Phys. Chem. **62** [1958] 1596/7).
[9] Naumann, D.; Lehmann, E. (J. Fluorine Chem. **5** [1975] 307/21).
[10] Fischer, J.; Rudzitis, E. (J. Am. Chem. Soc. **81** [1959] 6375/7).
[11] Weinstock, B.; Malm, J. G.; Weaver, E. E. (J. Am. Chem. Soc. **83** [1961] 4310/7).
[12] Weinstock, B.; Malm, J. G. (J. Inorg. Nucl. Chem. **2** [1956] 380/94).

[13] Gillespie, R. J.; Spekkens, P. H. (J. Chem. Soc. Dalton Trans. **1977** 1539/46).
[14] Kuz'min, A. I.; Shpanko, V. I.; Zviadadze, G. N.; Sukhoverkov, V. F.; Dzevitskii, B. E. (Zh. Neorg. Khim. **24** [1979] 2127/33; Russ. J. Inorg. Chem. **24** [1979] 1178/82).
[15] Schack, C. J.; Dubb, H. E.; Quaglino, J., Jr. (Chem. Ind. [London] **1967** 545/6).

10.16.2 Uses. Handling. Safety Factors

BrF_5, a fuming, colorless liquid (m.p. 212.7 K, b.p. 313.9 K), is used as a fluorinating agent and as the oxidizer component in some rocket propellants.

BrF_5 can be stored or handled in iron, copper, nickel, Inconel, Monel or Kel-F cylinders. Dry glass is attacked slowly, quartz is practically not attacked [1 to 3]; see also the corresponding section of BrF_3, p. 55. BrF_5 requires the "oxidizer" label for rail shipment and the "corrosive" label for air shipment [2].

Solutions of BrF_5 in acetonitrile, stored in a capped plastic vessel, were found to decompose spontaneously and violently [4], though they were reported to be stable at room temperature [5, 6].

BrF_5 is classified as a dangerous compound (highest hazard grade in the vapor hazard index) [7]. The greatest hazard is inhalation. The lethal effects are due to local corrosive destruction of the pulmonary membranes, resulting in the gas exchange to fail; see e. g. [8]. The threshold limit value (TLV) is 0.1 ppm (0.7 mg/m^3) and the short time exposure limit (STEL) is 0.3 ppm (2 mg/m^3), as published by the American Conference of Governmental Industrial Hygienists [2, 9].

References:

[1] Kwasnik, W. (in: Brauer, G.; Handbuch der Präparativen Anorganischen Chemie, Vol. 1, Enke, Stuttgart 1975, pp. 170/1).
[2] Dagani, M. J.; Bard, H. J.; Benya, T. J.; Sanders, D. C. (Ullmann's Encycl. Ind. Chem. 5th Ed. A**4** [1985] 405/29, 425).
[3] Stein, L. (in: Gutmann, V.; Halogen Chemistry, Vol. 1, Academic, New York 1967, pp. 168/74).
[4] Stein, L. (Chem. Eng. News **62** No. 28 [1984] 4).
[5] Meinert, H.; Gross, U. (Z. Chem. [Leipzig] **9** [1969] 190).
[6] Gross, U.; Meinert, H. (Z. Chem. [Leipzig] **11** [1971] 431/2).
[7] Pitt, M. J. (Chem. Ind. [London] **1982** 804/6).
[8] Dost, F. N.; Reed, D. J.; Finch, A.; Wang, C. H. (AD-681161 [1968] 1/113; C.A. **71** [1969] No. 11377).
[9] Sax, N. J. (Dangerous Properties of Industrial Materials, 6th Ed., Van Nostrand Reinhold, New York 1984, pp. 519/20).

10.16.3 Molecular Properties and Spectra

Molecular and Electronic Structure

The square-pyramidal structure of BrF_5 with bromine in the base of the pyramid and the fluorine atoms at the pyramid corners with C_{4v} symmetry in gas and liquid phase was deduced from or confirmed by analysis of vibrational spectra [1 to 4], ^{19}F NMR spectra [5 to 7], microwave spectra [8], electron-diffraction data [9, 10], and from mass-spectrometric and electron-deflection studies [11]. The BrF_5 molecule is described by the theory of electron pair repulsion as an octahedron with one edge occupied by lone-pair electrons. The FBrF angle is less than 90° due to the repulsion of the lone-pair electrons [12 to 16].

Thus, the structure of the BrF_5 molecule is a square pyramid with the bromine atom below the base, four equal equatorial (eq) fluorine atoms at the base, and one axial (ax) fluorine atom at the top of the pyramid. Recommended values for the bond lengths and angles in the gas phase, determined by combining electron diffraction [10] and microwave data [17 to 20], are [10]:

$$r(BrF_{eq}) = 1.768(1)\ \text{Å},\ r(BrF_{ax}) = 1.699(6)\ \text{Å},\ \text{angle}\ F_{ax}BrF_{eq} = 85.1(4)°$$

For a preliminary analysis, see [9]. Values obtained only from rotational constants agree quite well with one another [18].

The structure of the crystalline phase was investigated at 153 K by single-crystal photographic X-ray diffraction methods. Bond distances of $r(BrF_{eq}) = 1.85$ Å, 1.84 Å and $r(BrF_{ax}) = 1.72$ Å, and bond angles $F_{ax}BrF_{eq} = 82°$, 88° were reported [21].

The calculated bond orders [22, 23] agree with the sequence of experimentally determined bond lengths. Bond polarities determined from atomic constants are given in [24].

An ab initio MO SCF calculation using small basis sets (3-21 G, 3-21 $G^{(*)}$) deals with the geometry, the electric dipole moment, and the hydrogenation energy [25]. A number of semiempirical methods (SCC Xα [26], DVM Xα [27], various NDO procedures [28 to 30], EHT [31]) were employed to calculate various molecular properties.

Ionization Potentials

From the He I photoelectron spectrum the following adiabatic (ad) and vertical (vert) ionization energies E_i (in eV) were obtained [32]. The orbital assignments are based on semiempirical MO calculations (SCC Xα [26]; DVM Xα [27]) and correct the original assignments [32] which were obtained in analogy to ClF_5:

E_i(ad)	13.17(2)	14.35(2)	–	–	16.92(2)	–
E_i(vert)	13.59(2)	14.87(4)	15.69(2)	16.23(2)	17.46(2)	20.17(5)
assignment ...	a_1	a_2, b_1	e, a_1	b_1, b_2	e	e, a_1

Dipole Moment. Polarizability

The dipole moment of gaseous BrF_5 was determined from dielectric measurements to be 1.51 ± 0.10 D [16]. For liquid BrF_5 the dipole moment was reported to be 1.68 D [34].

The average polarizability was calculated with the δ-function model to be 6.384×10^{-24} cm^3 [33].

Nuclear Quadrupole Coupling

The bromine quadrupole coupling constants in the ground state [8, 19, 20] (with the most precise values given in [8]) and in the excited vibrational state $v_5 = 1$ [35] were determined from the rotational transitions $J = 4 \leftarrow 3$, $5 \leftarrow 4$, $6 \leftarrow 5$ as follows:

	$^{79}BrF_5$		$^{81}BrF_5$	
eqQ in MHz	-280.9 ± 0.3	-277.7 ± 0.5	-233.3 ± 0.3	-233.5 ± 0.5
state	ground state	$v_5 = 1$	ground state	$v_5 = 1$
Ref.	[8]	[35]	[8]	[35]

For solid BrF_5, |eqQ| was determined to be 220.07 MHz at 77 K [36].

Nuclear Magnetic Resonance

The ^{19}F NMR spectrum consists of a doublet with components of equal intensity and a quintuplet with relative intensities of 1:4:6:4:1 [37]. Chemical shifts δ and spin-coupling constants $J(F_{ax}-F_{eq})$ are compiled below (δ is defined positive at lower field with $CFCl_3$ as external reference):

δ in ppm F_{ax}	F_{eq}	T in K	$J(F_{ax}-F_{eq})$ in Hz	Ref.
275.9	139.0	300	75	[38]
273.8	136.1	214	76.3	[39]

For earlier NMR measurements, see [5 to 7, 40, 41].

^{19}F NMR studies of solid BrF_5 revealed the longitudinal relaxation time T_1 and the relaxation time of the dipole energy T_D of the fluorine atoms between 120 and 200 K [42]; see **Fig. 5**.

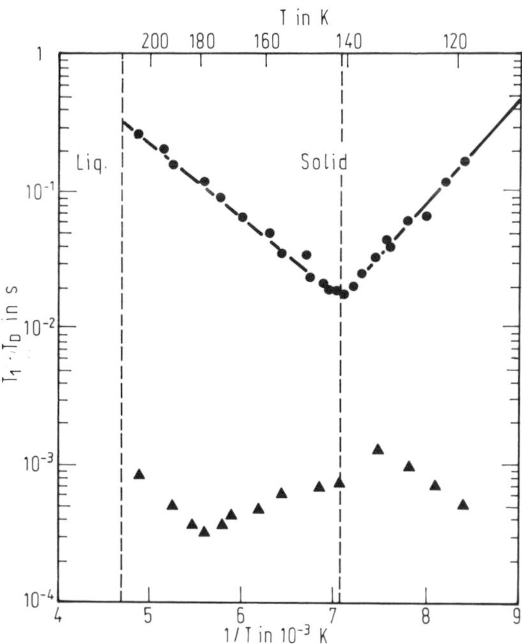

Fig. 5. Relaxation of the ^{19}F nuclei in solid BrF_5. Longitudinal relaxation time T_1 (●) and relaxation time of the dipole energy T_D at 10 MHz as a function of temperature [42, p. 103].

Spin-Rotation Interaction

From nuclear magnetic resonance studies on solid BrF_5 at 110 K, the spin-rotation interaction constants $c_{\parallel} = 2 \times 10^3$ Hz and $c_{\perp} = 14 \times 10^3$ Hz were determined [42].

Constants of Molecular Rotation and Vibration

Rotational Constants. The BrF_5 molecule is an oblate, symmetric top with the rotational constants C<A=B. The rotational constant B was obtained from rotational transitions in the ground state [8, 17 to 20] (with refined values from [35]) and two excited states ($v_5 = 1$ and $v_9 = 1$) [17, 18]. The rotational constant C_0 was calculated from the observed rotational constant C_5 and C_9 given in [18] using a correction from an anharmonic force field [10] or using the structural data [43] given in [9]. Less precise values for B_0 and C_0 are given in [2]. The following values are for ground-state BrF_5 (in MHz):

	B_0	B_0	B_0	C_0	C_0
$^{79}BrF_5$	3099.2612(8)	3099.267(7)	3099.267(3)	2147.12	2140.5
$^{81}BrF_5$	3096.8196(8)	3096.819(6)	3096.831(3)	2147.99	–
Ref.	[19]	[8]	[20]	[10]	[43]

Rotational constans for the $v_5 = 1$ state and $v_9 = 1$ state are given in [17, 18].

Centrifugal Distortion Constants. The following centrifugal distortion constants (in kHz) for the ground state were determined:

constant	$^{79}BrF_5$	$^{81}BrF_5$	Ref.
D_{JK}	−0.51(2)	−0.49(2)	[20]
	−0.455(2)	−0.450(2)	[19]
D_J	0.574(8)	0.561(8)	[20]
	0.5432(8)	0.5416(8)	[19]
D_K	0.0328	0.0322	[10]*)

constant	$^{79}BrF_5$	$^{81}BrF_5$	Ref.
$\lvert R_6 \rvert$	0.071(5)	0.069(5)	[20]
	0.078(2)	0.078(2)	[19]
	0.081(10)	–	[44]
	0.0773	0.0771	[10]*)

*) Calculated from a quadratic force field.

Less precise values for D_{JK} and D_J are given in [8].

For the excited state $v_5 = 1$, the constants $D_{JK} = 146(2)$ and 157(2) and $D_J = -61(1)$ and −65(1) were determined for $^{79}BrF_5$ and $^{81}BrF_5$, respectively [35].

Fundamental Vibrations. BrF_5 has 12 normal modes of vibration among which three are doubly degenerate: $3A_1 + 2B_1 + B_2 + 3E$. All modes are Raman-active, and the A_1 and E modes are infrared-active. Selected fundamental frequencies for BrF_5 in the gaseous, liquid, and solid state or trapped in an Ar matrix are listed in Table 17.

Table 17
Fundamental Vibrations of BrF_5 in cm^{-1}.
ν Denotes stretching, δ out-of-plane bending, δ′ in-plane bending, and F′ indicates the axial fluorine atom.

ν_i	approximate mode description	IR gas	IR liquid	IR Ar matrix	Raman liquid	Raman solid (180 K)
$\nu_1(A_1)$	ν BrF′	683	672	681	682	684
$\nu_2(A_1)$	sym. ν BrF	587	–	582	570	583, 562

Table 17 (continued)

ν_i	approximate mode description	IR gas	IR liquid	IR Ar matrix	Raman liquid	Raman solid (180 K)
$\nu_3(A_1)$	sym. δ FBrF	369	369	366	365	366, 373
$\nu_4(B_1)$	antisym. ν BrF	–	544	–	535	524, 539
$\nu_5(B_1)$	antisym. δ FBrF	–	–	–	281	–
$\nu_6(B_2)$	antisym. δ' FBrF	–	–	–	312	313, 320
$\nu_7(E)$	antisym. ν BrF	644	640	636	–	–
$\nu_8(E)$	δ FBrF′	415	414	415	414	415, 423
$\nu_9(E)$	antisym. δ' FBrF	–	–	240	237	235, 245
Ref.		[3]	[45]	[46]	[3]	[47]

Wave numbers (and intensities [48]) obtained from force fields are given in [10, 20, 48].

Coriolis Coupling Constants. Experimentally derived constants are given as:

	$^{79}BrF_5$	$^{81}BrF_5$	Ref.
ξ_9	-0.251 ± 0.004	-0.249 ± 0.004	[18]
$\lvert\xi^{y}_{5,9}\rvert$	0.781 ± 0.015	0.785 ± 0.015	[17]

Constants arising from the couplings $A_1 \times E$, $B_1 \times B_2$, $B_1 \times E$, and $E \times E$ were calculated [10, 49].

Mean amplitudes of vibration were calculated for different temperatures [10, 50, 51]. Values in Å are given at 300 K as [10]:

BrF_{eq}	BrF_{ax}	$F_{ax}F_{eq}$	$F_{eq}F_{eq}$(short)	$F_{eq}F_{eq}$(long)
0.0468	0.0430	0.0783	0.0915	0.0595

Force Constants

Force constants for BrF_5 with C_{4v} symmetry and an $F_{ax}BrF_{eq}$ angle less than 90° were calculated with a generalized valence force field [51] using the spectroscopic results for liquid [3] and solid [47] BrF_5 and the structure parameters for crystalline BrF_5 [21]. Nonzero valence force constants f in 10^2 N/m are derived with R denoting the BrF_{ax} and r the BrF_{eq} bonds, α the $F_{eq}BrF_{eq}$ and β the $F_{eq}BrF_{ax}$ angles [51]:

$f_r = 3.49$	$f_R = 4.07$	$f_\alpha = 0.21$	$f_\beta = 0.73$	$f_{rr} = 0.15$	$f'_{rr} = 0.06$
$f_{rR} = -0.01$	$f_{r\alpha} = 0.01$	$f'_{r\alpha} = -0.01$	$f_{r\beta} = 0.01$	$f_{R\beta} = 0.03$	$f_{\alpha\alpha} = -0.06$
$f'_{\alpha\alpha} = -0.05$	$f_{\alpha\beta} = 0.04$	$f'_{\alpha\beta} = 0.04$	$f'_{\beta\beta} = 0.15$		

Symmetry force constants are given in [10]. Constants of a modified Urey-Bradley force field are reported in [56]. Other force constants calculated with different methods are based on obsolete geometrical assumptions ($F_{eq}BrF_{ax}$ angle > 90° [1, 52, 53]; angle = 90° [3, 54, 55]).

Microwave Spectrum

Rotational spectra of BrF_5 were recorded in its ground vibrational state and in the excited vibrational states $v_5=1$ and $v_9=1$ at dry ice temperature. For both isotopic species, $^{79}BrF_5$ and $^{81}BrF_5$, the following transitions were observed and analyzed:

transition $J=n+1 \leftarrow n$	approximate frequency in GHz	Ref.	transition $J=n+1 \leftarrow n$	approximate frequency in GHz	Ref.
ground state			excited state $v_5=1$		
n=3, 4, 5	24, 31, 37	[8]	n=3, 4, 5	23, 29, 35	[17, 35]
n=8	55	[44]	excited state $v_9=1$		
n=10, 21, 22	70, 136, 142	[19]	n=3, 4, 5	25, 31, 37	[17, 35]
n=16	105	[20]	n=6, 7	44, 50	[18]

A K-type doubling, predicted for molecules with C_{4v} symmetry and obscured by the quadrupole structure in the spectrum, was observed for the transitions $J=9 \leftarrow 8$ [44] and $J=17 \leftarrow 16$ [20].

The vibrational states $v_5=1$ and $v_9=1$ are nearly degenerate, and an unusually strong Coriolis interaction perturbs both excited state spectra [35]. A detailed analysis of the Coriolis resonance was given and the q^- doubling of the $kl=-1$ levels, predicted for molecules with C_{4v} symmetry, was demonstrated for the $v_9=1$ state [17, 35]. A splitting of the $kl=3$ transitions for $J=7 \leftarrow 6$ and $J=8 \leftarrow 7$ for the $v_9=1$ state was reported [18]. Frequencies and intensities of "forbidden" $\Delta J=+1$, $\Delta K=\pm 4$ transitions were presented [43].

Infrared and Raman Spectra

The fundamental frequencies of BrF_5 are given on p. 120.

Infrared studies of gaseous BrF_5 were performed in the regions of 400 to 700 cm^{-1} [1, 4] and 215 to 4000 cm^{-1} [2, 3, 57], of liquid BrF_5 in the region 200 to 1100 cm^{-1} [45, 58], of solid BrF_5 at 77 K in the region of 400 to 750 cm^{-1} [47], and of BrF_5 in an argon matrix in the region 33 to 800 cm^{-1} [46, 59]. A solution of BrF_5 in CH_3CN was also investigated [60]. Frequencies in the overtone and combination band regions are given for gaseous BrF_5 [3, 57] and for liquid BrF_5 [58]. Selected Q-branch maxima in cm^{-1} are: 1364.3 ($2\nu_1$, $^{79}BrF_5$), 1361.3 ($2\nu_1$, $^{81}BrF_5$), and 1258 ($2\nu_7$) [3]. Raman studies were performed on liquid BrF_5 [1, 3, 4, 47], on solid BrF_5 at 180 and 10 K [47], and on matrix-trapped BrF_5 [59].

Ultraviolet Absorption

UV absorption of BrF_5 in gaseous mixtures of Br_2 and F_2 was recorded at 293 K in the range of 250 to 200 nm. The absorption cross section increases from about 0.3×10^{19} cm^2 to about 4.3×10^{19} cm^2 in this range [61].

References:

[1] Stephenson, C. V; Jones, E. A. (J. Chem. Phys. **20** [1952] 1830/4).
[2] McDowell, R. S.; Asprey, L. B. (J. Chem. Phys. **37** [1962] 165/7).
[3] Begun, G. M.; Fletcher, W. H.; Smith, D. F. (J. Chem. Phys. **42** [1965] 2236/42).
[4] Burke, T. G.; Jones, E. A. (J. Chem. Phys. **19** [1951] 1611).

[5] Gutowsky, H. S.; McCall, D. W.; Slichter, C. P. (J. Chem. Phys. **21** [1953] 279/92).
[6] Muetterties, E. L.; Phillips, W. D. (J. Am. Chem. Soc. **81** [1959] 1084/8).
[7] Muetterties, E. L.; Phillips, W. D. (J. Am. Chem. Soc. **79** [1957] 322/6).
[8] Whittle, M. J.; Bradley, R. H.; Brier, P. N. (Trans. Faraday Soc. **67** [1971] 2505/9).
[9] Robiette, A. G.; Bradley, R. H.; Brier, P. N. (J. Chem. Soc. D **1971** 1567/8).
[10] Heenan, R. K.; Robiette, A. G. (J. Mol. Struct. **54** [1979] 135/44).

[11] Falconer, W. E.; Jones, G. R.; Sunder, W. A.; Vasile, M. J.; Muenter, A. A.; Dyke, T. R.; Klemperer, W. (J. Fluorine Chem. **4** [1974] 213/34).
[12] Theilacker, W. (Z. Naturforsch. **3b** [1948] 231/7).
[13] Mellish, C. E.; Linnett, J. W. (Trans. Faraday Soc. **50** [1954] 657/64).
[14] Mellish, C. E.; Linnett, J. W. (Trans. Faraday Soc. **51** [1955] 1311).
[15] Searcy, A. W. (J. Chem. Phys. **31** [1959] 1/4).
[16] Rogers, M. T.; Pruett, R. D.; Thompson, H. D.; Speirs, J. L. (J. Am. Chem. Soc. **78** [1956] 44/5).
[17] Brier, P. N.; Jones, S. R.; Baker, J. G.; Gheorghiou, C. (J. Mol. Spectrosc. **64** [1977] 415/28).
[18] Gheorghiou, C.; Brier, P. N.; Baker, J. G.; Jones, S. R. (J. Mol. Spectrosc. **72** [1978] 282/92).
[19] Suzeau, P.; Jurek, R.; Chanussot, J. (Compt. Rend. B **276** [1973] 777/9).
[20] Bradley, R. H.; Brier, P. N.; Whittle, M. J. (J. Mol. Spectrosc. **44** [1972] 536/48).

[21] Burbank, R. D.; Bensey, F. N., Jr. (J. Chem. Phys. **27** [1957] 982/3).
[22] Wiebenga, E. H.; Kracht, D. (Inorg. Chem. **8** [1969] 738/46).
[23] Havinga, E. E.; Wiebenga, E. H. (Recl. Trav. Chim. **78** [1959] 724/38).
[24] Lakatos, B. (Z. Elektrochem. **61** [1957] 944/9).
[25] Dobbs, K. D.; Hehre, W. J. (J. Computat. Chem. **7** [1986] 359/78).
[26] Grodzicki, M.; Männing, V.; Trautwein, A. X.; Friedt, J. M. (J. Phys. B **20** [1987] 5595/625).
[27] Gutzev, G. L.; Smolyar, A. E. (Chem. Phys. Letters **71** [1980] 296/9).
[28] Scharfenberg, P. (Kem. Kozlem. **56** [1981] 167/88 from C.A. **96** [1982] No. 58094).
[29] Smolyar, A. E.; Charkin, O. P.; Klimenko, N. M. (Zh. Strukt. Khim. **15** [1974] 993/1003; J. Struct. Chem. [USSR] **15** [1974] 885/93).
[30] Deb, B. M.; Coulson, C. A. (J. Chem. Soc. A **1971** 958/70).

[31] Berry, R. S.; Tamres, M.; Ballhausen, C. J.; Johansen, H. (Acta Chem. Scand. **22** [1968] 231/46).
[32] DeKock, R. L.; Higginson, B. R.; Lloyd, D. R. (Faraday Discussions Chem. Soc. **54** [1972] 84/92).
[33] Lippincott, E. R.; Nagarajan, G.; Stutman, J. M. (J. Phys. Chem. **70** [1966] 78/84).
[34] Meinert, H. (Z. Chem. [Leipzig] **7** [1967] 41/57).
[35] Brier, P. N.; Jones, S. R.; Baker, J. G. (J. Mol. Spectrosc. **60** [1976] 18/30).
[36] Kuz'min, A. I.; Shpanko, V. I.; Zviadadze, G. N.; Sukhoverkhov, V. F.; Dzevitskii, B. E. (Zh. Neorg. Khim. **24** [1979] 2127/33; Russ. J. Inorg. Chem. **24** [1979] 1178/82).
[37] Stein, L. (in: Gutmann, V.; Halogen Chemistry, Vol. 1, Academic, New York 1967, pp. 168/74).
[38] Gillespie, R. J.; Schrobilgen, G. J. (Inorg. Chem. **15** [1976] 22/31).
[39] Keller, N.; Schrobilgen, G. J. (Inorg. Chem. **20** [1981] 2118/29).
[40] Gutowsky, H. S.; Hoffman, C. J. (J. Chem. Phys. **19** [1951] 1259/67).

[41] Bantov, D. V.; Dzevitskii, B. E.; Konstantinov, Yu. S.; Sukhoverkhov, V. F. (Izv. Sibirsk. Otd. Akad. Nauk SSSR Ser. Khim. Nauk **1968** No. 1, pp. 81/3; C.A. **70** [1969] No. 24519).
[42] Weulersse, J.-M. (CEA-R-4868 [1977] 1/213, 102/8; C.A. **89** [1978] No. 206985).
[43] Ghoshal, S.; Ghosh, P. N. (J. Mol. Spectrosc. **110** [1985] 364/8).

[44] Jones, S. R.; Brier, P. N.; Brookbanks, D. M.; Baker, J. G. (J. Mol. Spectrosc. **47** [1973] 351/2).
[45] Surles, T.; Quaterman, L. A.; Hyman, H. H. (J. Fluorine Chem. **3** [1973] 293/306).
[46] Frey, R. A.; Redington, R. L.; Aljibury, A. L. K. (J. Chem. Phys. **54** [1971] 344/55).
[47] Rousson, R.; Drifford, M. (J. Chem. Phys. **62** [1975] 1806/11).
[48] Krohn, B. J.; Person, W. B.; Overend, J. (J. Chem. Phys. **67** [1977] 5091/8).
[49] Venkateswarlu, K.; Mathew, M. P. (Curr. Sci. **37** [1968] 252/3; C.A. **69** [1968] No. 31641).
[50] Cyvin, S.; Brunvoll, J.; Robiette, A. G. (J. Mol. Struct. **3** [1969] 259/61).

[51] Mahmoudi, S.; Wendling, E. (Rev. Chim. Miner. **15** [1978] 254/67).
[52] Ramaswamy, K.; Muthusubramanian, P. (J. Mol. Struct. **7** [1971] 45/50).
[53] Khanna, R. K. (J. Sci. Ind. Res. [India] B **21** [1962] 352/6).
[54] Curtis, E. C. (Spectrochim. Acta A **27** [1971] 1989/97).
[55] Christe, K. O.; Curtis, E. C.; Schack, C. J.; Pilipovich, D. (Inorg. Chem. **11** [1972] 1679/82).
[56] Pillai, M. G. K.; Pillai, P. P. (Can. J. Chem. **46** [1968] 2393/7).
[57] Stein, L. (J. Am. Chem. Soc. **81** [1959] 1273/6).
[58] Haendler, H. M.; Bukata, S. W.; Millard, B.; Goodman, E. I.; Littman, J. (J. Chem. Phys. **22** [1954] 1939).
[59] Prochaska, E. S.; Andrews, L.; Smyrl, N. R.; Mamantov, G. (Inorg. Chem. **17** [1978] 970/7).
[60] Gross, U.; Meinert, H. (Z. Chem. [Leipzig] **11** [1971] 431/2).

[61] Buben, S. N.; Chaikin, A. M. (Kinet. Katal. **21** [1980] 1591/2; C.A. **94** [1981] No. 73863).

10.16.4 Crystallographic Properties

The structure of the orthorhombic crystalline phase was investigated at 153 K with photographic X-ray diffraction methods (single crystal, precession camera). The crystal data are: a = 6.422 Å, b = 7.245 Å, c = 7.846 Å, space group $Cmc2_1$-C_{2v}^{12} (No. 36). There are four molecules in the orthorhombic unit cell, with the bromine and three fluorine atoms on special positions and one fluorine atom in a general position; the final R value was 0.184 [1]. Bond distances and angles are given on p. 118.

Two solid phases possibly exist with a transition at about 143 K as indicated by the behavior of the longitudinal relaxation time T_1 of the ^{19}F nuclei as a function of temperature; see Fig. 5, p. 119 [2]. However, the same author did not mention this possible phase transition of BrF_5 in a more recent review [3].

References:

[1] Burbank, R. D.; Bensey, F. N., Jr. (J. Chem. Phys. **27** [1957] 982/3).
[2] Weulersse, J.-M. (CEA-R-4868 [1977] 1/213, 102/8; C.A. **89** [1978] No. 206985).
[3] Martin, D.; Rousson, R.; Weulersee, J.-M. (Chem. Non-Aqueous Solvents B **5** [1978] 157/95, 188).

10.16.5 Mechanical and Thermal Properties

Mechanical Properties

The **density** of solid, orthorhombic BrF_5, $\rho = 3.182$ g/cm^3, was calculated from crystallographic data [1] at 153 K. The density of liquid BrF_5 fits the equation $\rho = 2.5509 - 3.484 \times 10^{-3} \cdot t - 3.45 \times 10^{-6} \cdot t^2$ with the temperature t in °C and ρ in g/cm^3, valid from −15 to 76 °C (ρ = 2.4616 g/cm³ at 25 °C). The deviations for ρ are ±0.0003 g/cm³ for values above 0 °C and ±0.0006 g/cm³ below 0 °C [2]. Another equation, $\rho = 2.5474 - 0.00348 \cdot t$, was derived from measurements between 0 and 25 °C (ρ = 2.4604 g/cm³ at 25 °C) [3]. The differences between [1] and [2] may be due to impurities.

The **surface tension** γ of liquid BrF_5 was measured as 24.3, 23.5, 22.4, and 21.6 dyn/cm at 282.4, 287.9, 300.2, and 305.8 K, respectively [4].

The **viscosity** η of gaseous BrF_5 increases from about 168.5 cP to about 304.6 cP between 338.0 and 421.9 K. Using these data, the parameters σ = 5.494 Å and ε/k = 205 K were derived for the Stockmayer (3-6-12) potential [5].

The dynamic viscosity of liquid BrF_5 follows the empirical equation $\eta(P) = 1.11 \times 10^{-4} \exp(1195/T)$ between 275.5 and 302.1 K with a precision of ±2%. Applying Eyring's theory of absolute reaction rates to viscosity, the enthalpy, entropy, and free energy of activation for viscous flow were computed to be $\Delta H^* = 9.96$ kJ/mol, $\Delta S^* = -5.5\ J \cdot K^{-1} \cdot mol^{-1}$, and $\Delta G^* = 11.55$ kJ/mol, respectively [4].

Thermal Properties

A **melting point** of 212.65 ± 0.09 K was determined from the freezing curve of ten BrF_5 samples of about 99.6 mol% purity [6].

The **boiling points** of 313.91 ± 0.05 K [6] and of 314.45 K [3, 7] were extrapolated from the vapor pressures.

The **enthalpy of vaporization** was derived from vapor-pressure data to be 28.0 kJ/mol [4] and 30.59 kJ/mol [6]. The **Trouton constant** was calculated to be 97.5 $J \cdot K^{-1} \cdot mol^{-1}$ [6]. The **critical temperature** and **pressure** were estimated as 486 K and 47 bar [8].

The **vapor pressure** of BrF_5 is expressed by the Antoine equation

$$\log(p/\text{Torr}) = 6.4545 + 0.001101\ (T - 273.15) - 895/(T - 67.16)$$

which is valid for the temperature range from 298.223 up to 419.141 K. Vapor pressures predicted by this equation are about 1% lower than the former values in "Brom" 1931, p. 339. Noncondensable impurities may have caused the higher vapor pressures [7]. A vapor pressure of 2247 Torr was measured at 348.2 K [9, 10], compared to 2249.2 Torr calculated by the above equation [7]. The equation $\log(p/\text{Torr}) = 7.9727 - 1598.2/T$ was derived from an empirical correlation and is valid for the entire liquid range [6].

Thermodynamic Functions. Enthalpy and Free Energy of Formation

The functions C_p, S°, $-(G° - H°_{298})/T$ (in $J \cdot mol^{-1} \cdot K^{-1}$), $H° - H°_{298}$, $\Delta_f H°$, $\Delta_f G°$ (in kJ/mol) and the logarithms of the stability constant K_f were calculated for the ideal gas state at 0.1 MPa for temperatures in K up to 6000 K [11]. For previous results, see [12 to 15]. Excerpted values from the JANAF Tables are given below [11]:

T	$C_p°$	S°	$\frac{-(G° - H°_{298})}{T}$	$(H° - H°_{298})$	$\Delta_f H°$	$\Delta_f G°$	$\log K_f$
0	0	0	infinite	−19.180	−413.581	−413.581	infinite
100	47.334	243.616	398.823	−15.521	−418.586	−397.662	207.717
298.15	101.404	323.682	323.682	0	−428.718	−351.379	61.560
500	119.343	381.232	335.986	22.623	−441.858	−291.719	30.476

T	C_p^o	S°	$-\frac{(G^o - H^o_{298})}{T}$	$(H^o - H^o_{298})$	$\Delta_f H^o$	$\Delta_f G^o$	$\log K_f$
800	127.270	439.429	364.603	59.861	−436.812	−202.879	13.247
1000	129.277	468.067	382.531	85.537	−433.265	−144.806	7.564
1500	131.332	520.954	420.429	150.789	−424.606	−2.472	0.086
2000	132.069	558.852	450.520	216.665	−416.586	137.013	−3.578
2500	132.413	588.364	475.246	282.795	−408.878	274.522	−5.736
3000	132.601	612.524	496.173	349.053	−400.567	410.440	−7.146
4000	132.789	650.700	530.260	481.761	−378.889	677.733	−8.850
5000	132.876	680.342	557.422	614.598	−348.483	938.551	−9.805
6000	132.923	704.572	579.989	747.499	−310.068	1192.468	−10.381

The standard heat capacity and the standard entropy at 298 K and 1 atm were estimated as $C_p^o = 98.9$ and $S^o = 333\ J \cdot mol^{-1} \cdot K^{-1}$ from simple expressions containing the boiling point together with empirical parameters, characteristic for pentafluorides [16].

The enthalpy of formation $\Delta_f H^o_{298.15} = -428.71 \pm 2.1$ kJ/mol [11] was derived from the enthalpy of reaction for $5 F_2(g) + Br_2(g) \rightarrow 2 BrF_5(g)$ which was measured in an adiabatic calorimeter [17]. An earlier value of −444.17 kJ/mol was based on gaseous Br_2 adjusted to the liquid Br_2 reference state at 298.15 K [17].

References:

[1] Burbank, R. D.; Bensey, F. N., Jr. (J. Chem. Phys. **27** [1957] 982/3).
[2] Banks, A. A.; Maddock, J. J. (J. Chem. Soc. **1955** 2779/81).
[3] Stein, L.; Vogel, R. C.; Ludewig, W. H. (J. Am. Chem. Soc. **76** [1954] 4287/9).
[4] Rogers, M. T.; Garver, E. E. (J. Phys. Chem. **62** [1958] 952/4).
[5] Ostorero, J. (CEA-N-1293 [1970] 1/36; C.A. **74** [1971] No. 130644).
[6] Rogers, M. T.; Speirs, J. L. (J. Phys. Chem. **60** [1956] 1462/3).
[7] Long, R. D.; Martin, J. J.; Vogel, R. C. (Chem. Eng. Data Ser. **3** [1958] 28/34).
[8] Long, R. D. (Diss. Univ. Michigan 1955, 132 pp.; Diss. Abstr. **15** [1955] 1581).
[9] Fischer, J.; Bingle, J. (J. Am. Chem. Soc. **77** [1955] 6511/2).
[10] Fischer, J.; Bingle, J.; Vogel, R. C. (J. Am. Chem. Soc. **78** [1956] 902/4).

[11] Chase, M. W., Jr.; Davies, C. A.; Downey, J. R., Jr.; Frurip, D. J.; McDonald, R. A.; Syverud, A. N. (JANAF Thermochemical Tables, 3rd Ed., 1985; J. Phys. Chem. Ref. Data Suppl. **14** No. 1 [1985] 423).
[12] Stephenson, C. V.; Jones, E. A. (J. Chem. Phys. **20** [1952] 1830/4).
[13] McDowell, R. S.; Asprey, L. B. (J. Chem. Phys. **37** [1962] 165/7).
[14] Khanna, R. K. (J. Sci. Ind. Res. [India] B **21** [1962] 352/6).
[15] Kudchadker, A. P.; Kudchadker, S. A.; Agarwal, P. M. (Indian J. Chem. **9** [1971] 722/4).
[16] Stølevik, R. (Acta Chem. Scand. **43** [1989] 758/62).
[17] Stein, L. (J. Phys. Chem. **66** [1962] 288/91).

10.16.6 Electrical, Magnetic, and Optical Properties

The **dielectric constant** $\varepsilon = 7.76$ of liquid BrF_5 was reported for 298 K [1]. The constant decreases from 8.33 to 7.91 between 261.5 and 297.7 K and is represented by the equation $\varepsilon = 8.20 - 0.0117\ t$ with t in °C. The values are close to the values calculated from the gas phase dipole moment [2]. Selected values for ε of gaseous BrF_5 and for the **molar polarization** P from [2] are as follows:

T in K	374.9	402.4	430.8
ε	1.005525	1.004910	1.004378
P in cm^3/mol	56.6	54.0	51.5

The **specific conductivity** was determined at 298 K to be 9.1×10^{-8} S/cm [3 to 5]. Conductivity measurements of BrF_5 in acetonitrile gave a value of 4.74×10^{-7} S/cm [6]. A positive temperature coefficient was observed for pure liquid BrF_5 [3].

The measured (Gouy balance) molar **magnetic susceptibility** for liquid BrF_5 at room temperature, $\chi_M = -45.1 \times 10^{-6}$ cm^3/mol, is slightly smaller than $\chi_M = -48.1 \times 10^{-6}$ cm^3/mol calculated from the ionic susceptibilities [7].

The **refractive index** n of liquid BrF_5 is 1.3529 at 298 K, measured with natrium D light ($D_1 = 5889.965$ Å, $D_2 = 5895.93$ Å). The molar refraction is then 15.41 cm^3/mol [8]. For BrF_5 vapor at 298 K and 1 atm, the refractive index n = 1.000951 was determined with light of the mercury green line (λ = 5461 Å) resulting in a molar refraction of 15.48 cm^3/mol [9].

References:

[1] Meinert, H. (Z. Chem. [Leipzig] **7** [1967] 41/57).
[2] Rogers, M. T.; Pruett, R. D.; Thompson, H. B.; Speirs, J. L. (J. Am. Chem. Soc. **78** [1956] 44/5).
[3] Rogers, M. T.; Speirs, J. L.; Panish, M. B. (J. Am. Chem. Soc. **78** [1956] 3288/9).
[4] Quarterman, L. A.; Hyman, H. H.; Katz, J. J. (J. Phys. Chem. **61** [1957] 912/7).
[5] Surles, T.; Quarterman, L. A.; Hyman, H. H. (J. Fluorine Chem. **3** [1973] 293/306).
[6] Gross, U.; Meinert, H. (Z. Chem. [Leipzig] **11** [1971] 431/2).
[7] Rogers, M. T.; Panish, M. B.; Speirs, J. L. (J. Am. Chem. Soc. **77** [1955] 5292/3).
[8] Stein, L.; Vogel, R. C.; Ludewig, W. H. (J. Am. Chem. Soc. **76** [1954] 4287/9).
[9] Rogers, M. T.; Malik, J. G.; Speirs, J. L. (J. Am. Chem. Soc. **78** [1956] 46/7).

10.16.7 Chemical Behavior

10.16.7.1 Thermal Decomposition

BrF_5 is stable at moderate temperatures. Above 823 K BrF_5 dissociates via $BrF_5 \rightarrow BrF_3 + F_2$ [1]. pVT measurements showed dissociation up to 673 K to be less than 0.1%, corresponding to a value of $\Delta G_{670} > 75$ kJ/mol [2].

References:

[1] Stein, L. (J. Phys. Chem. **66** [1962] 288/91).
[2] Bernstein, R. B.; Katz, J. J. (J. Phys. Chem. **56** [1952] 885/8).

10.16.7.2 Electrolysis

An electrolysis of BrF_5 was performed with platinum electrodes at 298 K. Ohm's law was not followed. The observed cathodic process is a stepwise reduction of BrF_4^+ to bromine; an anodic process could not be analyzed.

Reference:

Meinert, H.; Gross, U. (Z. Chem. [Leipzig] **12** [1972] 150/1).

10.16.7.3 Reactions with Electrons

Mass spectra of BrF_5 at 50 eV [1] and 70 eV [2] show a dominant BrF_4^+ peak. Other ions (with intensities relative to BrF_4^+ (100) in parentheses) are: BrF_3^+ (26.4), BrF_2^+ (38.5), BrF^+ (18.8), Br_2^+ (11.7), and Br^+ (17.4). Ions containing oxygen (BrO^+, $BrOF^+$ et al.) were also noted. Appearance potentials (given in eV) were determined by the vanishing current method: $AP(BrF_4^+) = 14.0 \pm 0.3$; $AP(BrF_3^+) = 15.5 \pm 0.2$; $AP(BrF_2^+) = 16.1 \pm 0.2$; $AP(BrF^+) = 11.9 \pm 0.3$; $AP(Br^+) = 10.1 \pm 0.3$ [1].

References:

[1] Irsa, A. P.; Friedman, L. (J. Inorg. Nucl. Chem. **6** [1958] 77/90).
[2] Beattie, W. H. (Appl. Spectrosc. **29** [1975] 334/7).

10.16.7.4 Fluorine Exchange Reactions

F_2. The fluorine exchange between BrF_5 and F_2 was investigated qualitatively between 298 and 673 K [1] and quantitatively between 454 and 530 K [2] in metal vessels with the ^{18}F-tracer method. No exchange took place below 373 K. Complete exchange, however, occurred within a few minutes near 573 K. More details concerning the mechanism of exchange are given in "Fluorine" Suppl. Vol. 2, 1980, pp. 101/2.

XeF_6. One broad signal was observed in the ^{19}F NMR spectrum of XeF_6 dissolved in BrF_5. A rapid fluorine exchange was concluded [3].

HF, ClF, ClF_3, BrF_3. The exchange of ^{18}F between HF and BrF_5 was complete within ten minutes in the liquid and within three minutes in the vapor phase. An activated complex $HBrF_6$ with an IF_7-like structure was discussed [4]. No rapid exchange takes place between BrF_5 and HF, ClF, ClF_3, and BrF_3 based on ^{19}F NMR spectra of the corresponding binary mixtures [5]. For the exchange with HF, see also "Fluorine" Suppl. Vol. 3, 1982, p. 192.

$NaHF_2$. 16% exchange between gaseous BrF_5 and solid $NaHF_2$ was proven with the ^{18}F-tracer method [4].

CsF. The reaction of BrF_5 with $Cs^{18}F$ was studied between 298 and 373 K. Labeled products were $CsBr^{18}F_6$ which amounted to 30% of the total activity and $Br^{18}F_5$ [6].

NiF_2. The exchange of ^{18}F between BrF_5 and solid NiF_2 was studied at 299, 526, and 594 K: increasing the temperature reduced the extent of exchange [1].

References:

[1] Bernstein, R. B.; Katz, J. J. (J. Phys. Chem. **56** [1952] 885/8).
[2] Adams, R. M.; Bernstein, R. B.; Katz, J. J. (J. Chem. Phys. **22** [1954] 13/21).
[3] Seppelt, K.; Rupp, H. H. (Z. Anorg. Allgem. Chem. **409** [1974] 331/7).
[4] Rogers, M. T.; Katz, J. J. (J. Am. Chem. Soc. **74** [1952] 1375/7).
[5] Hamer, A. N. (J. Inorg. Nucl. Chem. **9** [1959] 98/9).
[6] Gross, U.; Meinert, H. (Z. Chem. [Leipzig] **11** [1971] 349/50).

10.16.7.5 Reactions with Elements

Xe, Kr. The formation of XeBr and KrBr exciplexes instead of XeF and KrF exciplexes was observed in the reactions of BrF_5 with metastable $Xe(^3P_2)$ and $Kr(^3P_2)$ atoms, studied with the flowing afterglow technique [1, 2].

Rn. BrF_5 oxidizes radon according to $Rn+BrF_5 \rightarrow BrF_3+RnF_2$ [3, 4]. The reaction was performed with BrF_5 as solvent [3], in mixtures of BrF_5 with F_2 [5], or in HF saturated with KF [6]. The postulated formation of higher radon fluorides [5] was questioned [7].

H_2. The hydrogenation enthalpy for the reaction $BrF_5+3\,H_2 \rightarrow HBr+5\,HF$ was calculated with the ab initio SCF MO method using the 3-21 $G^{(*)}$ basis set to be -778.2 kJ/mol. A value of -1000 kJ/mol was derived from experimental heats of formation [8].

O_2. $BrOF_3$ and BrO_2F were found mass spectrometrically as products of the reaction of BrF_5 with O_2 [9].

O_3. At 268 K BrF_5 reacts with ozone in the presence of bromine to give bromylfluoride according to $BrF_5+2\,Br_2+10\,O_3 \rightarrow 5\,BrO_2F+10\,O_2$ [10].

Br_2. BrF_5 reacts moderately with Br_2 at 423 K and very rapidly at 573 K forming BrF and BrF_3 [11].

C. The reaction of BrF_5 with graphite has been relatively little studied compared to the equivalent reaction of BrF_3 (see p. 77). Like BrF_3, BrF_5 forms fluorinated graphite intercalation compounds (FGIC) with graphite by fluorination of the carbon and insertion of unreacted BrF_5 between the graphite interlayers. The reaction can be performed with gaseous or liquid BrF_5 [12 to 15, 18], at room [15] and higher temperatures (up to 373 K) [13], and under varying BrF_5 pressure (up to 540 hPa) [14]. A mixture of BrF_5 with BrF_3, F_2, and HF was used at 298 K [16, 17]. Another procedure starts with a solution of BrF_5 in anhydrous HF, to which graphite is added at 77 K, followed by warming to 295 K [22]. A study of the intermediate products indicates that the intercalation reaction precedes fluorination. A reaction scheme was proposed, in which BrF_5 molecules are first weakly captured by van der Waals bonds between the graphite layers, then react with carbon to give fluorinated graphite and BrF_3 molecules [15]. FGIC of varying composition are obtained depending of reaction time and amounts of BrF_5; for example, a composition of $C_{9.7}BrF_{9.2}$ was found after ten days with excess of gaseous BrF_5 at 298 K [15], and of $C_{12.4}BrF_{8.9}H_{1.2}$ from a BrF_5-HF solution at 295 K after 30 days [22]. A composition of $C_{26}BrF_{12.1}$ was achieved with gaseous BrF_5 at 363 to 373 K [13]. The composition of the products depends also on the carbon particle size and the method of graphite preparation [19]. Bromine evolution according to the scheme $C+BrF_5 \rightarrow C\cdot F(BrF_3)_y+Br_2$ [13] was observed when heating the intercalation product above 623 K [12].

Si. Silicon is spontaneously etched by BrF_5 vapor. Physisorbed BrF_5 appears to act as a precursor to etching as the effective activation energy was found to be negative in the range studied (up to 420 K). The rate equation in this region, expressed as the etch rate R in Å/min, was found to be $R=1.4\times10^{-17}\,n\,T^{1/2}\exp(14.2\,kJ\cdot mol^{-1}/kT)$ with the density n of BrF_5 in cm^{-3}, and T in K (BrF_5 pressure was 8.1 Torr) [20].

Pt. A mixture of BrF_5 and Br_2 dissolves metallic platinum quite rapidly. A crystalline compound of the composition $PtBr_2F_{10}$ was isolated. Pure BrF_5 has no effect on metallic platinum [21].

References:

[1] Velazco, J. E.; Kolts, H. E.; Setser, D. W. (NBS-SP-526 [1978] 359/63; C.A. **90** [1979] No. 14354).
[2] Velazco, J. E.; Kolts, H. E.; Setser, D. W. (J. Chem. Phys. **65** [1976] 3468/80).
[3] Stein, L. (J. Am. Chem. Soc. **91** [1969] 5396/7).
[4] Stein, L. (Radiochim. Acta **32** [1983] 163/71).
[5] Avrorin, V. V.; Krasikova, R. N.; Nefedov, V. D.; Toropova, M. A. (Radiokhimiya **23** [1981] 879/83; Soviet Radiochem. **23** [1981] 708/11).
[6] Stein, L. (Science **168** [1970] 362/4).
[7] Stein, L. (Inorg. Chem. **23** [1984] 3670/1).
[8] Dobbs, K. D.; Hehre, W. J. (J. Computat. Chem. **7** [1986] 359/78).
[9] Irsa, A. P.; Friedman, L. (J. Inorg. Nucl. Chem. **6** [1958] 77/90).
[10] Schmeisser, M.; Pammer, E. (Angew. Chem. **69** [1957] 781).

[11] Stein, L. (J. Am. Chem. Soc. **81** [1959] 1273/6).
[12] Selig, H.; Sunder, W. A.; Vasile, M. J.; Stevie, F. A.; Gallagher, P. K.; Ebert, L. B. (J. Fluorine Chem. **12** [1978] 397/412).
[13] Sukhoverkhov, V. F.; Nikoronov, Yu. I.; Zhuzhgov, E. L. (Zh. Neorg. Khim. **30** [1985] 1391/4; Russ. J. Inorg. Chem. **30** [1985] 793/5).
[14] Danilenko, A. M.; Nazarov, A. S.; Yakovlev, I. I. (Zh. Neorg. Khim. **33** [1988] 42/7; Russ. J. Inorg. Chem. **33** [1988] 23/6).
[15] Danilenko, A. M.; Nazarov, A. S.; Yakovlev, I. I. (Izv. Akad. Nauk SSSR Ser. Khim. **1988** No. 5, pp. 953/7; Bull. Acad. Sci. USSR Div. Chem. Sci. **1988** 827/31).
[16] Hamwi, A.; Daoud, M.; Cousseins, J. C. (Synth. Metals **26** [1988] 89/98).
[17] Hamwi, A.; Daoud, M.; Cousseins, J. C.; Yazami, R. (J. Power Sources **27** [1989] 81/7).
[18] Danilenko, A. M.; Nazarov, A. S.; Yakovlev, I. I. (Zh. Neorg. Khim. **33** [1988] 884/8; Russ. J. Inorg. Chem. **33** [1988] 496/9).
[19] Danilenko, A. M.; Nazarov, A. S.; Yakovlev, I. I.; Potapova, O. G.; Fadeeva, V. P. (Izv. Akad. Nauk SSSR Neorg. Mater. **26** [1990] 1441/5; Inorg. Mater. [USSR] **26** [1990] 1228/31).
[20] Ibbotson, D. E.; Mucha, J. A.; Flamm, D. L.; Cook, J. M. (J. Appl. Phys. **56** [1984] 2939/42).

[21] Chernyaev, I. I.; Nikolaev, N. S.; Ippolitov, E. G. (Dokl. Akad. Nauk SSSR **130** [1960] 1041/3; Proc. Acad. Sci. USSR Chem. Sect. **130/135** [1960] 167/9).
[22] Danilenko, A. M.; Nazarov, A. S.; Yakovlev, I. I.; Fadeeva, V. P. (Izv. Akad. Nauk SSSR Neorg. Mater. **25** [1989] 1303/6; Inorg. Mater. [USSR] **25** [1989] 1100/3).

10.16.7.6 Reactions with Inorganic Compounds

Reactions with Oxides and Salts of Oxoacids

H_2O. The hydrolysis of BrF_5 at room temperature is violent, often explosive, and accompanied by formation of Br_2, O_2, and HF; see "Brom" 1931, p. 339. At low temperatures (about 210 K), the reaction with equimolar amounts or excess BrF_5 yields BrO_2F [1, 2]. It was suggested that the reaction proceeds via the intermediate formation of $BrOF_3$ (which hydrolyzes to give BrO_2F and HF), but no $BrOF_3$ was observed [1], neither was it found mass spectrometrically in partially hydrolyzed BrF_5 [3]. On the other hand, the product BrO_2F_3 formed via the reaction $2\,BrF_5 + 2\,H_2O \rightarrow BrF_3 + BrO_2F_3 + 4\,HF$ (another suggested route is $BrF_5 + H_2O_2 \rightarrow BrO_2F_3 + 2\,HF$) was identified in the mass spectrum [3].

The reaction of BrF_5 with small amounts of water in a nickel vessel at 353 K liberates oxygen in 100% yield according to $BrF_5+H_2O \rightarrow BrF_3+2\,HF+1/2\,O_2$ [4]. Water and oxide impurities can thus be removed from fluoride melts by adding BrF_5 [5]. The fact that the ^{16}OH bond in H_2O is broken in preference to the ^{18}OH bond in the reaction of H_2O with BrF_5 allows the separation of the oxygen isotopes [6].

BrF_5 is moderately hydrolyzed in acetonitrile to give $HBrO_3$ and HF [7].

A survey of the reactions with other oxygen compounds is given in Table 18.

Table 18
Reactions of BrF_5 with Oxides and Salts of Oxoacids.

reactant	products	conditions, remarks
N_2O_5	FNO_2, BrO_2F	reaction at 213 K [8]
$NaNO_2$	NaF, Br_2, FNO_2	with excess $NaNO_2$ at 298 K for 2 h [9]
KNO_2	$KBrF_4$, FNO_2	with excess BrF_5 [9]
$LiNO_3$	LiF, $BrNO_3$, FNO_2	molar ratio BrF_5:$LiNO_3$=1:3, at 298 K [10]
	O_2, $BrOF_3$	with excess BrF_5 at 273 K for 20 days [11]
$NaNO_3$	$NaBrOF_4$, $BrOF_3$	at 273 K [11]
	$NaBrF_4$, $NaBrO_2F_2$	at 298 K [11]
MNO_3	$MBrOF_4$, FNO_2	M=K at 373 K, M=Rb at 298 K, M=Cs at 242 K 100% conversion to $MBrOF_4$ [11]
BrO_2	BrO_2F, Br_2	at 213 to 223 K [12]
$KBrO_3$	$KBrF_4$, BrO_2F, O_2	at 223 K [12]
	$KBrO_2F_2$, BrF_3, O_2	at 298 K for 6 to 8 h, by-product $KBrF_4$ [13]
	BrO_2F, $KBrOF_4$, $KBrO_2F_2$	in contrast to [12, 13], there is no or slow reaction with pure reagents, but reaction is promoted by small amounts of HF [1, 14]
	$KBrOF_4$	after addition of F_2 at 353 K for 16 h, by-products BrF_3 and $KBrF_4$ [15]
$KBrO_4$	$KBrOF_4$	70% yield after addition of F_2 at 353 K for 95 h [16]
$CsBrO_4$	$CsBrOF_4$	after addition of F_2 at 298 K for 30 h [16]
I_2O_5	BrO_2F, IF_5	with excess BrF_5 at 225 to 243 K [1]
$CsIO_4$	$CsIO_2F_4$, $BrOF_3$	at 293 K for 24 h [17]
IO_4^-	$IO_2F_4^-$, $BrOF_3$	at 292 K [18]
SO_3	$BrF_5 \cdot SO_3$	at 313 K, addition compound is a viscous, yellow liquid [12]
	$S_2O_5F_2$, BrF_3, Br_2, O_2	reaction with liquid SO_3 [19]
Cs_2SO_4	$CsBrOF_4$, $CsSO_3F$	at 298 K [9]
$BaSO_4$	SO_2F_2, O_2	at 623 and 823 K, oxygen extraction for isotopic analysis [20]
MCO_3	MF_2, BrF_3, CO_2, O_2	at 398 K [21]
	MF_2, BrF_3, CF_4, O_2	at 973 K; used to extract oxygen from MCO_3 for isotopic analysis (M=Mg, Ca, Sr, Ba, Mn, Co, Ni, Cd, Pb) [21]

Table 18 (continued)

reactant	products	conditions, remarks
$KAlSi_3O_8$	KF, AlF_3, SiF_4, O_2, BrF_3	used for oxygen extraction from several minerals, e.g. quartz, feldspars, magnetite, hematite, ilmenite, garnet, and olivin, for isotopic analysis [22]
CrO_3	$CrOF_3 \cdot 0.25\ BrF_5$	at 298 K [23]
$K_2Cr_2O_7$	$KCrOF_4 \cdot 0.5\ BrF_5$	at 298 K [23]
UO_2	UF_6	above 550 K; reaction rates determined [24]; see also "Uran" Erg.-Bd. C8, 1980, p. 78
PuO_2	PuF_4	above 550 K [24]
RuO_4	$RuF_5 \cdot BrF_5$	formation of complex below 273 K suggested [25]

Reactions with Fluorides and Fluorine Compounds

A survey of the reactions of BrF_5 with fluorine compounds is given in Table 19.

Table 19
Reactions of BrF_5 with Fluorides and Fluorine Compounds.

reactant	products	conditions, remarks
KrF^+, $Kr_2F_3^+$	BrF_6^+, KrF_2	in AsF_5 or SbF_5; in SbF_5, BrF_4^+ is also formed [26 to 30]
IOF_3	BrO_2F, IF_5	excess BrF_5 between 210 and 238 K [1]
IO_2F	BrO_2F, IF_5	excess BrF_5 at 225 K [1]
IO_2F_3	BrO_2F, $BrOF_3$, IOF_5	excess BrF_5 at 298 K [1]
SF_4, SF_5Cl, S_2F_{10}	SF_6	at 673 K for 15 h [31]
BF_3	BrF_4BF_4	ionic complex as intermediate suggested in reactions between 183 and 223 K [32]
$(C_6F_5)SiF_3$	SiF_4, $(C_6F_5)BrF_4$	at 273 K for 24 h in CH_3CN as solvent [33]
$XeOTeF_5^+$ AsF_6^-	TeF_6, $Xe_2F_3^+$, $XeF_2BrOF_2^+$, $FXeFXeFOTeF_5^+$	dissolution of reactant in BrF_5 at 225 K; products analyzed by ^{19}F NMR and Raman spectroscopy [34]
AsF_5	BrF_4AsF_6	marginally stable at 176 K [35]
$IO_2F_3 \cdot AsF_5$	IOF_5, $BrOF_3 \cdot AsF_5$	excess BrF_5 at 298 K [36]
SbF_5	$BrF_4Sb_2F_{11}$	ionic solid complex [12, 37, 38]
	$BrF_5 \cdot k\ SbF_5$	addition compounds, k=0.66, 1, 2, 3, 7, detected during DTA of the BrF_5-SbF_5 system [39, 40]
$O_2^+SbF_6^-$	violet species	at 77 K oxygen fluoride radicals, e.g. O_2F, assumed [41]
$IO_2F_3 \cdot SbF_5$	IOF_5, $BrOF_3 \cdot SbF_5$	excess BrF_5 at 298 K [36]
$C_6F_3 \cdot BrF_3$	$C_{18.1}F_{9.0} \cdot BrF_5$	at 295 K in an inert atmosphere for 5 to 10 days [42]
MF	$MBrF_6$	in BrF_5 at 373 K: M=Li, Na, no reaction; M=K 49.4%; M=Rb 90%; M=Cs 100% conversion; $M=NH_4^+$ at 298 K in BrF_5 [43]
		in CH_3CN at 298 K:CsF:BrF_5=1.4:1 in 4 to 5 h [7]

Table 19 (continued)

reactant	products	conditions, remarks
	$MBrF_6 \cdot x\,MF$	in equimolar mixtures of BrF_5 and ClF_3 at 298 K: M=Rb, x=0.5; M=Cs, x=0.1 [43]
	$MClF_4 \cdot x\,MF$	in equimolar mixtures of BrF_5 and ClF_3 for M=K: x=1 at 298 K, x=0.7 at 373 K [43]
CsF	$CsBrF_6$, $Cs_2Br_3F_{17}$	detected at 293 K in the system BrF_5-HF-CsF [44]
$CsClOF_4$	$CsBrF_6$, $ClOF_3$	at 293 K for 12 h [45]
SnF_4	$(BrF_4)_2SnF_6$	reaction proceeds near the boiling point of BrF_5 [46, 47]
CrF_4	$CrOF_3 \cdot 0.25\,BrF_3$	in a mixture of BrF_3 and BrF_5, oxygen from the SiO_2 container; no reaction with pure BrF_5 [48]
PtF_6	?	no oxidation of BrF_5 at 298 and 373 K [27, 49]; a black viscous liquid and F_2 are decomposition products of PtF_6 [49]

Miscellaneous

BrF_5 fluorinates nitrides at room temperature, for example Li_3N, Mg_2N_3, Ca_3N_2, BN, and Si_3N_4, to give N_2F_2. The highest yield measured was 20 to 25% N_2F_2 for Mg_2N_3. The reactions of BrF_5 with TiN and Ba_3N_2 proceed violently even at low temperatures [50].

Excess BrF_5 ignites CsN_3 producing N_2 and an equimolar mixture of $CsBrF_4$ and $CsBrF_6$ [9].

BrF_5 forms a yellowish liquid with FNO when combined at 77 K, but no stable solid compound was observed below 209 K [51]. Vapor pressure measurements indicate some association of BrF_5 with FNO_2 [52].

BrF_5 fluorinates sulfides, for example Ag_2S via $4\,BrF_5 + Ag_2S \rightarrow 4\,BrF_3 + 2\,AgF + SF_6$, in a nickel vacuum system at 573 to 623 K [53].

References:

[1] Gillespie, R. J.; Spekkens, P. H. (J. Chem. Soc. Dalton Trans. **1977** 1539/46).
[2] Jacob, E. (Z. Anorg. Allgem. Chem. **433** [1977] 255/60).
[3] Sloth, E. N.; Stein, L.; Williams, C. W. (J. Phys. Chem. **73** [1969] 278/80).
[4] O'Neil, J.; Epstein, S. (J. Geophys. Res. **71** [1966] 4955/61).
[5] Green, G. L.; Hunt, J. B.; Sutula, R. A. (J. Inorg. Nucl. Chem. **35** [1973] 4305/7).
[6] Ustinov, V. I.; Sukhoverkhov, V. F.; Podzolko, L. G. (Zh. Fiz. Khim. **52** [1978] 610/4; Russ. J. Phys. Chem. **52** [1978] 344/7).
[7] Meinert, H.; Gross, U. (Z. Chem. [Leipzig] **9** [1969] 190).
[8] Schmeisser, M.; Taglinger, L. (Chem. Ber. **94** [1961] 1533/9).
[9] Christe, K. O.; Wilson, W. W.; Schack, C. J. (J. Fluorine Chem. **43** [1989] 125/9).
[10] Wilson, W. W.; Christe, K. O. (Inorg. Chem. **26** [1987] 1573/80).

[11] Wilson, W. W.; Christe, K. O. (Inorg. Chem. **26** [1987] 916/9).
[12] Schmeisser, M.; Pammer, E. (Angew. Chem. **69** [1957] 781).
[13] Tantot, G.; Bougon, R. (Compt. Rend. C **281** [1975] 271/3).
[14] Gillespie, R. J.; Spekkens, P. (J. Chem. Soc. Dalton Trans. **1976** 2391/6).
[15] Bougon, R.; Bui Huy, T.; Charpin, P.; Tantot, G. (Compt. Rend. C **283** [1976] 71/4).

[16] Christe, K. O.; Wilson, R. D.; Curtis, E. C.; Kuhlmann, W.; Sawodny, W. (Inorg. Chem. **17** [1978] 533/8).
[17] Christe, K. O.; Wilson, R. D.; Schack, C. J. (Inorg. Chem. **20** [1981] 2104/14).
[18] Christe, K. O. (unpublished results, given in [45] as ref. 27).
[19] Gross, U.; Meinert, H.; Grimmer, A. R. (Z. Chem. [Leipzig] **10** [1970] 441/3).
[20] Ustinov, V. I.; Grinenko, V. A. (Zh. Fiz. Khim. **56** [1982] 2730/3; Russ. J. Phys. Chem. **56** [1982] 1677/8).

[21] Sharma, T.; Clayton, R. N. (Geochim. Cosmochim. Acta **29** [1965] 1347/53).
[22] Clayton, R. N.; Mayeda, T. K. (Geochim. Cosmochim. Acta **27** [1963] 43/52).
[23] Clark, H. C.; Sadana, Y. N. (Can. J. Chem. **42** [1964] 702/4).
[24] Galkin, N. P.; Veryatin, U. D.; Zuev, V. A. (Radiokhimiya **22** [1980] 750/3; Soviet Radiochem. **22** [1980] 578/81).
[25] Hepworth, M. A.; Robinson, P. L. (J. Inorg. Nucl. Chem. **4** [1957] 24/9).
[26] Gillespie, R. J.; Schrobilgen, G. J. (Inorg. Chem. **15** [1976] 22/31).
[27] Christe, K. O.; Wilson, W. W.; Wilson, R. D. (Inorg. Chem. **23** [1984] 2058/63).
[28] Gillespie, R. J.; Schrobilgen, G. J. (Inorg. Chem. **13** [1974] 1230/5).
[29] Gillespie, R. J.; Schrobilgen, G. J. (J. Chem. Soc. Chem. Commun. **1974** 90/2).
[30] McKee, D. E. (LBL-1814 [1973] 65; C.A. **80** [1974] No. 66301).

[31] Bains-Sahota, S. K.; Thiemens, M. H. (Anal. Chem. **60** [1988] 1084/6).
[32] Bardin, V. V.; Furin, G. G.; Yakobson, G. G. (J. Fluorine Chem. **23** [1983] 67/86).
[33] Breuer, W.; Frohn, H. S. (J. Fluorine Chem. **34** [1987] 443/51).
[34] Keller, N.; Schrobilgen, G. J. (Inorg. Chem. **20** [1981] 2118/29).
[35] Christe, K. O.; Sawodny, W. (Inorg. Chem. **12** [1973] 2879/86).
[36] Bougon, R.; Bui Huy, T.; Carpin, P.; Gillespie, R. J.; Spekkens, P. H. (J. Chem. Soc. Dalton Trans. **1979** 6/12).
[37] Meinert, H.; Gross, U. (Z. Chem. [Leipzig] **10** [1970] 226/7).
[38] Surles, T.; Perkins, A.; Quarterman, L. A.; Hyman, H. H.; Popov, A. I. (J. Inorg. Nucl. Chem. **34** [1972] 3561/4).
[39] Kuz'min, A. I.; Shpanko, V. I.; Zviadadze, G. N.; Sukhoverkhov, V. F.; Dzevitskii, B. E. (Zh. Neorg. Khim. **24** [1979] 2127/33; Russ. J. Inorg. Chem. **24** [1979] 1178/82).
[40] Sukhoverkhov, V. F.; Shpanko, V. I.; Takanova, N. D. (Zh. Neorg. Khim. **22** [1977] 2534/8; Russ. J. Inorg. Chem. **22** [1977] 1371/4).

[41] Christe, K. O.; Wilson, R. D.; Goldberg, I. B. (J. Fluorine Chem. **7** [1976] 543/9).
[42] Danilenko, A. M.; Nazarov, A. S.; Yakovlev, I. I. (Zh. Neorg. Khim. **34** [1989] 1693/6; Russ. J. Inorg. Chem. **34** [1989] 958/61).
[43] Whitney, E. D.; MacLaren, R. O.; Fogle, C. E.; Hurley, T. J. (J. Am. Chem. Soc. **86** [1964] 2583/6).
[44] Sukhoverkhov, V. F.; Takanova, N. D.; Uskova, A. A. (Zh. Neorg. Khim. **18** [1973] 3333/9; Russ. J. Inorg. Chem. **18** [1973] 1774/7).
[45] Christe, K. O.; Wilson, W. W.; Wilson, R. D. (Inorg. Chem. **19** [1980] 1494/8).
[46] Sukhoverkhov, V. F.; Dzevitskii, B. Z. (Dokl. Akad. Nauk SSSR **170** [1966] 1099/102; Dokl. Chem. Proc. Acad. Sci. USSR **166/171** [1966] 983/6).
[47] Sukhoverkhov, V. F.; Dzevitskii, B. Z. (Dokl. Akad. Nauk SSSR **177** [1967] 611/4; Dokl. Chem. Proc. Acad. Sci. USSR **172/177** [1967] 1089/91).
[48] Clark, H. C.; Sadana, Y. N. (Can. J. Chem. **42** [1964] 50/6).
[49] Gortsema, F. G.; Toeniskoetter, R. H. (Inorg. Chem. **5** [1966] 1925/7).
[50] Shamir, J.; Binenboym, J. (Inorg. Nucl. Letters **2** [1966] 117/8).

[51] Christe, K. O. (Inorg. Chem. **11** [1972] 1215/9).

[52] Whitney, E. D.; MacLaren, R. O.; Hurley, T. J.; Fogle, C. E. (J. Am. Chem. Soc. **86** [1964] 4340/2).
[53] Ding, T.; Li, H.; Zhang, G.; Li, Y.; Li, J. (Certif. Ref. Mater. Proc. ISCRM'89, Beijing 1989, pp. 292/7; C.A. **112** [1989] No. 228928).

10.16.8 Reactions with Organic Compounds

BrF_5 reacts more or less violently with organic compounds; see "Brom" 1931, p. 339. There were no experimental data until 1960 on the preparation of organic fluorine compounds with BrF_5 apart from several patents with unsubstantiated claims [1]. BrF_5 reacts explosively with dioxane and pyridine [2]. Fluorination of octafluoronaphthalene with BrF_5, performed at 203 to 223 K, yields perfluorinated and oxidated dihydronaphthalenes [3]. Nucleophilic fluorine-aryl substitution on BrF_5 gives arylbrominetetrafluorides $RBrF_4$ [4, 5]. The reaction is performed below 273 K in an anhydrous, inert solvent, such as CH_2Cl_2, in the presence of a nitrogen base, such as acetonitrile or pyridine, which activates the axial fluorine by coordination of the equatorial fluorine atoms. Transferred aryl groups were C_6F_5 [4], C_6H_5, *o*-, *m*-, *p*-F-C_6H_4, *o*-, *m*-, *p*-CF_3-C_6H_4 [5]. By-products were fluorinated or brominated arylic compounds [5].

References:

[1] Musgrave, W. K. R. (Advan. Fluorine Chem. **1** [1960] 1/28, 16/7).
[2] Meinert, H.; Gross, U. (Z. Chem. [Leipzig] **9** [1969] 190).
[3] Bardin, V. V.; Furin, G. G.; Yakobson, G. G. (J. Fluorine Chem. **23** [1983] 67/86).
[4] Breuer, W.; Frohn, H. J. (J. Fluorine Chem. **34** [1987] 443/51).
[5] Breuer, W.; Frohn, H. J. (J. Fluorine Chem. **47** [1990] 301/15).

10.16.9 Bromine Pentafluoride as Solvent

In contrast to BrF_3, the self-ionization of BrF_5 is very small and its complexing power is low [1, 2]. It is useful as a solvent in certain cases, for example as a medium for other fluoride-containing compounds, such as xenon fluorides, in NMR studies [3, 4] or as a solvent for fluorinating reactions, for example in the preparation of $BrOF_3$ from $KBrO_4$ and O_2AsF_6 [5].

References:

[1] Meinert, H.; Gross, U. (J. Fluorine Chem. **2** [1973] 381/6).
[2] Martin, D.; Rousson, R.; Weulersse, J. M. (Chem. Non-Aqueous Solvents B **5** [1978] 157/95, 178/82).
[3] Schrobilgen, G. J.; Holloway, J. H.; Granger, P.; Brevard, C. (Inorg. Chem. **17** [1978] 980/7).
[4] Seppelt, K.; Rupp, H. H. (Z. Anorg. Allgem. Chem. **409** [1974] 331/7).
[5] Bougon, R.; Bui Huy, T. (Compt. Rend. C **283** [1976] 461/3).

10.16.10 Binary Systems with BrF_5

Br_2. The two compounds form an azeotrope with a F/Br ratio of 2.6 and a boiling point of 307 K [1].

XeF_2. A congruently melting adduct $XeF_2 \cdot 2\,BrF_5$ and an incongruently melting adduct $XeF_2 \cdot 9\,BrF_5$ were found in the system BrF_5-XeF_2. The maximum solubility of XeF_2 in BrF_5 does not exceed 11.2 moles per 1000 g BrF_5 [2]. Increasing association of XeF_2 with BrF_5 towards lower temperatures was concluded from the temperature dependency of the chemical shift in the ^{129}Xe NMR spectra [3]. According to [4], XeF_2 is soluble in BrF_5 without adduct formation.

HF. The two compounds are completely miscible in the liquid state and form an azeotrope; see "Fluorine" Suppl. Vol. 3, 1982, p. 192. The volatile azeotrope between 291 and 339 K was investigated by the rectification method. The temperature dependence of the vapor pressure of the azeotrope is given by $\log(p/Torr) = (8.17 \pm 0.01) - (1546 \pm 3)/T$. Its lowest HF concentration is about 54 mol% at about 308 K. With increasing temperature, the HF concentration slightly increases, while at temperatures below 308 K, the HF concentration appreciably increases. The azeotrope probably disappears at 283 ± 5 K [5].

BrF_3. The two compounds are completely miscible. The liquid-vapor phase diagram is shown in **Fig. 6**. The conductivity increases by adding BrF_3 from about 10^{-7} S/cm for pure BrF_5 to 8×10^{-3} S/cm for pure BrF_3 at 298 K [7]. The dependence of the refractive index n_D^{25} and the density d (in g/cm^3) on the mole fraction x of BrF_5 at 298 K are given by the equations [8]

$$n_D^{25} = 1.4536 - 0.1243\,x + 0.0124\,x^2 + 0.0112\,x^3$$
$$d = 2.8030 - 0.3884\,x + 0.0641\,x^2 - 0.0183\,x^3$$

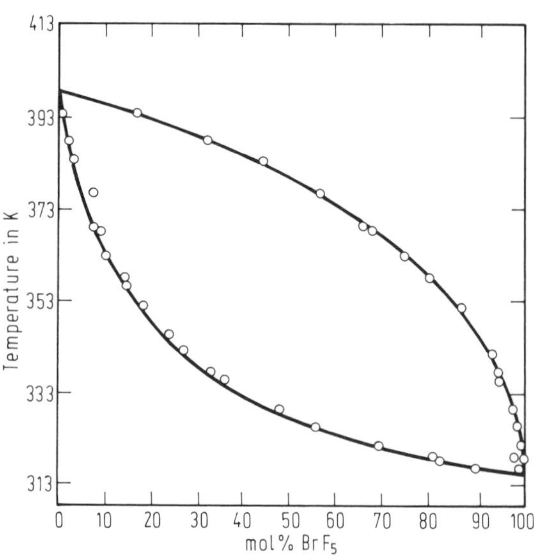

Fig. 6. Liquid-vapor equilibria for the BrF_5-BrF_3 system; data adjusted to 1 atm (from [6]).

References:

[1] Yosim, S. J. (J. Phys. Chem. **62** [1958] 1596/7).
[2] Prusakov, V. N.; Sokolov, V. B.; Chaivanov, B. B. (Zh. Fiz. Khim. **45** [1971] 1102/5; Russ. J. Inorg. Chem. **45** [1971] 616/8).
[3] Schrobilgen, G. J.; Holloway, J. H.; Granger, P.; Brevard, C. (Inorg. Chem. **17** [1978] 980/7).

[4] Meinert, H.; Gross, U. (Z. Chem. [Leipzig] **12** [1972] 150/1).
[5] Ezhov, V. K. (Zh. Neorg. Khim. **21** [1976] 2097/9; Russ. J. Inorg. Chem. **21** [1976] 1154/5).
[6] Long, R. D.; Martin, J. J.; Vogel, R. C. (Chem. Eng. Data Ser. **3** [1958] 28/34).
[7] Quarterman, L. A.; Hyman, H. H.; Katz, J. J. (J. Phys. Chem. **61** [1957] 912/7).
[8] Stein, L.; Vogel, R. C.; Ludewig, W. H. (J. Am. Chem. Soc. **76** [1954] 4287/9).

10.17 Pentafluorobromine(1+), BrF_5^+

CAS Registry Number: –

A weak signal of BrF_5^+ was observed in mass spectra of BrF_5 (see also p. 128) at ionization energies of 70 eV or more [1]. The ion was also found in the mass spectrum of partly hydrolyzed BrF_5 [2]. An earlier mass spectrum of BrF_5 at 50 eV did not show the molecular ion [3]. The ion was observed in the mass spectrum of BrF_7 [4]; see p. 144.

References:

[1] Falconer, W. E.; Jones, G. R.; Sunder, W. A.; Vasile, M. J.; Muenter, A. A.; Dyke, T. R.; Klemperer, W. (J. Fluorine Chem. **4** [1974] 213/34).
[2] Sloth, E. N.; Stein, L.; Williams, C. W. (J. Phys. Chem. **73** [1969] 278/80).
[3] Irsa, A. P.; Friedman, L. (J. Inorg. Nucl. Chem. **6** [1958] 77/90).
[4] Fogle, C. E.; Rewick, R. T.; United Aircraft Corp. (U.S. 3615206 [1971] 4 pp.; C.A. **76** [1972] No. 5447).

10.18 Bromine Hexafluoride, BrF_6

CAS Registry Numbers: $^{79}BrF_6$ *[56783-23-4]*, $^{81}BrF_6$ *[56783-24-5]*

The radical BrF_6 forms during γ irradiation of solid solutions of about 5% BrF_5 in SF_6 [1, 2] or TeF_6 [3] at 77 K.

The ESR spectrum in the range 105 to 110 K shows two superimposed septets of the Br isotopes, thereby confirming the composition BrF_6 [1, 2]. The septets indicate that the BrF_6 radical has on the average an octahedral symmetry, while the large ESR line width may be caused by a fluctuating structure [2]. The equivalence of pairs of F atoms in all directions of the magnetic field and the octahedral structure of BrF_6 are demonstrated by the hyperfine (hf) splitting of the anisotropic spectra measured at 27 K, when the rotation of the radical is frozen [3, 4].

The exceedingly large Br hyperfine interaction indicates s character and accordingly a totally symmetric representation of the semioccupied orbital. The ground state of BrF_6 is $^2A_{1g}$ in O_h symmetry [1 to 4]. Thus the F2p contributions to the semioccupied ns orbital of Br point to each other and to the central Br atom [3, 4]. The orbitals are overlapped out of phase [4]; the isotropic hyperfine interactions arise solely via polarization and not via direct F2s contributions to the semioccupied orbital [3].

The hf interaction of $^{79}BrF_6$ is constant from 4 to about 30 K and then drops increasingly when the temperature is raised to about 135 K. This change is compatible with a low-frequency vibration, whose amplitude increases with temperature bringing about a decrease in the bromine ns character and hence its hf interaction [4]. Table 20 lists g values and hf coupling constants a. Earlier values are given in [1, 2]. The ratio a_{81}/a_{79} is in satisfactory agreement with expectations [1, 2]. Anisotropic spectra at low temperature exhibit two different sites, A and B, of BrF_6 in the SF_6 matrix [3]. The observed behavior of site B is anomalous [5].

Table 20
g Values of BrF_6 and Hyperfine Coupling Constants a in G [3].

conditions	$^{79}BrF_6$		$^{81}BrF_6$		both isotopomers
	g	a_{Br}	g	a_{Br}	a_F
SF_6, 110 K	2.0158	4160	2.0158	4485	88.5(10)$^{c)}$
27 K	2.0147$^{a)}$	4175$^{a)}$	2.0147$^{a)}$	4501$^{a)}$	296(10)$^{d)}$, −18(2)$^{e)}$
	2.0148$^{b)}$	4191$^{b)}$	2.0148$^{b)}$	4519$^{b)}$	–
TeF_6, 27 K	2.0148	4191	2.0149	4483	–

$^{a)}$ Site A. — $^{b)}$ Site B. — $^{c)}$ Isotropic value. — $^{d)}$ Anisotropic; parallel value. —
$^{e)}$ Anisotropic; perpendicular value.

Spin densities were calculated for the atoms in BrF_6 from the hf coupling constants. They range from 0.4 to 0.54 for the s orbital of Br [1, 2, 4] and from 0.15 to 0.19 for the F2p orbital [3, 4].

The observed temperature dependence of the hyperfine interaction is compatible with a modulation by an intramolecular vibration. Vibrational wave numbers for $^{79}BrF_6$ of 163 cm^{-1} in SF_6 (site A) and 185 cm^{-1} in TeF_6 were calculated. The values of the frequencies suggest that a bending mode causes modulation of the bromine hf interaction. A dominant contribution of the F_{1u} bending mode in O_h symmetry was proposed; the vibration can be ascribed to ν_4 of BrF_6. A similar value was found for ν_4 of BrF_6^- [5].

References:

[1] Boate, A. R.; Morton, J. R.; Preston, K. F. (Inorg. Chem. **14** [1975] 3127/8).
[2] Nishikida, K.; Williams, F.; Mamantov, G.; Smyrl, N. (J. Chem. Phys. **63** [1975] 1693/4).
[3] Boate, A. R.; Morton, J. R.; Preston, K. F. (J. Phys. Chem. **80** [1976] 2954/9).
[4] Boate, A. R.; Morton, J. R.; Preston, K. F. (J. Magn. Resonance **29** [1978] 243/9).
[5] Boate, A. R.; Morton, J. R.; Preston, K. F.; Strach, S. J. (J. Chem. Phys. **71** [1979] 388/91).

10.19 Hexafluorobromine(1+), BrF_6^+

CAS Registry Number: *[51063-27-5]*

Formation. The synthesis of BrF_6^+ salts requires one to use the strongly oxidizing krypton-fluorine compounds. The preparation of BrF_6AsF_6 free from BrF_4^+ salts succeeds by decomposition of $KrFAsF_6$ in BrF_5 solution at ambient temperature [1] in yields ranging from 5.3 to 7% with respect to the applied KrF_2 [2]. The corresponding decomposition of $KrFSb_2F_{11}$ leads only to $BrF_4Sb_2F_{11}$ [3]. BrF_6AsF_6 and $BrF_6Sb_2F_{11}$ result when $Kr_2F_3EF_6$ with E=As, Sb is decomposed in BrF_5 at ambient temperature and after the volatiles are evaporated [4]. The yields do not exceed 20% [1]; the solid residue also contains the more volatile BrF_4^+ salts [1, 2]. The formation of BrF_6AuF_6 from BrF_3 or BrF_5 by $KrF_2 \cdot AuF_5$ was mentioned [5]. $BrF_4Sb_2F_{11}$ is oxidized by KrF_2 to $BrF_6SbF_6 \cdot x\, SbF_5$ ($x<1$) at 298 K in BrF_5 [2].

Failed attempts to oxidize BrF_5 to BrF_6^+ salts include reactions with PtF_6 in HF solution [6], with PtF_6 and F_2 by heating under pressure [7], with PtF_6 exposed to unfiltered UV irradiation [8], and with F_2 in the presence of AsF_5 at about 500 K and pressures up to 200 atm [1].

A BrF_6^+ peak was observed at m/z 193 in the mass spectrum of BrF_7 [9]; see p. 144.

^{19}F NMR Spectrum. The ^{19}F NMR signals obtained on about 0.5 molar solutions of BrF_6^+ salts in HF at 299 K are double quartets arising from interactions with the nuclear quadrupole moments of the Br atoms and exhibit low field shifts with respect to external CCl_3F:

anion	δ in ppm	$J(^{79}BrF)$ in Hz	$J(^{81}BrF)$ in Hz	Ref.
$Sb_2F_{11}^-$	339.4	1575	1697	[1, 4]
AsF_6^-	337.4	1587	1709	[1]

The well-resolved spin-spin coupling confirms the expected octahedral symmetry of BrF_6^+ in solution. The observed isotope coupling ratio of 1.078 is in excellent agreement with the gyromagnetic ratio of 1.0779 of the Br isotopes [1, 4]. The center of the fluorine resonance of $^{79}BrF_6^+$ is shifted 0.14±0.02 ppm to low field of the signal of $^{81}BrF_6^+$. The difference was attributed to different vibrational amplitudes for the Br isotopes [1].

Vibrational Spectra. The vibrational spectra of BrF_6^+ agree with an octahedral ion of point group O_h $(1\,A_{1g}+1\,E_g+2\,F_{1u}+1\,F_{2g}+1\,F_{2u})$[1, 2]. The Raman bands ν_1, ν_2, and ν_5 were assigned on the basis of relative intensities and the polarization of ν_1 and depolarization of the other bands in HF solution [1]. Vibrational bands of BrF_6^+ in salts in cm^{-1} are:

anion	method	$\nu_1(A_1)$ ν_s in plane	$\nu_2(E_g)$ ν_s out of phase	$\nu_3(F_{1u})$ ν_{as}	$\nu_4(F_{1u})$ δ_s out of plane	$\nu_5(F_{2g})$ δ_s in plane	Ref.
$Sb_2F_{11}^-$	Raman	658	668	–	–	405	[1]
AsF_6^- and $Sb_2F_{11}^-$	IR	–	–	775	430	–	[2]
AuF_6^-	IR	–	–	663, 669 m	–	–	[5]
AuF_6^-	Raman	651 and 657 m		–	–	409 w	[5]

A BrF bond length of 1.69 Å was estimated for BrF_6^+ [10] from a set of force constants [2].

Mean Amplitudes of Vibration. Values were calculated from the vibrational data in [2] between 0 and 1000 K by the method of characteristic vibrations assuming $\nu_6=\nu_5\cdot 2^{-1/2}$. Results at 300 K in Å are $u_{Br-F}=0.0405$, $u_{F\cdots F(short)}=0.072$, and $u_{F\cdots F(long)}=0.054$. The values are quite low in comparison to those of other Br-F compounds and indicate strong bonds [11].

Coriolis Coupling Constants. Force Constants. Calculated Coriolis coupling constants are $\zeta_1=0.380$ and $\zeta_2=0.120$ [12]. Force constants in the general valence force field (in mdyn/Å) were calculated [12] from the vibrational data in [2]:

f_r	f_{rr}	$f_{rr'}$	$f_{r\alpha}-f_{r\alpha'}$	$f_\alpha-f_{\alpha''\alpha'}$	$f_{\alpha\alpha}-f_{\alpha''\alpha}$	$f_{\alpha'\alpha}-f_{\alpha''\alpha'}$
4.896	−0.025	0.079	0.203	0.544	0.043	0.043

(f_r=bond stretch; f_{rr}, $f_{rr'}$=cis and trans bond stretch interaction; f_α=angle deformation; $f_{\alpha\alpha}$=angle deformation interaction; $f_{r\alpha}$, $f_{r\alpha'}$=bond stretch-angle deformation interaction). Results of modified Urey-Bradley and orbital valence force field calculations are also given [12]. Stretching force constants given in [2] agree with those in [12].

Quantum Chemical Calculation. Energy levels and the AO population of BrF_6^+ were calculated by the semiempirical NDDO-2(α, β) method [13].

Chemical Behavior. BrF_6AsF_6 and $BrF_6Sb_2F_{11}$ can be stored in Teflon vessels at room temperature without noticeable decomposition [2]. The strongly irritating and toxic nature of BrF_6AsF_6 was mentioned [14]. Thermal decomposition of BrF_6AuF_6 at 520 K yields BrF_5 and AuF_3 [5]. The salts BrF_6AsF_6 and $BrF_6Sb_2F_{11}$ are more powerful oxidants than F_2 at ambient temperature. The reaction with gaseous O_2 and Xe is rapid and yields O_2^+ and XeF^+ [1]. AgCl is attacked at room temperature, quartz reacts with formation of the corresponding O_2^+ salts [2]. Reduction of BrF_6AsF_6 by NOF yields $NOBrF_6$, $NOAsF_6$, and F_2 [1].

References:

[1] Gillespie, R. J.; Schrobilgen, G. J. (Inorg. Chem. **13** [1974] 1230/5).
[2] Christe, K. O.; Wilson, R. D. (Inorg. Chem. **14** [1975] 694/6).
[3] McKee, D. E. (LBL-1814 [1973] 1/93, 65; C.A. **80** [1974] No. 66301).
[4] Gillespie, R. J.; Schrobilgen, G. J. (J. Chem. Soc. Chem. Commun. **1974** 90/2).
[5] Sokolov, V. B.; Prusakov, V. N.; Ryzhkov, A. V.; Drobyshevskii, Yu. V.; Khoroshev, S. S. (Dokl. Akad. Nauk SSSR **229** [1976] 884/7; Dokl. Chem. Proc. Acad. Sci. **226/231** [1976] 525/8).
[6] Christe, K. O.; Wilson, W. W.; Wilson, R. D. (Inorg. Chem. **23** [1984] 2058/63).
[7] Gortsema, F. P.; Toeniskoetter, R. H. (Inorg. Chem. **5** [1966] 1925/7).
[8] Christe, K. O. (unpublished observation, ref. 10 in Plenary Main Sect. Lect. 24th Intern. Congr. Pure Appl. Chem., Hamburg 1973 [1974], Vol. 4, pp. 115/41).
[9] Fogle, C. E.; Rewick, R. T.; United Aircraft Corp. (U.S. 3615206 [1971] 4 pp.; C.A. **76** [1972] No. 5447).
[10] Mahjoub, A. R.; Hoser, A.; Fuchs, J.; Seppelt, K. (Angew. Chem. **101** [1989] 1528/9).
[11] Baran, E. J.; Lavat, A. E. (Indian J. Pure Appl. Phys. **20** [1982] 152/3).
[12] Goel, R. K.; Gupta, S. K. (Indian J. Pure Appl. Phys. **18** [1980] 718/22).
[13] Smolyar, A. E.; Charkin, O. P.; Klimenko, N. M. (Zh. Strukt. Khim. **15** [1974] 993/1003; J. Struct. Chem. [USSR] **15** [1974] 885/93).
[14] Fogle, C. E.; Breazeale, J. D. (AD-A022099 [1976] 1/112, 66; C.A. **85** [1976] No. 80474).

10.20 Hexafluorobromate(1−), BrF_6^-

CAS Registry Numbers: *[19702-39-7, 56713-48-5]*

Formation. Alkali salts of BrF_6^- are commonly prepared by solvolysis of alkali fluorides in excess BrF_5. The reaction is complete for CsF [1, 2], about 90% for RbF, and about 50% for KF, while the reaction of NaF is insignificant and LiF does not react [1]. Reaction with $Cs^{18}F$ yields $CsBr^{18}FF_5$ [3]. A white solid was obtained upon the reaction with NH_4F; it decomposed above 273 K and presumably was NH_4BrF_6 [1]. The composition of a compound formed by BrF_5 and FNO was not reproducible; FNO_2 does not react at low temperatures [4].

The reactions of BrF_5 with CsF [5] and AgF [6] in CH_3CN solution yield $CsBrF_6$ and $AgBrF_6$. This approach failed with BaF_2, while NH_4F reacted violently [6]. The reaction of CsF and gaseous BrF_5 was used to prepare samples for vibrational spectroscopy [7]. The formation of $CsBrF_6$ from BrF_5 and CsN_3 [8] or $CsClOF_4$ [9] involves CsF as an intermediate.

The metathetical reaction of $(CH_3)_4NF$ and $CsBrF_6$ gives a better yield of $(CH_3)_4NBrF_6$ than the reaction of $(CH_3)_4NF$ and BrF_5 [10]. Formation of R_4NBrF_6 with $M=CH_3$, C_2H_5

by metathetical reactions of $AgBrF_6$ and R_4NCl was reported [6]. BrF_6AsF_6 is reduced by FNO to $NOBrF_6$ and $NOAsF_6$ with release of F_2 [11]. Fluorination of KBr and RbBr with F_2 at elevated temperature probably yielded mixtures of BrF_6^- and BrF_4^- salts [12].

$KBrF_6$ forms when $KBrO_3$ is fluorinated by BrF_5 in the presence of F_2 (molar ratio 1:5:2) under autogenous pressure at temperatures exceeding 350 K. The reaction involves fluorination of the intermediate $KBrOF_4$ [13].

Uses. Only the Cs salt of BrF_6^- has been practically used. Applications include the quantitative determination of O_2 in oxides and minerals [14], the oxidative removal of Rn from dry air [15, 16], and acting as an ignition aid for hypergolic reactions involving mixtures of oxidizing compounds with granular fuels made from amines or carboranes [17].

Structural Investigations. The octahedral structure of BrF_6^- in CH_3CN solution is known from the vibrational spectrum; the distortion of the ion, if any, is apparently rather small [18]. Fluxional behavior was found on the slower NMR time scale [10, 18]; see below for details.

The BrF_6^- ion in solid $CsBrF_6$ is nearly octahedral, as demonstrated by diffraction studies [19]. The structural data agree well with each other; neutron diffraction at 5 K gave similar results:

conditions	Br-F distance in Å	angle in degree
X-ray diffraction of a single crystal at 113 K	1.854(1)	90.8(1), 89.2(1)
neutron diffraction of the powder at 298 K	1.847(1)	91.3, 88.7

The structure of BrF_6^- in solids is the same as in solution. The octahedral symmetry assigns the lone electron pair to a Br s orbital. The resulting shielding of the nuclear charge also explains the large Br-F bond length compared to an estimated value of 1.69 Å in BrF_6^+. The experimentally determined structure disagrees with the structure predicted by the VSEPR model [19] to be a distorted octahedral BrF_6^- ion because of the sterically active Br lone pair [18]. Earlier X-ray powder patterns showed that $MBrF_6$ salts are isomorphous for M=K, Rb, and Cs. Vibrational spectra of the anions were assigned in the lower D_{3d} symmetry [20]; see below.

^{19}F NMR Spectrum. The BrF_6^- signal in saturated CH_3CN solutions of the tetramethyl ammonium and caesium salts has a downfield shift of $\delta=94$ ppm with respect to CCl_3F. The relatively large line width of 80 Hz and the missing spin-spin coupling in the temperature range 213 to 313 K indicate fluxional behavior of BrF_6^- on the NMR time scale [10, 18].

Vibrational Spectra. The vibrational spectra indicate an ionic structure of the BrF_6^- compounds; they were assigned to the point group O_h (cf. p. 139) [10, 18, 21]. The assignment is confirmed by the stronger relative intensity of the ν_1 band and the lower relative intensity of the ν_2 band in solution [18]. Assigned bands in cm^{-1} are:

compound	$\nu_1(A_{1g})$ ν_s in plane Raman	$\nu_2(E_g)$ ν_s out of phase Raman	$\nu_3(F_{1u})$ ν_{as} IR	$\nu_5(F_{2g})$ δ_s in plane Raman	Ref.
solid $(CH_3)_4NBrF_6$	562 (59)	451 (100)	490 vs, br	239 (9)	[10]
solid $CsBrF_6$	563 vs	449 s	—	241 w	[21]
$CsBrF_6$ in CH_3CN	565 p	450 dp	—	240 dp	[18]

An earlier publication assigned vibrational spectra of solid alkali salts of BrF_6^- in the point group D_{3d}; wave numbers at 298 K in cm^{-1} are [20]:

compound	$\nu_1(A_{1g})$ Raman	$\nu_2(E_g)$ Raman	$\nu_4(A_{2u}, E_u)$ IR	$\nu_5(A_{1g}, E_g)$ Raman	$\nu_6(A_{2u},E_u)$ IR
$KBrF_6$	568 s	454 vs	204, 184	250 s	138
$RbBrF_6$	568 s	456 vs	203, 182	250 s	144
$CsBrF_6$	562 s	451 vs	193, 176	243 s	156

IR absorptions at 123 K are also given; they deviate by 8 cm^{-1} or less from the data at 298 K [20]. The wave numbers of Raman spectra in solution and in solids are in excellent agreement, but the relative intensities of the ν_1 and ν_2 bands are different. The change of the point group in vibrational spectra of BrF_6^- from O_h in solution to D_{3d} in the solid would be a consequence of the site symmetry in the crystal or a slight compression due to crystal packing effects [18]. Early unassigned vibrational data of BrF_6^- salts are given in [6, 7].

Force Constants. Sets of constrained force constants were calculated for octahedral BrF_6^- by a simple valence force field from vibrational data in [20]; values in mdyn/Å are [22]:

f_r	f_{rr}	$f_{rr'}$	$f_{r\alpha}-f_{r\alpha''}$	$f_\alpha-f_{\alpha\alpha'''}$	$f_{\alpha\alpha}-f_{\alpha\alpha''}$	$f_{\alpha\alpha'}-f_{\alpha\alpha'''}$
1.9185	0.2172	0.8218	0.2807	0.2606	0.0770	0.0429

The designations used therein are: f_r = bond stretch, f_{rr} = cis bond stretch cross term, $f_{rr'}$ = trans bond stretch cross term, $f_{r\alpha}$ = in-plane bond stretch-angle deformation cross term with the angle adjacent to the bond, $f_{r\alpha''}$ = in-plane bond angle cross term with one shared bond, $f_{\alpha\alpha}$ = out-of-plane deformation cross term with the angle sharing one bond, $f_{\alpha\alpha'}$ = in-plane deformation cross term with the angle sharing one bond, $f_{\alpha\alpha''}$ = out-of-plane deformation cross term with no shared bond, and $f_{\alpha\alpha'''}$ = in-plane deformation cross term with no shared bond. A second set of similar force constants is also given in [22]; the origin of the used fundamentals is uncertain however.

Quantum Chemical Calculations. The energy levels and the AO population of BrF_6^- were calculated by the semiempirical NDDO-2(α, β) method [23].

Chemical Behavior. The tetramethyl ammonium and alkali salts of BrF_6^- are white, crystalline solids [1, 10]. $CsBrF_6$ is a strong irritant and toxic [24]. While $(CH_3)_4NBrF_6$ decomposes thermally at about 490 K [10], $CsBrF_6$ is stable at 573 K [1]. Fluorine exchange between $CsBrF_6$ and BrF_5 was observed at 298 and 373 K in solution and between solid $CsBrF_6$ and a saturated BrF_5 solution. An associative mechanism is favored because of the moderate rates and the observation of a heterogeneous reaction [3]. Fluorine exchange between BrF_6^- and BrF_5 is fast on the ^{19}F NMR time scale [2].

The alkali salts of BrF_6^- are powerful fluorinating agents. They react violently with water [1]. $CsBrF_6$ oxidizes Rn [15]. $CsBrF_6$ dissolves in HF to form BrF_5 via $BrF_6^- + HF \rightleftharpoons BrF_5 + HF_2^-$. The equilibrium is shifted so far to the right side that the Raman spectrum does not show BrF_6^- bands [25]. Metathetic reactions of BrF_6^- salts are described under the preparation of the compounds. The neutralization of basic $CsBrF_6$ by acidic $BrF_2Sb_2F_{11}$ in the melt yields BrF_5 and $CsSbF_6$ [26]. $KBrF_6$ in CH_3CN solution is slowly decomposed to $KBrF_4$ and F_2. The reaction seems to be catalyzed by trace amounts of moisture; hydrolysis of this solution leads to $KBrOF_4$ and $KBrO_2F_2$ contaminated with $KBrF_4$. A mixture of $KBrOF_4$ and $KBrO_2F_2$ is also obtained when $KBrF_6$ and $KBrO_3$ react in a CH_3CN solution [27]. $CsBrF_6$ reacts with oxides and minerals above 480 K and is thereby reduced to $CsBrF_4$ with liberation of O_2 [14]. Slow reduction to BrF_4^- was also found during storage of $(CH_3)_4NBrF_6$, probably as a result of cation oxidation [10].

References:

[1] Whitney, E. D.; MacLaren, R. O.; Fogle, C. E.; Hurley, T. J. (J. Am. Chem. Soc. **86** [1964] 2583/6).
[2] Muetterties, E. L. (Advan. Chem. Coord. Compounds **1961** 509/19).
[3] Gross, U.; Meinert, H. (Z. Chem. [Leipzig] **11** [1971] 349/50).
[4] Whitney, E. D.; MacLaren, R. O.; Hurley, T. J.; Fogle, C. E. (J. Am. Chem. Soc. **86** [1964] 4340/2).
[5] Meinert, H.; Gross, U. (Z. Chem. [Leipzig] **9** [1969] 190).
[6] Meinert, H.; Gross, U. (Z. Chem. [Leipzig] **11** [1971] 469).
[7] Beaton, S. P.; Sharp, D. W. A.; Perkins, A. J.; Sheft, I.; Hyman, H. H.; Christe, K. (Inorg. Chem. **7** [1968] 2174/6).
[8] Christe, K. O.; Wilson, W. W.; Schack, C. J. (J. Fluorine Chem. **43** [1989] 125/9).
[9] Christe, K. O.; Wilson, W. W.; Wilson, R. D. (Inorg. Chem. **19** [1980] 1494/8).
[10] Wilson, W. W.; Christe, K. O. (Inorg. Chem. **28** [1989] 4172/5).

[11] Gillespie, R. J.; Schrobilgen, G. J. (Inorg. Chem. **13** [1974] 1230/5).
[12] Bode, H.; Klesper, E. (Z. Anorg. Allgem. Chem. **313** [1961] 161/9).
[13] Bougon, R.; Bui Huy, T.; Charpin, P.; Tantot, G. (Compt. Rend. C **283** [1976] 71/4).
[14] Sukhoverkhov, V. F.; Ustinov, V. I.; Grinenko, V. A. (Zh. Anal. Khim. **26** [1971] 2166/71; J. Anal. Chem. [USSR] **26** [1971] 1934/8).
[15] Stein, L. (Science **175** [1972] 1463/5, Noble Gases Symp., Las Vegas 1973, pp. 376/85, CONF-730915).
[16] Stein, L. (Can. 952290 [1974] 11 pp.; C.A. **81** [1974] No. 175632, U.S. 3784674 [1974] 3 pp.).
[17] Iwanciow, B. L.; Lawrence, W. J.; United Aircraft Corp. (U.S. 3797238 [1974] 5 pp.; C.A. **81** [1974] No. 15274).
[18] Christe, K. O.; Wilson, W. W. (Inorg. Chem. **28** [1989] 3275/7).
[19] Mahjoub, A. R.; Hoser, A.; Fuchs, J.; Seppelt, K. (Angew. Chem. **101** [1989] 1528/9).
[20] Bougon, R.; Charpin, P.; Soriano, J. (Compt. Rend. C **272** [1971] 565/8).

[21] Shamir, J.; Yaroslavky, I. (Israel J. Chem. **7** [1969] 495/7).
[22] Geldard, J. F.; McDowell, H. K. (Spectrochim. Acta A **43** [1987] 439/45).
[23] Smolyar, A. E.; Charkin, O. P.; Klimenko, N. M. (Zh. Strukt. Khim. **15** [1974] 993/1003; J. Struct. Chem. [USSR] **15** [1974] 885/93).
[24] Fogle, C. E.; Breazeale, J. D. (AD-A022099 [1976] 112 pp., 66; C.A. **85** [1976] No. 80474).
[25] Surles, T.; Quarterman, L. A.; Hyman, H. H. (J. Fluorine Chem. **3** [1973] 293/306).
[26] Meinert, H.; Gross, U.; Grimmer, A.-R. (Z. Chem. [Leipzig] **10** [1970] 226/7).
[27] Gillespie, R. J.; Spekkens, P. (J. Chem. Soc. Dalton Trans. **1976** 2391/6).

10.21 Bromine Heptafluoride and Heptafluorobromine(1+), BrF_7 and BrF_7^+

CAS Registry Number: BrF_7 *[32825-95-9]*

Fluorination of BrF_5 by excess F_2 in a nickel reactor in the presence of $KBrF_6$ or $CsBrF_6$ at 370 to 610 K and pressures of 3.45 to 32.8 MPa is said to yield BrF_7. The pressure and temperature dependence of the reaction could not be clearly resolved. The product was separated with low yield by fractional condensation and purified by fractional codistillation with He in the temperature range 146 to 163 K [1]. However, the isolation of BrF_7 claimed in [1] must be regarded with some skepticism, the compound apparently does not exist. An attempt to prepare BrF_7 by the displacement reaction of BrF_6AsF_6 and FNO failed [2]. UV photolysis of BrF_5-F_2 mixtures at 213 to 233 K did not yield BrF_7 [3].

An IR band at 735 cm^{-1} was assigned to BrF_7. The mass spectrum of BrF_7 differs from that of BrF_5; relative intensities of the observed ions with respect to the intensity of BrF^+ are given below [1]:

ion	BrF_7^+	BrF_6^+	BrF_5^+	BrF_4^+	BrF_3^+	BrF_2^+	BrF^+
m/z	212	193	174	155	136	117	98
rel. intensity	21.0	22.8	12.0	43.2	37.2	113.0	100

The boiling point of BrF_7 was said to be ~273 K. Hydrolysis of BrF_7 gave Br_2, $HBrO_3$, O_2, and HF [1].

Calculations using the semiempirical CNDO/2 and INDO methods predict for BrF_7 a structure with C_{2v} symmetry. The most stable structure derived is a tetragonal pyramid with two additional F atoms below the base of the pyramid. Internuclear distances are 1.79 Å for the bonds of the tetragonal pyramid and 1.76 Å for the F atoms below its base. The Br atom is 0.50 to 0.51 Å away from the base of the pyramid and the angle F-Br-F between the F atoms below the base of the pyramid is 83.5°. Other calculated values for this molecular structure applying the CNDO/2 and INDO methods are dipole moments of 2.677 and 2.255 D, ionization potentials of 7.86 and 9.20 eV, and electronic transition energies of 7.75 and 8.45 eV, respectively. Other structures considered were a pentagonal bipyramid or an "hourglass" structure; both were less stable by 10 to 12 eV [4].

The BrF_7^+ ion was identified in the mass spectrum of the product believed to be BrF_7 [1]; see above.

References:

[1] Fogle, C. E.; Rewick, R. T.; United Aircraft Corp. (U.S. 3615206 [1971] 4 pp.; C.A. **76** [1972] No. 5447).
[2] Gillespie, R. J.; Schrobilgen, G. J. (Inorg. Chem. **13** [1974] 1230/5).
[3] Pilipovich, D.; Rogers, H. H.; Wilson, R. D. (Inorg. Chem. **11** [1972] 2192/5).
[4] Deb, B. M.; Coulson, C. A. (J. Chem. Soc. A **1971** 985/70).

11 Compounds of Bromine with Fluorine and Hydrogen

11.1 $Br_2 \cdot HF$. $BrF \cdot HF$

CAS Registry Numbers: –

When bromine is condensed at 12 K with HF in excess argon, two 1:1 complexes, a hydrogen-bonded complex of the form $Br_2 \cdot HF$ and an anti-hydrogen-bonded complex of the form $HF \cdot Br_2$, are produced and identified on account of their infrared spectra. The hydrogen-bonded complex is predominantly formed. It has a strong absorption band at 3841 cm^{-1} (absorbance A = 1.0) and a weaker band at 3851 cm^{-1} (A = 0.2). The bands for the DF counterpart are at 2816 and 2824 cm^{-1}. A sharp band at 3892 cm^{-1} (A = 0.05) (2864 cm^{-1} for the DF analog) is due to the anti-hydrogen-bonded complex.

In analogous experiments with BrF instead of Br_2, a sharp doublet at 3857(2826) and 3845(2821) cm^{-1} was observed and assigned to the $BrF \cdot HF(DF)$ complex. Provided the $HF \cdot BrF$ complex is trapped its IR band is probably masked by a band at 3881 cm^{-1} due to the $N_2 \cdot HF$ complex.

Reference:

Hunt, R. D.; Andrews, L. (J. Phys. Chem. **92** [1988] 3769/74).

11.2 The $HBrF^-$ Ion

CAS Registry Number: *[53754-19-1]*

Compare the introductory remark in the chapter on $HBrCl^-$ (p. 232). More recently, an ab initio MRD (multi-reference double excitations)-CI calculation on the electronic structure and the hydrogen bond energy was performed [1].

$HBrF^-$ ions are observable in an ion cyclotron resonance (ICR) spectrometer [2]. $M^+ - HBrF^-$ ion pairs (M = K, Cs) were produced in Ar matrices at 15 K after reacting KF or CsF with HBr and CsBr with HF (DF gave the deuterium analog). Two types of linear $HBrF^-$ ions (point group $C_{\infty v}$) were identified (I and II; compare the corresponding section in the chapter on $HBrCl^-$, p. 232) [3]. A crystalline salt containing the $HBrF^-$ ion was prepared by treating solid $(C_2H_5)_4NBr$ with gaseous HF [4]. $HBrF^-$ in CH_3CN was obtained from the reaction of $(n\text{-}C_4H_9)_4NBr$ with HF [5]. Solutions in CH_3CN and CH_2Cl_2 were earlier prepared by dissolving the salt $(C_2H_5)_4NBrHF$ [4].

A hydrogen bond energy of $D(FH\text{-}Br^-)$ = 13 kcal/mol is said in [3] to be "well known from gas-phase ICR studies" [6]. A value of D = 17.0 kcal/mol was derived by correlating the shifts of the hydrogen bond stretching frequencies (relative to those of the free acids) and known hydrogen bond energies [2]; see also [7]. An empirical relation between $D(YH\text{-}X^-)$ and HX, and between HY acidities and X, Y electronegativities yielded D = 19 kcal/mol [7].

$D(BrH\text{-}F^-)$ = 65.0 kcal/mol was derived from the hydrogen bond energy (17.0 kcal/mol) by adding the difference of the HBr and HF gas phase acidities [2].

^1H and ^{19}F NMR spectra were recorded in a CH_3CN solution. The shifts δ^1H and $\delta^{19}F$ (positive towards low field) were measured against internal TMS and CF_4 standards [5]:

T in K	δ^1H in ppm	$\delta^{19}F$ in ppm	$^1J(HF)$ in Hz
307	$+8.76 \pm 0.05$	-84.4 ± 0.5	428 ± 2
233	$+8.88 \pm 0.03$	-83.2 ± 0.3	427.1 ± 0.2

The results of IR absorption spectra were partly inconsistent. Three bands were observed with the salt $(C_2H_5)_4NBrHF$ at 220 ± 3, 740 ± 10, and $\sim 2900\ cm^{-1}$ and were assigned to the fundamentals ν_1 (symmetric stretching), ν_2 (bending), and ν_3 (antisymmetric stretching) of a "type I" ion. $\nu_1 \approx 215\ cm^{-1}$ was measured in a CH_3CN solution, $\nu_3 \approx 3050\ cm^{-1}$ in a CH_2Cl_2 solution [4]. Three bands of matrix-isolation spectra (wave numbers of the deuterium analogs in parentheses) were at 744 (532) or 749 cm^{-1}, at 849 (609) or 872 cm^{-1}, and at 2801 (2124) or 2803 cm^{-1}. They were tentatively assigned to ν_2 (type I), ν_3 (type II), and ν_3 (type I) [3]. A more recent review listed only one of these bands, namely the one at 849 (609) cm^{-1} [8].

References:

[1] Sannigrahi, A. B.; Peyerimhoff, S. D. (Chem. Phys. Letters **164** [1989] 348/52).
[2] Larson, J. W.; McMahon, T. B. (Inorg. Chem. **23** [1984] 2029/33).
[3] Ault, B. S. (J. Phys. Chem. **83** [1979] 837/44).
[4] Evans, J. C.; Lo, G. Y.-S. (J. Phys. Chem. **70** [1966] 543/5).
[5] Fujiwara, F. Y.; Martin, J. S. (J. Am. Chem. Soc. **96** [1974] 7625/31).
[6] Beauchamp, J. L. (private communication to [3]).
[7] Larson, J. W.; McMahon, T. B. (J. Am. Chem. Soc. **109** [1987] 6230/6).
[8] Jacox, M. E. (J. Phys. Chem. Ref. Data **13** [1984] 945/1068, 961).

11.3 $(HBr)_n \cdot HF$, $n \geq 1$

CAS Registry Numbers: –

Data for complexes of the form $(HBr)_n \cdot HF$, $n \geq 1$, are given in Section "Hydrogen-Bonded Complexes" of HBr in "Bromine" Suppl. Vol. B1, 1990, p. 375.

12 Compounds of Bromine with Fluorine and Oxygen

Compounds composed of bromine, fluorine, and oxygen are described in this chapter in the following order: $BrOF^+$, BrO_2F, BrO_2F^+, BrO_3F, BrO_3F^+, $BrOF_2^+$, $BrO_2F_2^+$, $BrO_2F_2^-$, $BrOF_3$, BrO_2F_3, $BrO_2F_3^+$, $BrOF_4^-$, $BrOF_5$, and BrO_2F_5.

The study of these compounds started in the early 70's, but information is still limited: Compounds are identified mainly by IR, Raman, and ^{19}F NMR spectra, and a few data are available about the chemistry. BrO_2F, BrO_3F, and $BrOF_3$ are stable. $BrOF_2^+$, $BrO_2F_2^-$, and $BrOF_4^-$ can be stabilized as salts. The other ions, $BrOF^+$, BrO_2F^+, BrO_3F^+, $BrO_2F_2^+$, and $BrO_2F_3^+$, were identified only by mass spectrometry. The syntheses of BrO_2F_3 and $BrOF_5$ have been unsuccessful so far, while the existence of BrO_2F_5 (O_2BrF_5) is doubtful.

12.1 Fluorooxobromine(1+), $BrOF^+$

CAS Registry Number: –

The $BrOF^+$ ion was identified at m/z 114 in the mass spectra of BrO_3F [1], BrF_3, and BrF_5 [2]. The oxygen in the investigations of the bromine fluorides stems from an impurity or from reactions with the surface of the sampling system [2].

References:

[1] Appelman, E. H.; Studier, M. H. (J. Am. Chem. Soc. **91** [1969] 4561/2).
[2] Irsa, A. P.; Friedman, L. (J. Inorg. Nucl. Chem. **6** [1958] 77/90).

12.2 Bromine Fluoride Dioxide, BrO_2F

Other name: Bromyl fluoride

CAS Registry Numbers: BrO_2F *[22585-64-4]*, $Br^{18}O_2F$ *[64544-65-6]*, $^{79}BrO_2F$ *[67452-70-4]*

Preparation. Formation

Hydrolysis of BrF_5 yields BrO_2F almost quantitatively. BrF_5 and H_2O in a molar ratio of 5:1 are cocondensed in a steel vessel at 77 K. The reaction starts on warming at about 170 K and becomes complete in the molten mixture at about 210 K. The volatiles HF and BrF_5 are pumped from the mixture at a temperature as low as possible, and BrO_2F is isolated by sublimation at 258 K from the residue of the reaction. This method can also be used to prepare $Br^{18}O_2F$ [1]. Hydrolysis of BrF_5 by an equimolar amount of H_2O in HF solution at 210 K also yields BrO_2F free from $BrOF_3$ [2].

The reaction of pure $KBrO_3$ and BrF_5 (molar ratio 1:2.5) in the presence of some HF yields BrO_2F and a mixture of $KBrO_2F_2$ and $KBrOF_4$ at ambient temperature. The product was separated in 55% yield from its decomposition product Br_2 and the nonvolatile residue by fractional condensation [3]. The first paper on this reaction stated that the preparation of BrO_2F from $KBrO_3$ and BrF_5 in a 1:1 molar ratio at 223 K was accompanied by formation of $KBrF_4$ and O_2. BrO_2F thus formed was isolated by fractional condensation at 218 K [4]. This reaction yields BrO_2F at an acceptable rate only when carried out in a quartz vessel or in the presence of a trace of HF if metal or fluoropolymer vessels are used [5]. Earlier investigations had also shown that the reaction of equimolar quantities of $KBrO_3$ and BrF_5 did not yield BrO_2F in the absence of HF at 223 K [3, 6] and was very slow at ambient temperature [3].

BrO_2F is also prepared by reacting a mixture of Br_2 and BrF_5 with a stream of O_3 in O_2 at 268 K [4] or by hydrolyzing $KBrO_2F_2$ in HF below 273 K [2]. The product is isolated by fractional condensation after evaporation of more volatile compounds at low temperatures [2, 4].

Formation of BrO_2F from BrF_5 results in reactions with Cl_2O_6 [7], with BrO_2 at 213 to 223 K [4], with I_2O_5, IO_2F, IOF_3, and IO_2F_3 in the temperature range 238 to 298 K [2], and in the reaction with excess N_2O_5 above 213 K [8]. Since direct fluorination of pure BrO_2 is exceedingly vigorous [9], the reaction must be carried out in C_5F_{12} or Cl_2 solution at 223 K [10]. The reaction of a solution of $NaBrO_3$ in HF with F_2 yields BrO_2F together with Br_2 and bromine compounds below ambient temperature [11]. $MBrO_4$ (M=K, Cs) and F_2 form some BrO_2F as a by-product of $MBrOF_4$ in BrF_5 solution [12]. The reaction of excess NO_2F with BrO_2NO_3 in CCl_3F [8] or BrO_2PtF_6 at 195 K leads to BrO_2F [13]. The formation of BrO_2F from $KBrO_3$ and BrF_3 in Br_2 solution can not be used preparatively, because the resulting mixture with BrF_3 is difficult to separate [9]. NF_4BrO_4 decomposes to BrO_2F above 243 K [14].

Solid BrO_2F may be stored indefinitely in quartz ampules [9], although storage below 170 K is recommended [1]. Liquid BrO_2F at ambient temperature is colorless and quite stable in Kel-F vessels [15], but attacks glass [10]. Liquid and gaseous BrO_2F are quite stable when handled in a passivated steel apparatus [1]. Slow decomposition turns the colorless liquid yellow [9]. Solutions of BrO_2F in HF are stable at 195 K and decompose at ambient temperature; the BrF_5 solution is stable for several hours at ambient temperature [2]. Gaseous BrO_2F has a half-life of about 30 minutes at 288 K [1].

Physical Properties

Molecular Structure. Association. The similarity of the Raman spectra of solid, liquid, and dissolved BrO_2F indicates a basically unchanged, monomeric structure. The molecule forms a trigonal pyramid with each atom occupying one vertex [2]. The relatively low Br-F and high Br-O stretching frequencies were supposed to indicate a large contribution from the ionic resonance structure $BrO_2^+F^-$ [2, 15]. Large mean amplitudes of vibration and a small force constant of the Br-F bond can be explained in the same way [16]. However, a larger estimated Br-F force constant based on the same vibrational band was taken to indicate a normal bond strength [1]. The significant deviation of gas-phase spectra from those in the liquid, solid, and also in matrices affects mainly the Br-F stretching vibration and indicates association in condensed phases by intermolecular fluorine bridging [1, 2, 16]. Molecular association reduces the ν_2 value from 551 cm^{-1} in the gas to 490 cm^{-1} in the solid. The ν_5 shift of 33 cm^{-1} indicates a contribution from the Br-F vibration to this band. The shift of the other bands is very small [1].

^{19}F NMR Spectra. A sharp singlet of BrO_2F in BrF_5 solution at $\delta=210\pm4$ ppm downfield from CCl_3F is observed at 238 K; the chemical shift depends slightly on the temperature and the composition of the solution [2, 15]. The chemical shift of $\delta=205\pm2$ ppm in SO_2ClF at 150 to 195 K agrees with the shift in BrF_5 [2]. The signal of BrO_2F at $\delta=205$ ppm in HF solution [17] is replaced under different experimental conditions by a broad line at $\delta=186$ ppm which is attributed to BrO_2F undergoing rapid fluorine exchange with the solvent [2]. The chemical shift of BrO_2F confirms the Br(V) oxidation state [2, 15].

Vibrational Spectra. The vibrational spectra of BrO_2F were analyzed on the basis of a pyramidal molecule with C_s symmetry; they are essentially the same for the solid, liquid, and solution state [1, 2, 15, 16]. The six fundamentals ($\Gamma=4\,A'+2\,A''$) are all Raman- and IR-active; the assignment is confirmed by the observation of four polarized and two depolarized bands in the Raman spectrum of the liquid [2, 15]. The bands in the gas-phase spectra all have PQR structure, and their contours confirm the assignment to the individual fundamentals [1]. The IR frequencies of gaseous BrO_2F in cm^{-1} at 288 K are as follows [16]:

isotopomer	$\nu_1(A')$ $\nu_s(BrO)$	$\nu_2(A')$ $\nu(BrF)$	$\nu_3(A')$ $\delta_{sciss}(BrO_2)$	$\nu_4(A')$ $\delta_s(OBrF)$	$\nu_5(A'')$ $\nu_{as}(BrO_2)$	$\nu_6(A'')$ $\delta_{as}(OBrF)$
$^{79}Br^{16}O_2F$	921.0	551.9	385.8	~310	978.9	276, 271[a)]
$^{81}Br^{16}O_2F$	919.5	550.4	385.8	~310	976.2	276, 271[a)]
$^{79}Br^{18}O_2F$	876.3	551.5	370.4	~296	936.5	260[a)]
$^{81}Br^{18}O_2F$	874.8	550.1	370.4	~296	933.7	260[a)]
BrO_2F[b)]	908 (100) p	506 (36) p	394 (14) p	305 (21) p	953 (14) dp	271 (16) dp

[a)] From the spectrum of the solid at 186 K in [1]. – [b)] Raman bands of the liquid at 263 K [15].

Raman [2] and IR spectra [1] of the solid exhibit additional lines. IR spectra of BrO_2F in Ne, Ar, and N_2 matrices are reported in [1, 16]; the IR frequencies observed in a Ne matrix are closest to the gas-phase values [16]. The Raman spectrum of the HF solution is described in [2].

Mean Amplitudes of Vibration. Force Constants. The Raman fundamentals of liquid BrO_2F [15] and estimated structural parameters were used to calculate mean amplitudes of vibration from 0 to 700 K. Values at 298.16 K in Å are $u_{Br-O}=0.0373$, $u_{Br-F}=0.0508$, $u_{O\cdots O}=0.067$, and $u_{O\cdots F}=0.082$. The large value of u_{Br-F} and its considerable increase with temperature is consistent with an ionic contribution to the Br-F bond, while the small value of u_{Br-O} indicates a partial double bond [18].

A calculation of the valence force constants of BrO_2F was based on the gas-phase fundamentals [16] except for the value of ν_6 from the Raman spectrum of the solid [15] and on the estimated internuclear distances R(Br-F) = 1.76 Å, r(Br-O) = 1.577 Å and bond angles α(OBrF) = 102° and β(OBrO) = 115°. Obtained values are $f_r=6.984\pm0.08$, $f_R=2.750\pm0.04$, and $f_{rr}=-0.05\pm0.08$ mdyn/Å, $f_\alpha=1.125\pm0.04$, $f_\beta=1.453\pm0.08$, $f_{\alpha\alpha}=0.363\pm0.04$, and $f_{\alpha\beta}=0.49\pm0.07$ mdyn·Å·rad^{-2}, and $f_{R\alpha}=0.095\pm0.09$, $f_{r\alpha}=-0.07\pm0.11$, and $f_{r\alpha'}=0.07\pm0.11$ mdyn/rad [16]. Corrections of the anharmonic oxygen isotopic shifts are included; anharmonicity corrections for Br were not necessary because of the small isotopic shift. The potential energy distribution of the force constants is very characteristic and supports the approximate description of the vibrational modes except for ν_3 and ν_4, where a moderate amount of mutual mixing occurs. The force constant f_R indicates a weaker Br-F bond than expected [16].

The reported set of force constants differs significantly [16] from earlier sets in [18, 19] which were based on the Raman spectrum of liquid BrO_2F in [15]. A valence force constant of 3.1 mdyn/Å was estimated for the Br-F bond [1].

Thermal Properties. Solid, colorless BrO_2F sublimes between 248 and 253 K in vacuum and melts at 264 K [4] or 264.9 ± 0.5 K [1]. The approximate vapor pressure of the liquid at 273 K is 2 ± 1 Torr [1]. The liquid decomposes violently at 329 K [4]. Thermodynamic properties of $^{79}BrF^{16}O_2$ as an ideal gas at 1 atm were calculated from vibrational frequencies together with estimated structural parameters in the harmonic-oscillator, rigid-rotor approximation; results are as follows [16]:

T in K	C_p^o in $J \cdot mol^{-1} \cdot K^{-1}$	$H^o - H_0^o$ in kJ/mol	$-(G^o - H_0^o)/T$ in $J \cdot mol^{-1} \cdot K^{-1}$	S^o in $J \cdot mol^{-1} \cdot K^{-1}$
0	0	0	0	0
298.15	62.898	13.941	247.354	294.114
600	75.655	35.259	284.223	342.992
1000	80.140	66.613	316.323	382.932
1500	81.760	107.165	344.356	415.798
2000	82.358	148.214	365.305	439.412

Chemical Behavior

The thermal decomposition of pure BrO_2F or its solutions yields BrF_3, Br_2, and O_2. The speed of the reaction is moderate at ambient temperature and high at 329 K [1, 2, 4]. The thermodynamic instability is probably caused by labile Br-O bonds. The kinetic instability in the gas phase and in HF solution seems to result from the unsaturated Br coordination, while BrO_2F in the solid and in BrF_5 solution is stabilized by F bridges [1].

The reaction of BrO_2F with H_2O is violent [10]. Hydrolysis with a NaOH solution at low temperatures yields BrO_3^-, F^-, and H_2O [9]. BrO_2F is a weaker fluoride acceptor than $BrOF_3$ [3]. The reaction of BrO_2F with the Lewis acids BF_3 [20], AsF_5 [13, 20, 21], and SbF_5 [21] at low temperatures, either after melting the cocondensed reactants [13, 21] or in solution [20], yields the BrO_2^+ salts with the anions BF_4^-, AsF_6^-, or $Sb_2F_{11}^-$. The reaction of BrO_2F and BBr_3 is vigorous above 227 K, but only Br_2, BF_3, and B_2O_3 could be isolated [9]. An adduct formed by BrO_2F and BrF_5 in concentrated solution at room temperature decomposed even at 223 K under dynamic vacuum. The reaction of BrO_2F with the Lewis base KF (molar ratio 2:1) at ambient temperature proceeded only to a small extent and yielded a mixture of $KBrO_2F_2$ and $KBrF_4$ [2].

Fluorination of BrO_2F by excess KrF_2 in HF solution at 275 K slowly evolves $BrOF_3$ [2]. Reaction with PtF_6 (molar ratio 1:2.5) starts at 150 K and yields a mixture of $BrOF_2PtF_6$, O_2PtF_6, PtF_5, and some F_2 at ambient temperature [13]. Organic substances are ignited by BrO_2F [10].

References:

[1] Jacob, E. (Z. Anorg. Allgem. Chem. **433** [1977] 255/60).
[2] Gillespie, R. J.; Spekkens, P. H. (J. Chem. Soc. Dalton Trans. **1977** 1539/46).
[3] Gillespie, R. J.; Spekkens, P. (J. Chem. Soc. Dalton Trans. **1976** 2391/6).
[4] Schmeisser, M.; Pammer, E. (Angew. Chem. **69** [1957] 781).
[5] Naumann, D. (private communication, ref. 8 in [1]).
[6] Tantot, G.; Bougon, R. (Compt. Rend. C **281** [1975] 271/3).
[7] Weiss, R. (Diss. T.H. Aachen 1959, ref. 199 in [9]).
[8] Schmeisser, M.; Taglinger, L. (Chem. Ber. **94** [1961] 1533/9).
[9] Schmeisser, M.; Brändle, K. (Advan. Inorg. Chem. Radiochem. **5** [1963] 41/89, 73/5).
[10] Schmeisser, M.; Pammer, E. (Angew. Chem. **67** [1955] 156).

[11] Appelman, E. H.; Studier, M. H. (J. Am. Chem. Soc. **91** [1969] 4561/2).
[12] Christe, K. O.; Wilson, R. D.; Curtis, E. C.; Kuhlmann, W.; Sawodny, W. (Inorg. Chem. **17** [1978] 533/8).
[13] Adelhelm, M.; Jacob, E. (Angew. Chem. **89** [1977] 476/7; Angew. Chem. Intern. Ed. Engl. **16** [1977] 461).

[14] Christe, K. O.; Wilson, R. D. (Inorg. Nucl. Chem. Letters **15** [1979] 375/6).
[15] Gillespie, R. J.; Spekkens, P. (J. Chem. Soc. Chem. Commun. **1975** 314/6).
[16] Christe, K. O.; Curtis, E. C.; Jacob, E. (Inorg. Chem. **17** [1978] 2744/9).
[17] Christe, K. O.; Wilson, W. W.; Wilson, R. D. (Inorg. Chem. **19** [1980] 1494/8).
[18] Baran, E. J. (Spectrosc. Letters **9** [1976] 323/7).
[19] Bougon, R.; Joubert, P.; Tantot, G. (J. Chem. Phys. **66** [1977] 1562/5).
[20] Spekkens, P. (Diss. Hamilton 1977, ref. 18 in Gillespie, R. J.; Spekkens, P. H.; Israel J. Chem. **17** [1978] 11/9).
[21] Jacob, E. (Angew. Chem. **88** [1976] 189/90; Angew. Chem. Intern. Ed. Engl. **15** [1976] 158).

12.3 Fluorodioxobromine(1+), BrO_2F^+

CAS Registry Number: –

The ion BrO_2F^+ was observed at m/z 130 in the mass spectrum of BrO_3F [1]. The identification of BrO_2F^+ in effusion rate measurements of BrF_5-O_2 mixtures was attributed to formation of BrO_2F in the sampling system of the mass spectrometer. The ion was also observed in the mass spectra of BrF_3 (oxygen probably originated from reactions with the surface of the sampling system) [2] and partly hydrolyzed samples of BrF_3 and BrF_5 [3].

References:

[1] Appelman, E. H.; Studier, M. H. (J. Am. Chem. Soc. **91** [1969] 4561/2).
[2] Irsa, A. P.; Friedman, L. (J. Inorg. Nucl. Chem. **6** [1958] 77/90).
[3] Sloth, E. N.; Stein, L.; Williams, C. W. (J. Phys. Chem. **73** [1969] 278/80).

12.4 Bromine Fluoride Trioxide, BrO_3F

Other name: Perbromyl fluoride

CAS Registry Numbers: BrO_3F *[25251-03-0]*, $^{79}BrO_3F$ *[131085-58-0]*, $^{81}BrO_3F$ *[131085-59-1]*

Preparation. Formation

Fluorination of $KBrO_4$ by excess SbF_5 in anhydrous HF yields BrO_3F after warming the frozen mixture from 196 K to ambient temperature. The product is separated in a 97% yield by passing the volatiles through a NaF bed which absorbs HF and SbF_5 [1]. Other agents for fluorinating $KBrO_4$ in HF are AsF_5, BrF_5, and BrF_6AsF_6 [2]. The Br_2 content of BrO_3F can be lowered by treating it with F_2 at a low pressure and ambient temperature and distilling BrO_3F at 193 K from the formed BrF_3 [3]. The reaction of BrO_4^- and F_2 in BrF_5 yields BrO_3F as a by-product of $BrOF_4^-$ salts [4]. A sample of $^{79}BrO_3F$ was prepared by reacting $K^{79}BrO_4$ and SbF_5 in HF [5].

BrO_3F should be handled in Monel or Pt vessels, because glass, Kel-F, and Teflon are slowly attacked at room temperature [1, 6]. However, at low temperatures Kel-F, fluoroethene polymer, and Teflon are inert [2].

Physical Properties

Molecular Structure. The pseudotetrahedral structure and C_{3v} symmetry of gaseous BrO_3F is known from an electron-diffraction study at about 250 K. Internuclear distances in

Å and bond angles are: r(Br-O) = 1.581(1), r(Br-F) = 1.707(3), r(O···O) = 2.663, and r(F···O) = 2.577, angle OBrO = 114.9(3)°, and angle FBrO = 103.3(3)°. The Br-O bond length is consistent with a nominal bond order of 2, and Br and F are linked by a single bond. Structural parameters referring to the equilibrium atomic positions were also calculated [7]. The pseudotetrahedral geometry of BrO_3F is also consistent with the analysis of the vibrational spectra [3, 5].

Polarizability. The δ-function potential model of chemical bonding yields for BrO_3F an average polarizability of 7.41×10^{-24} cm^3. Parallel and perpendicular components of the polarizability were also calculated [8].

^{19}F NMR Spectrum. The ^{19}F NMR spectrum of liquid BrO_3F at 193 K consists of a single broad line with a low field chemical shift of δ = 274 ppm relative to $CClF_3$. The signal from a saturated HF solution at δ = 269 ppm is even broader. The chemical shift agrees with a bond between F and heptavalent Br. The line width is presumably due to a poorly resolved Br-F coupling which is caused by rapid Br quadrupole relaxation. The line width increases with temperature for the HF solution [2].

Rotational Constants. BrO_3F is a near-spherical prolate, symmetric top molecule. The ground-state rotational parameters $A_0 = 0.14822$ and $D_K^o = -8.37 \times 10^{-8}$ cm^{-1} were derived [5] from the molecular structure and harmonic force field given in [7]. Other ground-state rotational parameters were determined from a high-resolution FTIR spectrum of BrO_3F [5]:

isotopomer	B_0 in cm^{-1}	D_J^o in 10^{-8} cm^{-1}	D_{JK}^o	H_J^o in 10^{-13} cm^{-1}	H_{JK}^o	H_{KJ}^o
$^{79}BrO_3F$	0.14216640(14)	4.003(4)	6.389(26)	−0.12(4)	2.51(27)	−4.4(11)
$^{81}BrO_3F$	0.14213455(31)	4.005(5)	6.43(5)	*)	*)	*)

*) The sextic parameters were insignificant.

Values of the rotational distortion constants calculated from a force field are given in [9].

Rotational constants and centrifugal distortion constants were calculated for the v = 1 states of the A_1 vibrations. Including sextic terms did not improve the results significantly and was omitted. No clear indication was found for rotational disturbations of the excited states. Relatively large changes of 5 to 15% from the ground-state values were found for the $v_3 = 1$ state of the centrifugal distortion parameters. Data are listed in [5].

Vibrational Spectra. The vibrational spectra of gaseous BrO_3F are consistent with a tetrahedral molecule of C_{3v} symmetry ($\Gamma = 3\,A_1 + 3\,E$); assigned fundamentals in cm^{-1} are as follows [3]:

	$\nu_1(A_1)$ ν(BrO)	$\nu_2(A_1)$ ν(BrF)	$\nu_3(A_1)$ $\delta(BrO_3)$	$\nu_4(E)$ ν(BrO)	$\nu_5(E)$ $\delta(BrO_3)$	$\nu_6(E)$ τ
IR	875.5 s	606 vs	359.8 m	976.5 vs	382.0 s	286 w
Raman	875.2 vs, p	605.0 vs, p	~354 w, p	974 w, dp	~376 w	~296 vw

Polarization ratios obtained from the spectrum of molten BrO_3F and the Raman spectrum of the solid confirm the assignment of the bending modes [2].

Origins of the A_1 bands were obtained from the FTIR spectrum; values in cm^{-1} are [5]:

isotopomer	ν_1	ν_2	ν_3
$^{79}BrO_3F$	876.814360(18)	609.181431(8)	369.755555(31)
$^{81}BrO_3F$	876.33048(11)	607.62547(3)	359.8663(6)

Combination bands are listed in [3].

Mean Amplitudes of Vibration. Experimentally determined mean amplitudes of vibrations of BrO_3F at about 250 K in Å are $u_{Br-O}=0.037$, $u_{Br-F}=0.044$, $u_{O\cdots O}=0.064$, and $u_{F\cdots O}=0.078$, taken from the gas-phase electron diffraction study [7]. Values calculated from valence force fields are given in [8 to 11].

Force Constants. The most reliable set of force constants of BrO_3F [12] is based on experimental structural parameters [7] and vibrational data from [3]. Calculated values are as follows [7]:

$F_1(\nu_{Br-O})$ in mdyn/Å	$F_2(\nu_{Br-F})$	$F_3(\delta_{OBrO})$ in mdyn·Å	$F_4(\delta_{FBrO})$	$F_5(F_3\text{-}F_4$ interact.)
6.92(2)	3.22(3)	1.06(2)	0.81(3)	−0.19(2)

The stretching force constants indicate a Br–O double bond and a Br–F single bond [12]. Force constants were recently calculated using the L^0 matrix approximation [11]. The vibrational data from [3] were also used in earlier calculations of the force constants together with estimated structural parameters [8 to 10, 13]. The potential energy distribution of the vibrational frequencies was calculated in [13].

Coriolis Coupling Constants. Coriolis coupling constants were calculated from vibrational data [3] and estimated structural parameters. Negative values resulted where expected for BrO_3F [9]:

$\zeta_4=0.2146$	$-\zeta^z_{45}=0.1650$	$\zeta_{14}=0.2023$	$-\zeta_{15}=0.7471$	$\zeta_{16}=0.1653$	$\zeta_{45}=0.2706$
$\zeta_5=0.5925$	$\zeta^z_{46}=0.8952$	$-\zeta_{24}=0.1069$	$\zeta_{25}=0.4615$	$-\zeta_{26}=0.0889$	$\zeta_{46}=0.2503$
$-\zeta_6=0.3394$	$-\zeta^z_{56}=0.1569$	$\zeta_{34}=0.7449$	$\zeta_{35}=0.1584$	$\zeta_{36}=0.5432$	$\zeta_{56}=0.3610$

Thermal Properties. BrO_3F is colorless in the gas and liquid state and becomes a colorless solid at ~163 K. The vapor pressure of BrO_3F is ~6 Torr at 193 K [1] and 689 ± 3.5 Torr at 273.15 K. The vapor pressure is described by $\ln(p/Torr)=17.6986-3048.45/T$ between 188 and 292 K from which a normal boiling point of 275.5 K can be calculated. The heat of evaporation is $\Delta_{vap}H°=25.36\pm0.25$ kJ/mol [6].

Standard thermodynamic data were calculated from vibrational frequencies in [3] and estimated structural parameters for BrO_3F as an ideal gas at 1 atm in the rigid-rotor harmonic-oscillator approximation [6, 8]. Values in $J\cdot mol^{-1}\cdot K^{-1}$ are [8]:

T in K	$(H^\circ - H^\circ_0)/T$	$-(G^\circ - H^\circ_0)/T$	S°	C°_p
298.16	52.4343	247.9338	300.3681*)	76.5136
600	70.5079	290.8537	361.3616	96.1579
1000	82.4888	330.0373	412.5261	103.2519
1500	89.9234	365.0565	454.9795	105.8460
2000	94.0400	391.5387	485.5787	106.8037

*) A value of 298.7 ± 8.4 is given in [6].

A standard entropy of 207 ± 13 $J \cdot mol^{-1} \cdot K^{-1}$ was calculated for liquid BrO_3F at 298.16 K [6].

Thermodynamic data of formation at 298.15 K for $0.5\ Br_2(l) + F_2(g) + 1.5\ O_2(g) \rightarrow BrO_3F(l)$ were obtained by calorimetric measurements of the hydrolysis and were calculated for gaseous BrO_3F [6]:

	Δ_fH° in kJ/mol	Δ_fG° in kJ/mol	Δ_fS° in $J \cdot mol^{-1} \cdot K^{-1}$
$BrO_3F(l)$	112.05 ± 1.05	195.02 ± 3.85	−278.2 ± 12.6
$BrO_3F(g)$	137.40 ± 1.09	192.92 ± 3.85	−186.2 ± 12.6

Chemical Behavior

The mass spectrum of BrO_3F contains the ions BrO_3F^+, BrO_2F^+, $BrOF^+$, BrO_3^+, BrO_2^+, BrO^+, BrF^+, Br_2^+, and Br^+ [1]. Contact of BrO_3F with surfaces results in partial decomposition and produces some Br_2 [3]. A reaction of gaseous BrO_3F and $F(^2P)$ atoms was not observed in a flow apparatus; thus the bimolecular rate constant has to be smaller than $7 \times 10^{-16}\ cm^3 \cdot molecule^{-1} \cdot sec^{-1}$ [14]. The hydrolysis of BrO_3F via $BrO_3F(l) + H_2O(l) \rightarrow BrO_4^-(aq) + HF(aq) + H^+(aq)$ [1, 6] has an average enthalpy of $\Delta H = -134.64 \pm 0.21$ kJ/mol [6]. BrO_3F is not fluorinated by KrF_2 in a HF solution at room temperature. The reaction with the more powerful fluorinating agent $KrFAsF_6$ leads to the decomposition products BrF_5, Br_2, and O_2AsF_6. Adducts with the Lewis acids AsF_5 and SbF_5 do not form in HF solution [2].

References:

[1] Appelman, E. H.; Studier, M. H. (J. Am. Chem. Soc. **91** [1969] 4561/2).
[2] Spekkens, P. (Diss. Hamilton 1977, ref. 18 in [12]).
[3] Claassen, H. H.; Appelman, E. H. (Inorg. Chem. **9** [1970] 622/4).
[4] Christe, K. O.; Wilson, R. D.; Curtis, E. C.; Kuhlmann, W.; Sawodny, W. (Inorg. Chem. **17** [1978] 533/8).
[5] Bürger, H.; Pawelke, G.; Appelman, E. H. (J. Mol. Spectrosc. **144** [1990] 201/11).
[6] Johnson, G. K.; O'Hare, P. A. G.; Appelman, E. H. (Inorg. Chem. **11** [1972] 800/2).
[7] Appelman, E. H.; Beagley, B.; Cruickshank, D. W. J.; Foord, A.; Rustad, S.; Ulbrecht, V. (J. Mol. Struct. **35** [1976] 139/48).
[8] Nagarajan, G.; Redmon, M. J. (Monatsh. Chem. **103** [1972] 1406/26).
[9] Mohan, S.; Ravikumar, K. G.; Boopathy, T. J.; Chandraprakash, N. (Indian J. Phys. B **58** [1984] 56/60).
[10] Baran, E. J.; Aymonino, P. J. (Z. Naturforsch. **27b** [1972] 1568/9).

[11] Dublish, A. K.; Alix, A. J. P. (Acta Ciencia Indica Phys. **13** [1987] 97/125 from C.A. **110** [1989] No. 181739).
[12] Gillespie, R. J.; Spekkens, P. H. (Israel J. Chem. **17** [1978] 11/9).
[13] So, S. P.; Chau, F. T. (Z. Physik. Chem. [N.F.] **85** [1973] 69/75).
[14] Appelman, E. H.; Clyne, M. A. A. (J. Chem. Soc. Faraday Trans. I **71** [1975] 2072/84).

12.5 Fluorotrioxobromine(1+), BrO_3F^+

CAS Registry Number: –

The ion BrO_3F^+ was identified in the mass spectrum of BrO_3F.

Reference:

Appelman, E. H.; Studier, M. H. (J. Am. Chem. Soc. **91** [1969] 4561/2).

12.6 Difluorooxobromine(1+), $BrOF_2^+$

CAS Registry Numbers: $BrOF_2^+$ *[62521-26-0]*, $Br^{18}OF_2^+$ *[62521-27-1]*

Formation

The ion $BrOF_2^+$ was detected in the mass spectrum at m/z 133 during effusion measurements of BrF_5-O_2 mixtures. The ion probably forms from the intermediate $BrOF_3$ and was also observed in the mass spectra of BrF_3 and BrF_5, with oxygen probably originating from the surface of the sampling system [1].

Salts of the $BrOF_2^+$ result from $BrOF_3$ and Lewis acids. The salt with the anion AsF_6^- forms with cocondensed AsF_5 upon warming from 77 K [2] or from the reactants in HF or BrF_5 solution at ambient temperature. The corresponding reaction of $BrOF_3$ with BF_3 in HF is to be carried out with excess BF_3 at 200 K; $BrOF_3$ can be replaced by $KBrOF_4$ in this reaction. The salts are isolated by evaporation of the volatiles at low temperatures [3].

Dissolution of $EF_5 \cdot IO_2F_3$ (E = As, Sb) in excess BrF_5 yields $BrOF_2EF_6$ at ambient temperature. The salts are isolated by evaporation of the volatiles [3]. Warming of cocondensed BrO_2F and PtF_6 (molar ratio 1:2.5) to ambient temperature yields $BrOF_2PtF_6$ in a mixture with O_2PtF_6 and PtF_5. This reaction was used to prepare $Br^{18}OF_2PtF_6$ [2].

Spectra

^{19}F NMR Spectra. Signals of the cation in HF solutions of $BrOF_2AsF_6$ were observed at $\delta = 199$ ppm at 213 K and at $\delta = 202$ ppm at 283 K to low field of CCl_3F. A sharp singlet of $BrOF_2^+$ in $BrOF_2BF_4$ is found at $\delta = 192$ ppm at 195 K. Rapid F exchange of the cation with the solvent is suppressed by an excess of a Lewis acid [3].

Vibrational Spectra. The vibrational spectra of $BrOF_2^+$ can be interpreted on the basis of a species with pyramidal geometry and C_s symmetry; all 6 fundamentals ($\Gamma = 4\,A' + 2\,A''$) are active in both IR and Raman spectra [2, 3]. Bands are listed in Table 21. Spectra of solid $BrOF_2^+$ salts of AsF_6^-, BF_4^-, and SbF_6^- [3] demonstrate F bridging between cation and anion by the lowered anion symmetry.

Table 21
Fundamental Vibrations from Raman Spectra of $BrOF_2^+$ Salts in the Solid (I) and in HF Solution (II) in cm^{-1}.

anion		$\nu_1(A')$ $\nu(BrO)$	$\nu_2(A')$ $\nu_s(BrF_2)$	$\nu_3(A')$ $\delta_s(OBrF)$	$\nu_4(A')$ $\delta_s(BrF_2)$	$\nu_5(A'')$ $\nu_{as}(BrF_2)$	$\nu_6(A'')$ $\delta_{as}(OBrF)$	Ref.
AsF_6^-,	(I)	1059 (59)	647 (100)	366 and 360 (27)	290 (12)	632 sh	311 (37)	[3]
	(II)	1056 (79)	654 (100)	367 (22), 386 sh	290 sh	637 sh	311 (18)	
BF_4^-,	(I)	1051 (40)	656 (100), 666 (25)	373 (25)	296 (6)	640 sh, 627 (50)	322 (15)	[3]
	(II)	1052 (95) p	654 (100) dp	369 (25)	296 sh	633 sh	314 (17)	
SbF_6^-,	(I)	1061 (66)	656 (37)	367 (20)	278 (22)	636 sh	312 (21)	[3]
PtF_6^-,	(I)[a]	1062 vs	655 s	365 vs	290 s	630 s	315 m	[2]
	(II)[b]	1002 vs	–	355 vs	–	–	304 m	

[a] IR absorptions at 195 K. – [b] IR absorptions of $Br^{18}OF_2PtF_6$ at 195 K.

The polarization data were inconclusive and allowed only a tentative assignment of ν_4 and ν_6. Three IR bands of solid $BrOF_2AsF_6$ were also assigned [3].

Chemical Behavior

The $BrOF_2^+$ salts are white, crystalline solids except for the frequently yellow BF_4^- salt [3] and the yellow-brown $BrOF_2PtF_6$ [2]. Thermal decomposition at ambient temperature is fast for the $BrOF_2BF_4$ and slower for $BrOF_2AsF_6$; products are Br_2BF_4 and Br_2AsF_6. The slow decomposition of $BrOF_2SbF_6$ yields mainly BrF_2SbF_6. Rapid fluorine exchange of $BrOF_2^+$ is observed in ^{19}F NMR spectra of $BrOF_2AsF_6$ in HF solution. The exchange can be attributed to the reaction with HF_2^- and becomes slow on the NMR time scale when an excess of AsF_5 is added [3]. The reaction of $BrOF_2PtF_6$ with excess NO_2F at 195 K results in the release of BrO_2F and $BrOF_3$ [2].

References:

[1] Irsa, A. P.; Friedman, L. (J. Inorg. Nucl. Chem. **6** [1958] 77/90).
[2] Adelhelm, M.; Jacob, E. (Angew. Chem. **89** [1977] 476/7; Angew. Chem. Intern. Ed. Engl. **16** [1977] 461).
[3] Bougon, R.; Bui Huy,T.; Charpin, P.; Gillespie, R. J.; Spekkens, P. H. (J. Chem. Soc. Dalton Trans. **1979** 6/12).

12.7 Difluorodioxobromine(1+), $BrO_2F_2^+$

CAS Registry Number: –

The ion $BrO_2F_2^+$ was observed in the mass spectrum of partly hydrolyzed BrF_5 [1]. The attempted preparation by oxidizing BrO_2F with PtF_6 did not succeed [2].

References:

[1] Sloth, E. N.; Stein, L.; Williams, C. W. (J. Phys. Chem. **73** [1969] 278/80).
[2] Adelhelm, M.; Jacob, E. (Angew. Chem. **89** [1977] 476/7; Angew. Chem. Intern. Ed. Engl. **16** [1977] 461).

12.8 Difluorodioxobromate(1−), $BrO_2F_2^-$

CAS Registry Number: −

Formation

Reaction of $KBrO_3$ and BrF_5 in a Monel apparatus at ~340 K [1] or in a Kel-F apparatus at 298 K [2] yields $KBrO_2F_2$ with formation of BrF_3 and evolution of O_2. The by-product, $KBrOF_4$, can be reduced to less than 2% by carefully purifying and drying the starting materials and by using at least a fivefold excess of BrF_5 [1]. Presence of HF in the mixture of $KBrO_3$ and excess BrF_5 leads to BrO_2F as the principal product and a nonvolatile residue of $KBrO_2F_2$ containing ~10% of $KBrOF_4$. An approximately equimolar mixture of $KBrO_2F_2$ and $KBrOF_4$ results from the partial fluorination of $KBrO_3$ with a slight excess of $KBrF_6$ in CH_3CN at room temperature. $KBrO_2F_2$ can be isolated by extraction of $KBrOF_4$ with CH_3CN. The reaction of excess $KBrO_3$ and $KBrOF_4$ at ambient temperature yields $KBrO_2F_2$, the reaction being complete in CH_3CN and incomplete in BrF_5 [3].

The reaction of BrO_2F and a half molar amount of KF yields $KBrO_2F_2$ and $KBrF_4$ at ambient temperature [4]. Partial hydrolysis of $KBrF_6$ in CH_3CN always leads to mixtures of $KBrO_2F_2$, $KBrOF_4$, and $KBrF_4$, precluding the preparative use of this reaction [3]. Some $NaBrO_2F_2$ forms from $NaNO_3$ and BrF_5 at 298 K as a by-product of $NaBrOF_4$ [5]. The claimed formation of $BrO_2F_2^-$ salts from BrO_3^- in aqueous HF was withdrawn [6].

Vibrational Spectra. Force Constants

The VSEPR theory predicts for $BrO_2F_2^-$ a pseudotrigonal bipyramid of C_{2v} symmetry with the O atoms and the Br lone electron pair in the equatorial positions and the F atoms in the axial positions. The observation of the expected 9 fundamentals ($\Gamma = 4\,A_1 + A_2 + 2\,B_1 + 2\,B_2$) in the vibrational spectra agree with the predicted ion structure [1, 3]. The most reliable assignment [7] of the bands of solid $KBrO_2F_2$ is that in [1]; data (in cm^{-1}) are given below:

vibration	approximate description	Raman	IR
$\nu_1(A_1)$	$\nu_s(BrO_2)$	885 (100), 894 (17) sh	900 s, 885 sh
$\nu_2(A_1)$	$\delta_s(BrO_2)$	429 (17)	−
$\nu_3(A_1)$	$\nu_s(BrF_2)$	374 (52)	−
$\nu_4(A_1)$	$\delta_s(BrF_2)$	201 (4)	−
$\nu_5(A_2)$	τ	405 (13)	−
$\nu_6(B_1)$	$\nu_{as}(BrF_2)$	448 sh	440 s, br
$\nu_7(B_1)$	$\delta_{rock}(BrO_2)$	339 (9)	350 w, 330 w
$\nu_8(B_2)$	$\nu_{as}(BrO_2)$	912 (9)	914 sh
$\nu_9(B_2)$	$\delta_{sciss\ out\ of\ plane}(BrF_2)$	314 (11), 300 sh	315 vw, ~285 sh

The spectra measured at 4 K differ only slightly from those observed at ambient temperature. Band splitting was attributed to interanion coupling. The assignment of ν_5, ν_7, and ν_9 was rather tentative, but was confirmed by a calculation of the vibrations [1]. Earlier Raman data in [2, 3] agree with each other [3], but their assignment differs somewhat from that in [1].

Force constants of $BrO_2F_2^-$ were calculated from a simple valence force field with experimental vibrational data and estimated structural data r(Br−O) = 1.6, R(Br−F) = 2 Å, α(OBrF) = 90°, β(OBrO) = 120°, and γ(FBrF) = 180°; values in mdyn/Å are tabulated below:

f_r	f_R	f_γ/R^2	f_β/r^2	f_α/rR	f_{rr}	f_{RR}	$f_{\alpha\alpha}/rR$	$f'_{\alpha\alpha}/rR$
6.32	1.63	0.10	0.52	0.53	0.35	0.43	−0.03	0.12

The force constants indicate essentially double bond character of Br-O. The value of the Br-F bond is about half of the value of a covalent bond and agrees with a semi-ionic three-center four-electron pσ bond pair [1].

Chemical Behavior

The white, solid $KBrO_2F_2$ is stable at ambient temperature but very sensitive to moisture [2, 3] and has no measurable vapor pressure. It decomposes at ~360 K. Reaction with BrF_3 is fast and yields $KBrF_4$ at room temperature [2]. The reaction of mixed $KBrO_2F_2$ and $KBrOF_4$ with a deficit of the Lewis acid AsF_5 leads to complete conversion only for $KBrO_2F_2$ [3].

References:

[1] Bougon, R.; Joubert, P.; Tantot, G. (J. Chem. Phys. **66** [1977] 1562/5).
[2] Tantot, G.; Bougon, R. (Compt. Rend. C **281** [1975] 271/3).
[3] Gillespie, R. J.; Spekkens, P. (J. Chem. Soc. Dalton Trans. **1976** 2391/6).
[4] Gillespie, R. J.; Spekkens, P. H. (J. Chem. Soc. Dalton Trans. **1977** 1539/46).
[5] Wilson, W. W.; Christe, K. O. (Inorg. Chem. **26** [1987] 916/9).
[6] Mitra, G. (Z. Anorg. Allgem. Chem. **340** [1965] 110/2, **368** [1969] 336).
[7] Gillespie, R. J.; Spekkens, P. H. (Israel J. Chem. **17** [1978] 11/9).

12.9 Bromine Trifluoride Oxide, $BrOF_3$

Other name: Bromosyl fluoride

CAS Registry Number: *[61519-37-7]*

Preparation. Formation

A convenient preparation of $BrOF_3$ uses the slow oxygenation of BrF_5 by $LiNO_3$ in excess BrF_5 at 273 K. Separation from the by-products, FNO_2 and LiF, and from the solvent by fractional condensation gives $BrOF_3$ in 95% yield; yields from other alkali nitrates are much lower [1]. Oxygenation of excess BrF_5 by IO_2F_3 at ambient temperature leads to $BrOF_3$ as a by-product of BrO_2F. The attempted oxygenation of BrF_5 by hydrolysis in HF solution at low temperature did not yield $BrOF_3$ [2].

Displacement reactions of $BrOF_4^-$ salts in presence of stronger fluoride acceptors yield $BrOF_3$. The reaction of $KBrOF_4$ and O_2AsF_6 in BrF_5 at ambient temperature can be used preparatively. Condensation of the volatiles and evaporation of BrF_5 at ~240 K leaves a residue of solid $BrOF_3$ [3]. A similar synthesis uses the dissolution of $MBrOF_4$ in HF, which leads to equilibria containing $BrOF_3$ and MHF_2 for M=K [2] and Cs [4]. Isolation of $BrOF_3$ starts with evaporation of HF at 195 K. Addition of BrF_5 allows mechanical separation of the floating KHF_2. Evaporation of the solvent at 225 K leaves solid $BrOF_3$. The attempted separation of the products KHF_2 and $BrOF_3$ by evaporation of HF and then of $BrOF_3$ at ambient temperature shifts the equilibrium to the side of the reactants [2]. Thermal decomposition of NF_4BrOF_4 in BrF_5 solution at 293 K led to NF_3, F_2, and $BrOF_3$, which was isolated in 96% yield by fractional condensation of the volatiles [5, 6].

The fluorination of BrO_2F by excess KrF_2 in HF at 275 K formed $BrOF_3$ in a slow reaction [2]. The displacement reaction of $BrOF_2PtF_6$ and excess FNO_2 at 195 K formed $BrOF_3$ mixed with BrO_2F [7].

Physical Properties

Molecular Structure. Association. The $BrOF_3$ molecule is described as a pseudotrigonal bipyramid of C_s symmetry. Two F atoms occupy the axial positions, the equatorial ones are filled by F, O, and the bromine lone electron pair. Molecular association in condensed phases is indicated by the relatively high melting point and boiling point and the low vapor pressure, and also by ^{19}F NMR [4] and vibrational spectra [3, 4]. Association predominates in the solid, and in solution is stronger in the polar HF than in the rather nonpolar ClO_3F [4]. The proposed mechanism of association involves coordination of an axial F atom [2, 4] of one $BrOF_3$ molecule as a fourth equatorial ligand of a second one. The mechanism is based on the comparison of gas-phase spectra with those of condensed phases, where the stretching vibration of the axial Br-F bond shifts to lower frequencies [4], on the small change in Br-O and Br-F_{eq} frequencies, and on the observation of additional bands in the vicinity of fundamental vibrations [2, 4].

^{19}F NMR Spectra. The single signals of liquid $BrOF_3$ at 283 K are observed at $\delta=164$ [3] or $\delta=165.2$ ppm to low field of CCl_3F [4], whereas a shift of $\delta=152$ ppm at ambient temperature was given in [2]. Broad signals with shifts of $\delta=164$ and 162 ppm are found in SO_2ClF at 170 to 190 K and in SOF_2 at 137 K. The chemical shifts are consistent with F bonded to Br(V) [2]. The observed single signals are due to rapid intramolecular exchange of axial and equatorial F atoms [2, 4].

Chemical shifts in ClO_3F at 263 K are $\delta=163.2$ ppm for a 2.3 molar solution and $\delta=$ 169.7 ppm for a 0.025 molar solution. The signals shift to higher field with decreasing temperature and increasing concentration, possibly because the resonance of associated $BrOF_3$ is upfield from that of the monomer [4]. The signal of 10 mol% of $BrOF_3$ in CH_3CN is found at $\delta=147$ ppm at 283 K and is taken to indicate the acidic character of $BrOF_3$ in the basic solvent [3].

Rotational Constants. Rotational constants of $BrOF_3$ were calculated for a pseudotrigonal bipyramid of C_s symmetry with estimated structural parameters: values are A=0.189, B= 0.119, and C=0.086 cm^{-1} [4].

Vibrational Spectra. Nine vibrational bands ($\Gamma = 6\,A' + 3\,A''$) [2 to 4] are expected for $BrOF_3$ in C_s symmetry. All bands should be active in the IR and Raman spectra [3, 4], the A′ bands should be polarized in liquid samples [4]. Assigned fundamentals in cm^{-1} appear below:

vibration	approx. description	gas (IR) [4]	liquid (Raman) [3]	in HF (Raman) [4]
$\nu_1(A')$	$\nu(BrO)$	995 s[a)]	1002 (73) p	1008 s, p
$\nu_2(A')$	$\nu(BrF_{eq})$	625 s[b)]	621 (91) p	619 vs, p
$\nu_3(A')$	$\nu_s(BrF_{2,ax})$	531 mw	489 (100) p	502 vs, p
$\nu_4(A')$	$\nu_{sciss}(OBrF_{eq})$	345 ms	351 (38)	350 m, p
$\nu_5(A')$	$\delta_{sciss}(BrF_{2,ax})$[c)]	240 m[d)]	242 (5)	235 w
$\nu_6(A')$	$\delta_{sciss}(BrF_{2,ax})$[e)]	–	202 (21)	198 mw, p
$\nu_7(A'')$	$\nu_{as}(BrF_{2,ax})$	601 vs[f)]	–	–
$\nu_8(A'')$	$\delta_{wag}(OBrF_{eq})$	397 mw[g)]	396 (26)	394 mw, dp
$\nu_9(A'')$	$\tau(OBrF_{eq})$	–	334 (18)	330 sh, dp

[a)] Isotopic splitting 2.25 cm^{-1}. – [b)] Isotopic splitting 1.5 cm^{-1}. – [c)] Out of the BrF_3 plane. – [d)] In Ne matrix. – [e)] In the BrF_3 plane. – [f)] Isotopic splitting 2.7 cm^{-1}. – [g)] In the neat solid.

The assignment was confirmed by the observed polarization of the bands and also by calculations of band contours and isotopic splitting. The observation of ν_3 in the IR spectrum demonstrates that the $BrF_{2,ax}$ group is slightly bent [4]. Raman data in [2] are only partly assigned but agree with data in [4]. Additional unassigned bands are listed in [4]. IR bands of gaseous $BrOF_3$ are given in [3]. A Ne matrix was found to give IR frequency values closest to those observed in the gas phase. The positions of several bands under high resolution in Ne, Ar, and N_2 matrices were listed [4].

Force Constants. The force constants of $BrOF_3$ were calculated from a modified valence force field with vibrational data [4] and from the following estimated structural parameters: D(Br-O) = 1.56, R(Br-F_{eq}) = 1.72, r(Br-F_{ax}) = 1.81 Å, $\alpha(OBrF_{eq})$ = 120°, and $\beta(OBrF_{ax})$ = $\gamma(F_{eq}BrF_{ax})$ = 90°. Values are given in mdyn/Å for stretching, in mdyn·Å/rad² for deformation, and in mdyn/rad for interaction force constants [4]:

force constant	vibration	value
f_D	ν_1	7.68
f_R	ν_2	3.51
$f_r + f_{rr}$	ν_3	3.16
f_α	ν_4	1.21
$f_\beta + f_{\beta\beta'}$	ν_5	1.70
$f_\gamma + f_{\gamma\gamma'}$	ν_6	1.62

force constant	vibration	value
$f_{\beta\gamma} + f_{\beta\gamma''}$	ν_6	0.65
$f_r - f_{rr}$	ν_7	2.70
$f_\beta - f_{\beta\beta'}$	ν_8	1.23
$f_\gamma - f_{\gamma\gamma'}$	ν_9	1.15
$f_{r\beta} - f_{r\beta'}$	ν_9	0.2

The f_D value agrees well with expectations for a BrO double bond. The equatorial BrF bond is stronger than the axial BrF bonds, which are possibly of the semi-ionic three-center four-electron pσ type. The calculated potential energy distributions show that all fundamentals are highly characteristic, except for ν_5 and ν_6 which correspond to approximately equal mixtures of $(f_\beta + f_{\beta\beta'})$ and $(f_\gamma + f_{\gamma\gamma'})$ [4].

Thermal Properties. Colorless, solid $BrOF_3$ has a melting range of 268 to 273 K [2]. The colorless liquid is of moderate stability at ambient temperature [2, 3]. The vapor pressure of less than 5 Torr at 298 K indicates an associated liquid [3].

Thermodynamic functions were calculated for $BrOF_3$ as an ideal gas at 1 atm from the fundamental vibrations and estimated structural data in the harmonic-oscillator rigid-rotor approximation [4]:

T in K	C_p° in $J \cdot mol^{-1} \cdot K^{-1}$	$H^\circ - H_0^\circ$ in kJ/mol	$-(G^\circ - H_0^\circ)/T$ in $J \cdot mol^{-1} \cdot K^{-1}$	S° in $J \cdot mol^{-1} \cdot K^{-1}$
0	0	0	0	0
298.15	82.969	16.945	261.680	318.515
600	99.475	45.078	307.905	383.037
1000	104.696	86.144	349.222	435.366
1500	106.533	139.039	385.543	478.235
2000	107.207	192.497	412.743	508.992

Chemical Behavior

$BrOF_3$ is a powerful oxidizing agent, contact with moisture or organic material should be avoided [1]. Thermal decomposition at ambient temperature is slow and yields BrF_3

and O_2 [2, 4]. Fluorination in HF solution by KrF_2 slowly leads to BrF_5 and O_2 above 228 K [2]. $BrOF_3$ is a stronger fluoride acceptor than BrO_2F [8], reaction with KHF_2 during warming from 253 K to ambient temperature yielding the main product $KBrOF_4$ [2]. Reaction with the Lewis acids BF_3 and AsF_5 produces $BrOF_2^+$ salts [7, 9], see p. 155. $BrOF_3$ reacts rapidly with AgCl and AgBr [4].

References:

[1] Wilson, W. W.; Christe, K. O. (Inorg. Chem. **26** [1978] 916/9).
[2] Gillespie, R. J.; Spekkens, P. H. (J. Chem. Soc. Dalton Trans. **1977** 1539/46).
[3] Bougon, R.; Bui Huy, T. (Compt. Rend. C **283** [1976] 461/3).
[4] Christe, K. O.; Curtis, E. C.; Bougon, R. (Inorg. Chem. **17** [1978] 1533/9).
[5] Christe, K. O.; Wilson, W. W.; Wilson, R. D. (Inorg. Chem. **19** [1980] 1494/8).
[6] Christe, K. O.; Wilson, W. W. (Inorg. Chem. **25** [1986] 1904/6).
[7] Adelhelm, M.; Jacob, E. (Angew. Chem. **89** [1977] 476/7; Angew. Chem. Intern. Ed. Engl. **16** [1977] 461).
[8] Gillespie, R. J.; Spekkens, P. (J. Chem. Soc. Dalton Trans. **1976** 2391/6).
[9] Bougon, R.; Bui Huy, T.; Charpin, P.; Gillespie, R. J.; Spekkens, P. H. (J. Chem. Soc. Dalton Trans. **1979** 6/12).

12.10 Bromine Trifluoride Dioxide and Trifluorodioxobromine(1+), BrO_2F_3 and $BrO_2F_3^+$

CAS Registry Number: BrO_2F_3 *[22680-01-9]*

Attempts to prepare BrO_2F_3 failed; they included fluorination of BrO_2F by excess KrF_2 [1], fluorination of BrO_3F by $KrFAsF_6$ [2], and hydrolysis of BrF_5 in HF solution at low temperature [1].

A peak of low intensity was observed in the mass spectrum of BrF_5 which had been hydrolyzed with less than equimolar amounts of water; its m/z was consistent with the $BrO_2F_3^+$ ion [3].

References:

[1] Gillespie, R. J.; Spekkens, P. H. (J. Chem. Soc. Dalton Trans. **1977** 1539/46).
[2] Spekkens, P. (Diss. Hamilton 1977, ref. 18 in: Gillespie, R. J.; Spekkens, P. H.; Israel J. Chem. **17** [1978] 11/9).
[3] Sloth, E. N.; Stein, L.; Williams, C. W. (J. Phys. Chem. **73** [1969] 278/80).

12.11 Tetrafluorooxobromate(1−), $BrOF_4^-$

Other name: Tetrafluorobromate(V)

CAS Registry Number: *[71075-29-1]*

Formation

The most convenient synthesis of $MBrOF_4$ is that from MNO_3 and excess BrF_5. Maximum yields of 99.5 to 99.9% of $MBrOF_4$ result for M=K at 373 K, Rb at 298 K, and Cs at 242 K after evaporation of BrF_5 and NO_2F. The maximum yield for M=Na is 72% at 273 K; the product contains $NaBrF_4$ and $NaBrO_2F_2$. With $LiNO_3$, BrF_5 forms LiF and $BrOF_3$ exclusively. The reaction probably involves formation of the intermediate $M[NO_3 \cdot BrF_5]$ and its decomposition [1].

The preparation of $MBrOF_4$ from $MBrO_4$ in BrF_5 in presence of F_2 is quantitative for M=Cs at 298 K and also at 343 K; for M=K yields of $KBrOF_4$ are about 70% at 353 K. The conversion to $CsBrOF_4$ is low in the absence of F_2, even at 353 K; addition of HF does not catalyze the reaction [2]. The reaction of $KBrO_3$ with F_2 (molar ratio 1:2) in BrF_5 at 350 K leads to $KBrOF_4$ [3] only under carefully controlled conditions because $KBrF_4$ and $BrOF_3$ form more easily, especially in the absence of F_2 [2]. Formation of $KBrOF_4$ as a by-product of $KBrO_2F_2$ is favored by traces of H_2O in $KBrO_3$ [4] and of HF in BrF_5 [4, 5].

A mixture of $KBrOF_4$ and $KBrO_2F_2$ forms from $KBrF_6$ and a deficit of $KBrO_3$ in CH_3CN at ambient temperature, the products can be separated by extraction of $KBrOF_4$ with CH_3CN. Hydrolysis of $KBrF_6$ in CH_3CN yields the same products, contaminated by $KBrF_4$ [5]. The reaction of BrF_5 with Cs_2SO_4 leads directly to a product mixture containing $CsBrOF_4$ after evaporation of the volatiles. The reaction of BrF_5 with KNO_2 yields a mixture containing $KBrOF_4$ via several intermediates. Reaction paths were proposed [6]. Warming a mixture of $BrOF_3$ and KHF_2 from ~250 K yields $KBrOF_4$ as the main product [7]. A solution of NF_4BrOF_4 results from the metathesis reaction of $CsBrOF_4$ and NF_4SbF_6 in BrF_5 at 218 K [8].

Physical Properties

^{19}F NMR Spectrum. The saturated CH_3CN solution of $KBrOF_4$ exhibits a single, fairly broad ^{19}F NMR peak with a low field shift of $\delta=104$ ppm from CCl_3F [5].

Vibrational Spectra. Vibrational spectra of $BrOF_4^-$ salts were analyzed on the basis of a square pyramidal ion of C_{4v} symmetry with the BrF_4 unit as the base of the pyramid and the O atom in the apex. This geometry leads to 9 fundamentals ($\Gamma=3\,A_1+2\,B_1+B_2+3\,E$). They are all active in the Raman spectrum and the A_1 and E modes are IR active [2, 5]. Fundamental vibrations of $BrOF_4^-$ were deduced from the IR and Raman spectra of solid $CsBrOF_4$. Averaged frequencies in cm^{-1} are as follows [2]:

vibration	approx. description	position
$\nu_1(A_1)$	$\nu_s(BrO)$	930
$\nu_2(A_1)$	$\nu_s(BrF_4)$	500
$\nu_3(A_1)$	$\delta_s(BrF_4$ out of plane)	302
$\nu_4(B_1)$	$\nu_s(BrF_4$ out of phase)	417
$\nu_5(B_1)$	$\delta_{as}(BrF_4$ out of plane)	205
$\nu_6(B_2)$	$\delta_s(BrF_4$ in plane)	235
$\nu_7(E)$	$\nu_{as}(BrF_4)$	505
$\nu_8(E)$	$\delta(OBrF_4)$	395
$\nu_9(E)$	$\delta_{as}(BrF_4$ in plane)	179

Vibrational data of solid $BrOF_4^-$ salts with the cations Na^+ [1], K^+ [3, 5], Rb^+ [1], and NF_4^+ [8] were also published. The vibrations of the $BrOF_4^-$ anion are similar in all salts [1 to 3, 5, 8]. The observed number of bands usually exceeds the number of the fundamentals, especially in the well resolved Raman spectra obtained at low temperature [2, 5]. The splitting was attributed to a lower site symmetry of $BrOF_4^-$ in the solids which lifts the double degeneracy of the E modes [2, 3]. The splitting of the ν_3, ν_7, and ν_8 bands probably arises from Fermi resonance [2]. Dynamic coupling in the crystal lattice was used in [3] in order to explain the splitting of the ν_8 band.

Mean Amplitudes of Vibration. Values of $BrOF_4^-$ between 0 and 1000 K were calculated with measured vibrational and estimated structural data (see below) from [2]. The results in Å at 298.16 K are [9]: $u_{Br-F}=0.0521$, $u_{Br-O}=0.0373$, $u_{F\cdots F(short)}=0.118$, $u_{F\cdots F(long)}=0.069$, and $u_{F\cdots O}=0.074$. The values are similar to those obtained for BrO_2F. The strong increase of the Br–F vibrational amplitude with temperature indicates a partly ionic character of the bond [9]. Mean square amplitudes in [10] were calculated from the same molecular data. Nonlinear shrinkage constants are $\delta_{F...F}=0.0375$ and $\delta_{F...O}=0.0177$ Å [10].

Force Constants. Force constants of $BrOF_4^-$ were calculated in the modified valence force field approximation with the estimated structural parameters R(Br–O) = 1.56, r(Br–F) = 1.88 Å, and α(FBrF) = β(OBrF) = 90°. Results in mdyn/Å are:

f_R	f_r	f_{rr}	$f_{rr'}$	f_α	f_β
6.70	2.142	0.213	0.232	0.148	0.482

Values of zero were assumed for $f_{\alpha\alpha}$ and $f_{\beta\beta}$. The calculated potential energy distribution showed that all $BrOF_4^-$ vibrations were highly characteristic except for the E block where the vibrations are slightly mixed [2]. Force constants calculated from a modified Urey-Bradley force field in [10] were also based on $BrOF_4^-$ data in [2].

Constants of Molecular Rotation and Vibration. Data of $BrOF_4^-$ were calculated from molecular parameters in [2]. Coriolis coupling constants for the degenerate E × E coupling are:

ζ_{77}	ζ_{88}	ζ_{99}	ζ_{78}	ζ_{79}	ζ_{89}
0.4401	0.6388	−0.3645	−0.1984	−0.8730	−0.2932

Centrifugal distortion constants in kHz are $D_J=0.7249$, $D_K=0.3456$, and $D_{JK}=-0.9516$ [10].

Chemical Behavior

The alkali salts of $BrOF_4^-$ are stable, white, crystalline solids at ambient temperature which are very sensitive to moisture [1 to 3, 5]. Thermal decomposition of $MBrOF_4$ (M = Na, K, Rb, Cs) occurs in the temperature range 434 (M = Na) to 499 K (M = Rb) [1]. The white NF_4BrOF_4 is thermally much less stable, see p. 170.

Hydrolysis of $KBrOF_4$ is rapid [3, 5]. The displacement reaction of $KBrOF_4$ [7] or $CsBrOF_4$ [8, 11] with HF yields $BrOF_3$ and KHF_2 or $CsHF_2$. Liberation of $BrOF_3$ also results from $KBrOF_4$ and the Lewis acid AsF_5 [5] or O_2AsF_6 [12], while the reaction of $KBrOF_4$ with excess BF_3 in HF solution yields $BrOF_2BF_4$ and $KBrF_4$ [13]. Fluorination of $KBrOF_4$ in BrF_5 in the presence of F_2 requires a temperature exceeding 350 K and leads to $KBrF_6$ [3]. Reaction of $KBrOF_4$ with $KBrO_3$ in CH_3CN at ambient temperature produces $KBrO_2F_2$ and BrF_5 [5].

References:

[1] Wilson, W. W.; Christe, K. O. (Inorg. Chem. **26** [1987] 916/9).
[2] Christe, K. O.; Wilson, R. D.; Curtis, E. C.; Kuhlmann, W.; Sawodny, W. (Inorg. Chem. **17** [1978] 533/8).
[3] Bougon, R.; Bui Huy, T.; Charpin, P.; Tantot, G. (Compt. Rend. C **283** [1976] 71/4).
[4] Bougon, R.; Joubert, P.; Tantot, G. (J. Chem. Phys. **66** [1977] 1562/5).
[5] Gillespie, R. J.; Spekkens, P. (J. Chem. Soc. Dalton Trans. **1976** 2391/6).
[6] Christe, K. O.; Wilson, W. W.; Schack, C. J. (J. Fluorine Chem. **43** [1989] 125/9).
[7] Gillespie, R. J.; Spekkens, P. H. (J. Chem. Soc. Dalton Trans. **1977** 1539/46).

[8] Christe, K. O.; Wilson, W. W. (Inorg. Chem. **25** [1986] 1904/6).
[9] Baran, E. J. (Monatsh. Chem. **110** [1979] 715/9).
[10] Natarajan, A.; Somasundaram, S. (Acta Ciencia Indica Phys. **6** [1980] 38/41).
[11] Christe, K. O.; Curtis, E. C.; Bougon, R. (Inorg. Chem. **17** [1978] 1533/9).
[12] Bougon, R.; Bui Huy, T. (Compt. Rend. C **283** [1976] 461/3).
[13] Bougon, R.; Bui Huy, T.; Charpin, P.; Gillespie, R. J.; Spekkens, P. H. (J. Chem. Soc. Dalton Trans. **1979** 6/12).

12.12 Bromine Pentafluoride Oxide, $BrOF_5$

CAS Registry Number: –

An attempt to prepare $BrOF_5$ by UV irradiation of BrF_5 and excess O_2 at 213 to 233 K failed [1]. Other unsuccessful syntheses include heating a mixture of BrF_5 and O_2 under F_2 pressure at 480 K, hydrolyzing BrF_6AsF_6 in HF or reacting it with $KBrO_4$ [2], and fluorinating $BrOF_3$ with KrF_2 in HF solution [3].

References:

[1] Pilipovich, D.; Rogers, H. H.; Wilson, R. D. (Inorg. Chem. **11** [1972] 2192/5).
[2] Spekkens, P. (Diss. Hamilton 1977, ref. 18 in: Gillespie, R. J.; Spekkens, P. H.; Israel J. Chem. **17** [1978] 11/9).
[3] Gillespie, R. J.; Spekkens, P. H. (J. Chem. Soc. Dalton Trans. **1977** 1539/46).

12.13 Dioxygenylpentafluorobromate, O_2BrF_5

CAS Registry Number: *[12297-37-9]*

The formation of violet O_2BrF_5 from cocondensed equimolar amounts of BrF_3 and O_2F_2 at 130 K in 80 % yield was described in a patent. The reaction was accompanied by some gas evolution due to decomposition. The application of the compound as an oxidizer in solid propellant technology was proposed. Further information is given in "Fluorine" Suppl. Vol. 4,1986, p. 100.

Reference:

Grosse, A. V.; Streng, A. G. (U.S. 3341294 [1967] 1 p.; C.A. **67** [1967] No. 110198).

13 Compounds of Bromine with Fluorine, Oxygen, and Noble Gases

13.1 Fluoro(bromine trifluoride oxide-F)xenon(1+), $XeF_2 \cdot BrOF_2^+$

CAS Registry Number: *[77071-46-6]*

Formation

A colorless solution of $XeF_2 \cdot BrOF_2AsF_6$ and TeF_6 resulted from the bright yellow solution of $XeOTeF_5AsF_6$ in BrF_5 upon warming from 225 K to ambient temperature. The solution is stable for some hours, and solid $XeF_2 \cdot BrOF_2AsF_6$ can be isolated by evaporation of the volatiles at 225 K.

Spectra

NMR Spectra. NMR spectra (negative high field shifts) of $XeF_2 \cdot BrOF_2AsF_6$ were measured in BrF_5 solution at 214 K. The ^{19}F spectrum shows singlets at $\delta = -163.9$ ppm for F on Xe and at $\delta = 193.9$ ppm for F on Br from the reference CCl_3F. The latter shift is in the range where signals of F bonded to Br(V) are usually observed. There is no fluorine exchange with the solvent. The ^{129}Xe NMR spectrum consists of a triplet at $\delta = -1359$ ppm with respect to neat $XeOF_4$ measured at 298 K. The coupling constant of the AX_2 spin system is J(Xe-F) = 5680 Hz. Weak covalent F bridging in $FXeF \cdot BrOF_2^+$ is indicated by the low field shifts of the Xe and F resonances compared to those of pure XeF_2 in BrF_5. The bridging is exchange-averaged on the NMR time scale. Both an intermolecular and an intramolecular exchange mechanism were found to be consistent with the data.

Raman Spectrum. The Raman spectrum of solid $XeF_2 \cdot BrOF_2AsF_6$ was analyzed tentatively on the basis of a fluorine-bridged species $F\text{-}Xe\text{-}F' \cdots BrOF_2^+$ of C_1 symmetry ("end-on structure"). A structure with both F atoms on Xe coordinated to the Br atom ("edge-on structure") was also considered to be possible. The expected 15 fundamentals of the end-on structure have symmetry A and were all identified. The modes of the $BrOF_2$ group were assigned by comparison with the bands of $BrOF_2AsF_6$; values in cm^{-1} are shown below:

band, description	position	band, description	position
ν_1, $\nu(BrO)$	1051 (29), 1045 (17)	ν_8, $\delta(OBr \cdots F')$	229 (1)
ν_2, $\nu_s(BrF_2)$	648 (72)	ν_9, $\delta(FBr \cdots F')$	209 (1)
ν_3, $\nu(Xe\text{-}F)$	546 (100), 550 (57), 561 (46)	ν_{10}, $\delta_s(FXeF')$	157 (10)
		ν_{11}, $\delta(Xe\text{-}F' \cdots Br)$	127 (2)
ν_4, $\nu(Xe\text{-}F')$	468 (32)	ν_{12}, $\tau(Xe\text{-}F' \cdots Br)$	118 (4)
ν_5, $\nu(Br \cdots F')$	404 (12)	ν_{13}, $\nu_{as}(BrF_2)$	636 (40)
ν_6, $\delta_s(OBrF)$	375 (2)	ν_{14}, $\delta_{as}(OBrF)$	328 (8), 317 (4)
ν_7, $\delta(BrF_2)$	295 (10)	ν_{15}, $\delta_s(FXeF')$	142 (6)

Fluorine bridging in $XeF_2 \cdot BrOF_2^+$ is demonstrated by identification of the appropriate bands and also by the shift of $\nu(BrO)$ to lower wave numbers in the adduct in comparison to the band of the free ion. The bridging lowers the symmetry of XeF_2 from $D_{\infty h}$ to C_s which leads to a strong set of factor-group split lines of the terminal Xe-F stretching vibration ν_3. Anion and cation in $XeF_2 \cdot BrF_2OAsF_6$ are also linked by F bridging, resulting in a lowered symmetry of the anion.

Reference:

Keller, N.; Schrobilgen, G. J. (Inorg. Chem. **20** [1981] 2118/29).

14 Compounds of Bromine with Fluorine and Nitrogen

14.1 Dibromofluoroamine, NBr_2F

CAS Registry Number: –

Formation of NBr_2F possibly resulted from BrF and NaN_3 at 273 K. The product composition was deduced by analogy with the ClF + NaN_3 reaction. The erratic nature of the reaction prevented analysis of the product.

Reference:

Allied Chemical Corp. (unpublished results, ref. 12 in: Lawless, E. W.; Smith, I. C.; Inorganic High-Energy Oxidizers, Dekker, New York 1968, p. 89).

14.2 Bromodifluoroamine, $NBrF_2$

CAS Registry Number: *[15605-95-5]*

Preparation. Formation. $NBrF_2$ is obtained when NHF_2 is bubbled through an aqueous solution containing Br_2 and HgO. Fractional condensation of the escaping gas yields more than 90% of $NBrF_2$. The high yield in the presence of Hg(II) is due to the formation of $HgBr_2$ and an increased concentration of HBrO. Absence of HgO leads to N_2F_4 as the main product and only 30% of $NBrF_2$ [1].

The reaction of ground-state NF_2 radicals with Br atoms diluted by Ar was investigated at 293 K in a flow system. The overall reaction $Br + NF_2 + Ar \rightarrow NBrF_2 + Ar$ is of third order with a rate constant of $k = (1.0 \pm 0.2) \times 10^{-31}$ $cm^6 \cdot molecule^{-2} \cdot s^{-1}$ and an enthalpy of reaction of $\Delta_rH^\circ_{298} \leq 25$ kJ/mol. The concentration of $NBrF_2$ remained low due to the rapid secondary reaction $NBrF_2 + Br \rightarrow NF_2 + Br_2$ with a rate constant of $k \geq 2 \times 10^{-14}$ $cm^3 \cdot molecule^{-1} \cdot s^{-1}$, an activation energy of $E_a \leq 25$ kJ/mol, and a heat of formation of $\Delta_fH^\circ_{298} \geq -39$ kJ/mol at 293 K. The upper limit of Br-NF_2 dissociation was estimated to be $D^\circ_{298} \leq 220$ kJ/mol [2].

The intermediate formation of $NBrF_2$ during the reduction of NHF_2 to NH_3 by Br^- in aqueous solution was discussed [3]. The synthesis of $NBrF_2$ by the reactions of BrN_3 and F_2 [4] and of $KOCF_2NF_2$, KF, and Br_2 in CH_3CN at 233 K did not succeed [5].

Properties. Reactions. Pure $NBrF_2$ is stable at 77 K and in general more stable in glass than in metal containers. The colorless liquid boils at 237 K. Its thermal decomposition to Br_2 and N_2F_4 in a few hours upon warming is not affected by water or weak bases at 273 K. No reaction occurs between $NBrF_2$ and $AlBr_3$ at 200 K, however AlF_3 is formed in CCl_4 solution. Boron-nitrogen compounds could not be isolated after reaction with BBr_3. Mixtures of $NBrF_2$ and CO, SO_2, or C_2H_4 at 333 K did not react [6]. The application of $NBrF_2$ as an additive to oxidizers, e.g. H_2O_2, in rocket propellants was mentioned [1].

References:

[1] Marshall, M. D.; Bernauer, W. H.; Callery Chemical Co. (U.S. 3238013 [1966] 3 pp.; C.A. **64** [1966] 17105).
[2] Clyne, M. A. A.; Connor, J. (J. Chem. Soc. Faraday Trans. II **68** [1972] 1220/30).
[3] Yap, W. T.; Craig, A. D.; Ward, G. A. (J. Am. Chem. Soc. **89** [1967] 3442/6).
[4] Allied Chemical Corp. (unpublished results, ref. 12 in: Lawless, E. W.; Smith, I. C.; Inorganic High-Energy Oxidizers, Dekker, New York 1968, pp. 86/7).
[5] Fraser, G. W.; Shreeve, J. M. (Inorg. Chem. **6** [1967] 1711/5).
[6] Callery Chemical Co. (unpublished results, ref. 74 in: Lawless, E. W.; Smith, I. C.; Inorganic High-Energy Oxidizers, Dekker, New York 1968, pp. 86/7).

14.3 Bromotrifluoroammonium, $NBrF_3^+$

CAS Registry Number: *[64710-00-5]*

The heat of formation of $NBrF_3^+$ of $\Delta_fH° = 854$ kJ/mol was estimated by combining the core 1s binding energy and the heat of formation of the isoelectronic $CBrF_3$ with the "core replacement energy" of the N atom.

Reference:

Jolly, W. L.; Gin, C. (Intern. J. Mass Spectrom. Ion Phys. **25** [1977] 27/37).

14.4 Tetrafluoroammonium Tetrafluorobromate(III), NF_4BrF_4

CAS Registry Number: *[101756-83-6]*

Preparation. The metathetical reaction of $CsBrF_4$ and NF_4SbF_6 in BrF_5 solution at 218 K yields NF_4BrF_4. It is isolated by filtering the insoluble $CsSbF_6$ and evaporating the volatiles from the filtrate at 242 K. The resulting solid NF_4BrF_4 is stable at 251 K; thermal decomposition at 298 K yields NF_3 and BrF_5.

Vibrational Spectra. The vibrational bands confirm the ionic structure of NF_4BrF_4. The degeneracy of the ideally triply degenerate F_2 modes of tetrahedral NF_4^+ (see "Fluorine" Suppl. Vol. 4, 1986, p. 168) is removed in the solid, and these bands are split into their components. Fundamental vibrations of NF_4^+ in cm^{-1} with intensities in parentheses relative to a maximum value of ten were observed as follows:

	$\nu_1(A_1)$, ν_s	$\nu_2(E)$, δ_s	$\nu_3(F_2)$, ν_{as}	$\nu_4(F_2)$, δ_{as}
Raman	851(2.0)	448(0.8)	1182(0.2), 1158(0.2), 1149(0.2)	622(0.5), 608(1.0)
IR	–	–	1156(sh), 1147(vs)	618(mw), 608(w), 600(m)

The BrF_4^- ion in solid NF_4BrF_4 seems not to be square-planar, but slightly distorted from D_{4h} to C_{4v} symmetry. Fundamentals of BrF_4^- are given on p. 111; lattice modes and overtones are listed in the paper.

Reference:

Christe, K. O.; Wilson, W. W. (Inorg. Chem. **25** [1986] 1904/6).

15 Compounds of Bromine with Fluorine, Nitrogen, and Oxygen

15.1 Nitrosyl Tetrafluorobromate(III), $NOBrF_4$

CAS Registry Number: *[28888-57-5]*

Preparation. Formation. White, crystalline $NOBrF_4$ results quantitatively when excess gaseous FNO is contacted with liquid BrF_3 at ambient temperature in an exothermal reaction and the remaining FNO is evaporated [1]. The exothermal reaction to $NOBrF_4$, when passing NO through BrF_3 under an atmosphere of N_2 at 293 K, frees Br_2. The yield of the white, solid product is 60 to 70%; the reaction probably involves intermediate formation of FNO [2].

The reaction of molten FNO·3 HF and liquid BrF_3 in the temperature range 293 to 393 K is fast and yields $NOBrF_4$ [3]. Formation of $NOBrF_4$ by solvolysis was discussed for BrF_3 solutions of $NOMF_6$, where M=Nb or Ta, in order to explain the BrF_3 content of the salts after isolation [4].

Crude $NOBrF_4$ can be purified by sublimation at ~380 K in a slow stream of N_2; transparent octahedrons and needles are obtained [2]. Single crystals of $NOBrF_4$ grow during sublimation at ambient temperature [1]. Purification by crystallization from hot BrF_3 was mentioned [2].

Use. Contact of granules of $NOBrF_4$ and granules of boranes results in hypergolic combustion which is potentially useful in propellant systems [5].

Vibrational Spectra. The NO^+ stretch observed at 2300 cm^{-1} in the IR spectrum and at 2309 cm^{-1} in the Raman spectrum demonstrates the ionic structure of powdered $NOBrF_4$. A very weak Raman band at 183 cm^{-1} was not assigned [1]. BrF_4^- bands are given on p. 111.

Bulk Physical Properties. The density of $NOBrF_4$ was determined to be 2.83 g/cm^3 at 289 K [2]; a value of 2.85 g/cm^3 is given in [6]. X-ray densities of 2.89 [2] and 2.90 g/cm^3 [6] were calculated. The dissociation pressure of $NOBrF_4$ is 2 Torr at 298 K [1]. The solid melts at ~528 K [2].

The crystal structure of $NOBrF_4$ was investigated by X-ray powder diffraction at room temperature. The tetragonal unit cell with $a=6.28\pm0.04$ and $c=10.84\pm0.06$ Å contains four formula units. $NOBrF_4$ seems to be isostructural with $KBrF_4$ and $RbBrF_4$ [2].

The conductivity of BrF_3 rises linearly with the concentration of dissolved $NOBrF_4$ from $\sim6\times10^{-3}\ \Omega^{-1}\cdot cm^{-1}$ for pure BrF_3 to $\sim19\times10^{-3}\ \Omega^{-1}\cdot cm^{-1}$ for a 0.1 M $NOBrF_4$ solution at 293 K. The increase in conductivity begins to abate with a further increase of the $NOBrF_4$ concentration [2].

Reactions. The violent reaction of $NOBrF_4$ with cold water yields HF, Br_2, and nitrogen oxides [2, 6]. The reaction of $NOBrF_4$ with gaseous SiF_4 yields $(NO)_2SiF_6$ and BrF_3 because BrF_4^- reacts as a base with fluoride acceptors. The product also forms when glass is attacked by $NOBrF_4$ at ambient temperature with liberation of Br_2 and O_2; the reaction probably involves intermediate formation of HF by hydrolysis. The formation of $(NO)_2GeF_6$ from $NOBrF_4$ and GeF_4 is slow. The preparation of $(NO)_2SnF_6$ from $NOBrF_4$ in BrF_3 solution succeeds with SnF_4 and $(BrF_2)_2SnF_6$; the product is isolated by evaporation of the volatiles. Dissolving $TiCl_4$ in BrF_3 and adding solid $NOBrF_4$ yields a precipitate, which was identified as $(NO)_2TiF_6$ after filtration and drying at 420 K. The formation of $(NO)_2MnF_6$ from a hot BrF_3 solution of MnF_3 and excess solid $NOBrF_4$ is known from the minimum of conductivity at a reactant ratio of 1:2. However, a mixture of solid $(NO)_2MnF_6$ and MnF_3 resulted after evaporation of the volatiles [2]; see also "Mangan" C4, 1977, p. 271.

The reaction of $NOBrF_4$ with alcohol, ether, and acetone is violent [2].

References:

[1] Christe, K. O.; Schack, C. J. (Inorg. Chem. **9** [1970] 1852/8).
[2] Bouy, P. (Ann. Chim. [Paris] [13] **4** [1959] 853/90, 865/87).
[3] Seel, F.; Birnkraut, W.; Werner, D. (Angew. Chem. **73** [1961] 806).
[4] Clark, H. C.; Emeléus, H. J. (J. Chem. Soc. **1958** 190/5).
[5] Iwanciow, B. L.; Lawrence, W. J.; United Aircraft Corp. (U.S. 3797238 [1974] 5 pp.; C.A. **81** [1974] No. 15274).
[6] Chrétien, A.; Bouy, P. (Compt. Rend. **246** [1958] 2493/5).

15.2 Nitryl Tetrafluorobromate(III), NO_2BrF_4

CAS Registry Number: *[28888-58-6]*

Preparation. Contacting excess gaseous FNO_2 and liquid BrF_3 quantitatively yields white, powdery NO_2BrF_4. The reaction is started at ambient temperature and completed at ~230 K while shaking the reaction vessel. A viscous solution and later lumps are formed. Excess FNO_2 is evaporated at 208 K [1]. The attempt to isolate the product at ambient temperature from the formed viscous liquid after the reaction failed [2].

Properties. The ionic structure of NO_2BrF_4 is evident from the stretching vibration of NO_2^+ at 2384 cm^{-1} in the vibrational spectrum. Solid NO_2BrF_4 melts at 300 K under autogenous pressure. The dissociation $NO_2BrF_4(s) \rightarrow FNO_2(g) + BrF_3(g)$ can be described by $\log(p/Torr) = 11.0208 - (2508.9/T)$; the standard error of log p is 0.019 (the temperature range was not specified). Calculated thermodynamic data of dissociation are $\Delta H^\circ_{298} = 96.06$ kJ/mol, $\Delta G^\circ_{298} = 6.669$ kJ/mol, and $\Delta S^\circ_{298} = 299.87$ $J \cdot K^{-1} \cdot mol^{-1}$. The solubility of FNO_2 in BrF_3 and the enthalpy of sublimation or vaporization of BrF_3 were neglected in this calculation. The heat of formation of $\Delta_f H^\circ_{298} = -431.4$ kJ/mol was obtained with the $\Delta_f H^\circ$ values of FNO_2 and BrF_3.

The reaction of NO_2BrF_4 with an excess of the stronger Lewis acid SiF_4 at ambient temperature frees the weaker Lewis acid BrF_3 and leads to $(NO_2)_2SiF_6$ [1].

References:

[1] Christe, K. O.; Schack, C. J. (Inorg. Chem. **9** [1970] 1852/8).
[2] Aynsley, E. E.; Hetherington, G.; Robinson, P. L. (J. Chem. Soc. **1954** 1119/24).

15.3 Tetrafluoroammonium Perbromate, NF_4BrO_4

CAS Registry Number: *[25483-10-7]*

Preparation. The metathetical reaction of $CsBrO_4$ and NF_4SbF_6 yields NF_4BrO_4 in HF at 293 K. The second product is $CsSbF_6$ which is crystallized from the concentrated solution at 195 K and filtered. The solution slowly decomposes at ambient temperature with formation of NF_3; also formed O_2 and BrO_2F possibly result from the intermediate $FOBrO_3$. An explosion occurred during an attempt to isolate NF_4BrO_4 from the solution, even though it had never been warmed above 195 K.

Spectra. The ^{19}F NMR spectrum shows a sharp triplet of equal intensity due to NF_4^+ with $\delta = 217$ downfield of external CCl_3F and J(NF) = 227 Hz. The Raman spectrum confirms the ionic structure of NF_4BrO_4 by NF_4^+ and BrO_4^- bands.

Reference:

Christe, K. O.; Wilson, W. W.; Wilson, R. D. (Inorg. Chem. **19** [1980] 1494/8).

15.4 Nitrosyl Hexafluorobromate(V), $NOBrF_6$

CAS Registry Numbers: *[50859-39-7, 90993-20-7]*

The formation of $NOBrF_6$ via $BrF_6AsF_6 + 2\,FNO \rightarrow NOBrF_6 + NOAsF_6 + F_2$ was observed in excess liquid FNO at 195 K during the attempted synthesis of BrF_7. The white, solid $NOBrF_6$ formed can be sublimated under dynamic vacuum at 273 K after evaporating the formed F_2 and excess FNO [1]. An attempt to prepare $NOBrF_6$ from pure FNO and BrF_5 at 248 K failed. The addition of a small amount of $NOClF_4$ induced a reaction, but the formed solid did not yield reproducible analytical results [2].

The isolated $NOBrF_6$ was identified by its Raman spectrum [3].

References:

[1] Gillespie, R. J.; Schrobilgen, G. J. (Inorg. Chem. **13** [1974] 1230/5).
[2] Whitney, E. D.; MacLaren, R. O.; Hurley, T. J.; Fogle, C. E. (J. Am. Chem. Soc. **86** [1964] 4340/2).
[3] Gillespie, R. J.; Schrobilgen, G. J. (unpublished results, ref. 38 in [1]).

15.5 Nitryl Hexafluorobromate(V), NO_2BrF_6

CAS Registry Number: –

An attempt to synthesize NO_2BrF_6 from FNO_2 and BrF_5 at low temperature failed since the starting materials did not react.

Reference:

Whitney, E. D.; MacLaren, R. O.; Hurley, T. J.; Fogle, C. E. (J. Am. Chem. Soc. **86** [1964] 4340/2).

15.6 Tetrafluoroammonium Tetrafluorobromate(V), NF_4BrOF_4

CAS Registry Number: *[101652-54-4]*

Preparation. The metathetical reaction of $CsBrOF_4$ and NF_4SbF_6 in BrF_5 yields NF_4BrOF_4 at 218 K. Filtration of insoluble $CsSbF_6$ and evaporation of the volatiles from the filtrate at 218 K leaves a residue of white, solid NF_4BrOF_4. The salt slowly decomposes at 218 K; thermal decomposition at 298 K yields NF_3, $BrOF_3$, and F_2 [1]. The same products were obtained when the metathesis reaction was carried out at 293 K [2].

Vibrational Spectra. The vibrational spectra confirm the ionic structure of NF_4BrOF_4. The degeneracy of the ideally triply degenerate F_2 modes of the tetrahedral NF_4^+ ion (see also "Fluorine" Suppl. Vol. 4, 1986, p. 168) disappears in the solid, and these bands are split into their components [1]. Fundamental vibrations of NF_4^+ in cm^{-1} with intensities in parentheses relative to a maximum value of ten are observed at:

	$\nu_1(A_1)$, ν_s	$\nu_2(E)$, δ_s	$\nu_3(F_2)$, ν_{as}	$\nu_4(F_2)$, δ_{as}
Raman	853(2.7)	451(3)	1165(0.4), 1152(sh)	614(2.0), 605(sh)
IR	–	–	1165(s), 1149(vs)	614(w), 608(ms), 600(sh)

Fundamentals of $BrOF_4^-$ are given on p. 111; lattice modes and overtones are listed in [1].

References:

[1] Christe, K. O.; Wilson, W. W. (Inorg. Chem. **25** [1986] 1904/6).
[2] Christe, K. O.; Wilson, W. W.; Wilson, R. D. (Inorg. Chem. **19** [1980] 1494/8).

16 Compounds of Bromine with Chlorine

Compounds composed of bromine and chlorine described in this chapter are arranged in decreasing bromine content: Br_6Cl^-, Br_3Cl, Br_2Cl^+, Br_2Cl^-, BrCl, its hydrate, $BrCl^+$, $BrCl^-$, Br_2Cl_2, $Br_2 \cdot (Cl_2)_n$, $Br_2Cl_3^-$, $BrCl_2^+$, $BrCl_2^-$, $BrCl_2^{2-}$, $BrCl_3$, $BrCl_4^-$, and $BrCl_6^{5-}$. Whenever information is available, isomers are treated separately.

Out of all neutral and ionic species, BrCl is characterized best by its molecular properties and spectra. Little information on its chemistry is available, probably because it readily dissociates into bromine and chlorine. Chemical and physical properties of the single isomers $BrBrCl^-$ and $ClBrCl^-$ are known to some extent. Information on all other species is very limited, and even serious doubts may be raised on the existence of species, such as Br_6Cl^-, $BrCl_4^-$, and $BrCl_6^{5-}$.

General References:

Martin, D.; Rousson, R.; Weulersse, J.-M.; The Interhalogens, Chem. Non-Aqueous Solvents B **5** [1978] 157/95.

Mills, J. F.; Interhalogens and Halogen Mixtures as Disinfectants, in: Johnson, J. D.; Disinfection: Water Wastewater, Ann Arbor Sci., Stoneham, Mass., 1975, pp. 113/43.

Mills, J. F.; Schneider, J. A.; Bromine Chloride: an Alternative to Bromine, Ind. Eng. Chem., Prod. Res. Develop. **12** [1973] 160/5.

Downs, A. J.; Adams, C. J.; Chlorine, Bromine, Iodine and Astatine, in: Bailar, J. C.; Emeléus, H. J.; Nyholm, R; Trotman-Dickenson, A. F.; Comprehensive Inorganic Chemistry, Vol. 2, Pergamon, Oxford 1973, pp. 1107/594, 1476/563.

Meinert, H.; Interhalogenverbindungen, Z. Chem. [Leipzig] **7** [1967] 41/56, 47/8.

Popov, A. I.; Polyhalogen Complex Ions, in: Gutmann, V.; Halogen Chemistry, Vol. 1, Academic, London 1967, pp. 225/64.

Schmeisser, M.; Schuster, E.; Compounds of Bromine with Non-Metals, in: Jolles, Z. E.; Bromine and its Compounds, Benn, London 1966, pp. 179/252, 186/8.

Greenwood, N. N.; Compounds of Chlorine with Bromine and Iodine, in: Mellor's Comprehensive Treatise on Inorganic and Theoretical Chemistry, Suppl. II, Pt. I, Longmans Green, London 1956, pp. 476/513, 476/82.

16.1 Hexabromochlorate(1−), Br_6Cl^-

CAS Registry Number: −

Reportedly, a compound of the formula "$(C_2H_5)_4NBr_6Cl$" formed after exposing $(C_2H_5)_4NCl$ to excess Br_2 vapor. The product readily lost Br_2 again as recognized by the color of the vapor phase above a specimen and by the formation of $(C_2H_5)_4NBr_2Cl$ when stored over lime.

Reference:

Chattaway, F. D.; Hoyle, G. (J. Chem. Soc. **123** [1923] 654/62).

16.2 Tribromine Chloride, Br_3Cl

CAS Registry Number: *[12360-52-0]*

Cl_2 and excess Br_2, both diluted by a rare gas, were passed through a microwave discharge and condensed at 20 K. Among the IR bands observed in the matrix, those at 290 (in Ar), 289.5, 286.0 (in Kr), and 284.6, 283.8, 282.4, 280.2, 279.3, 278.4 cm^{-1} (in Xe) were assigned to the T-shaped Cl—Br(Br)—Br molecule. Isotopic frequency shifts and reasonable values for the force constants were the basis for the very tentative assignment.

Reference:

Nelson, L. Y.; Pimentel, G. C. (Inorg. Chem. **7** [1968] 1695/9).

16.3 Bromochlorobromine(1+) and Dibromochlorine(1+), $BrBrCl^+$ and $BrClBr^+$

CAS Registry Number: $BrBrCl^+$ *[56488-80-3]*

The possible existence of two structural isomers for Br_2Cl^+ was mentioned only once without giving an assignment. Thus, data are summarized under the general formula Br_2Cl^+.

CCl_2Br_2 was diluted with Ar and irradiated by 0.5 keV protons. The products were deposited on a CsI window kept at 15 K. IR bands at 432, 416, 362, and 339 cm^{-1} were tentatively assigned to the two isomeric Br_2Cl^+ species [1].

Solid $[Br_2^+]$ $[Sb_3F_{16}^-]$ was reacted with a 1.5-fold excess of Cl_2 via

$$2\,[Br_2^+]\,[Sb_3F_{16}^-] + Cl_2 \rightarrow 2\,Br_2Cl^+ + 2\,SbF_6^- + 4\,SbF_5$$

After removing the excess Cl_2, $[Br_2Cl^+]$ $[Sb_3F_{16}^-]$ was obtained as a solution in SbF_5. Separating the product from the solvent was impossible [2, 3].

The viscous solution of $[Br_2Cl^+]$ $[Sb_3F_{16}^-]$ in SbF_5 is dark red-brown and diamagnetic. Raman bands at 424 and 300 cm^{-1} were assigned to the symmetric and antisymmetric stretching modes. (Only the $BrBrCl^+$ isomer was considered as indicated by the mode description as Br—Cl and Br—Br stretching.) The bending mode could not be observed. UV-VIS bands of Br_2Cl^+ were observed at 370, 305, and 285 nm as shoulders on the 252 nm band, tentatively assigned to SbF_6^- [2, 3].

The Br_2Cl^+ cation is thermally stable at 25 °C as long as $[Br_2Cl^+]$ $[Sb_3F_{16}^-]$ is stored under an N_2 or Cl_2 atmosphere [3]. Specifically, no evidence was found for dissociation or disproportionation to Br_2^+. Br_2Cl^+ is not stable in protic acids [2].

References:

[1] Andrews, L.; Grzybowski, J. M.; Allen, R. O. (J. Phys. Chem. **79** [1975] 904/12, 910).
[2] Wilson, W. W.; Thompson, R. C.; Aubke, F. (Inorg. Chem. **19** [1980] 1489/93).
[3] Wilson, W. W.; Landa, B.; Aubke, F. (Inorg. Nucl. Chem. Letters **11** [1975] 529/34).

16.4 Bromochlorobromate(1−), $BrBrCl^-$

CAS Registry Number: $BrBrCl^-$ *[16871-87-7]*

There is evidence from other trihalide anions that $BrBrCl^-$ with the heavier halogen in the middle position is more stable than the isomeric $BrClBr^-$ species; see, for instance, a general review [1]. For the present description it is assumed that work on the Br_2Cl^- anion deals with $BrBrCl^-$ unless the $BrClBr^-$ isomer (see p. 178) is explicitely mentioned.

Formation

In general, Br_2Cl^- forms when adding a chloride anion to molecular bromine or a bromide anion to bromine chloride.

First evidence for Br_2Cl^- formation was provided by the heat of solution of Br_2 in concentrated HCl (or other chlorides) [2]. Later on, relatively stable Br_2Cl^- salts were isolated with alkali cations, such as Cs^+ and Rb^+ [3 to 6], but also with more complex onium cations, such as substituted ammonium [7], phosphonium [8, 9], sulfonium [8], and thiopyrylium [8] cations. In the gas phase, appreciable amounts of Br_2Cl^- were detected by mass spectrometry during drift-tube experiments on Xe–Cl_2 mixtures in an apparatus that was contaminated by Br_2 [51]. Matrix-isolated Br_2Cl^- anions were obtained by codeposition of KCl [10] or CsCl [11] vapor with Br_2 in excess Ar at about 15 K. Codeposition of KBr vapor with BrCl produced $BrBrCl^-$ besides some $BrClBr^-$ [10]. Surface reactions between solid alkali chlorides and Br_2 vapor likewise led to Br_2Cl^- anions [12, 13].

Experimental methods used to detect Br_2Cl^- formation in solution include UV-VIS spectroscopy [14 to 19] (very early study [20]), measuring the distribution of bromine in H_2O and CCl_4 in the presence of Cl^- [21 to 24], determining the solubility in aqueous systems containig Br_2 and Cl^- [6, 25 to 28], electrochemical techniques [29 to 32], calorimetry [33] (first study [2]), Br_2 vapor pressure measurements over Br_2-containing chloride solutions [34], and the freezing point depression in aqueous HCl containing Br_2 [35].

Several equilibrium constants for $Br_2 + Cl^- \rightleftharpoons Br_2Cl^-$ that were measured in (mostly acidified) aqueous solutions [14, 15, 19, 23, 25, 29, 33, 34] are compiled in "Bromine" Suppl. Vol. A, 1985, pp. 454/5. A review of these data shows that the equilibrium constant is very close to 1 L/mol at room temperature and only slightly depends on external conditions, as discussed below. Equilibrium constants from the most recent critical evaluation [18] and from a number of very early measurements (reviewed in [15, 29]) support the average value above, specifically 0.84 L/mol (298 K, value extrapolated for zero ionic strength, from UV-VIS spectrophotometry) [18], 1.38 L/mol (298 K, thermodynamic estimate) [36], 1.726 L/mol (271.6 K, from freezing point depression of aqueous HCl) [35], 1.39 L/mol (298 K, average value from distribution in the presence of Cu^{2+}, Mg^{2+}, Ca^{2+}, Sr^{2+}, Ba^{2+}, Al^{3+}; slightly smaller values in the presence of Cd^{2+}, Hg^{2+}) [21], 1.41 L/mol (303 K, from distribution) [24], 1.2 L/mol (293.9 K, average value from solubility of Br_2 in aqueous NaCl or HCl solutions as calculated by [15]) [28], and 1.3 L/mol (298 K, from distribution as calculated by [15]) [22]. Only a few values lie outside the expected range: 12.9 L/mol (298 K, from solubility of $PbCl_2$ in H_2O ($I=3$, 0.1 M $HClO_4$) containing Br_2) [26], 4.7 L/mol (298 K, from Br_2 solubility in aqueous NH_4Cl solutions) [53], and 5.3 L/mol (298 K, from spectrophotometry) [17].

In aqueous solutions, the $Br_2 + Cl^- \rightleftharpoons Br_2Cl^-$ equilibrium constant was reported to depend slightly on the ionic strength: $K_{298} = 1.25$ and 1.42 L/mol for $I = 1.14$ and 4 [34]. A reverse trend, e.g. $K_{298} = 0.83$ and 0.74 L/mol for $I = 1.3$ and 7.7, was noted when the experimental data were corrected for other feasible equilibria [18]. Early measurements showed no definite trend in K changing the bivalent cations (see above) [21]. However, equilibrium constants were reported to decrease in the order of the univalent cations $K^+ > Rb^+ > Cs^+$ [19, 23].

Similar to those of other trihalide anions, the equilibrium constants of Br_2Cl^- increases by five orders of magnitude when the solvent water is replaced by sulfolane, giving $K_{295} = 6.3 \times 10^4$ L/mol.

Most of the published $Br_2 + Cl^- \rightleftharpoons Br_2Cl^-$ equilibrium constants slightly decrease with increasing temperature. Heats of reaction are between −4.4 and −9.2 kJ/mol and entropies of reaction between −14.2 and −29.3 $J \cdot K^{-1} \cdot mol^{-1}$ (all data in aqueous solution) [15, 19, 23, 33, 35]; see also "Bromine" Suppl. Vol. A, 1985, p. 455. Free energies of reaction at 298 K of −0.75 and +0.17 kJ/mol were estimated for $Br_2(aq) + Cl^-(aq)$ and $Br_2(g) + Cl^-(aq)$, respectively [36]. A reinvestigation of the $Br_2 + Cl^-$ equilibrium in acidified aqueous solutions between 283 and 318 K led, after extrapolating of data to zero ionic strength, to equilibrium constants with numerical values close to those previously known, but with an opposite temperature effect: $\Delta_rH = 1.19 \pm 0.03$ kJ/mol, $\Delta_rS = 4.03 \pm 0.03$ $J \cdot K^{-1} \cdot mol^{-1}$ [18]. The change in sign with respect to the previous data is caused by taking four additional equilibria into account [18].

Only estimated free energies of reaction are available for $Br^-(aq) + BrCl(aq)$, namely, $\Delta_rG_{298} = -15.9$ kJ/mol, and $Br^-(aq) + BrCl(g)$, $\Delta_rG_{298} = -24.2$ kJ/mol [36].

Thermodynamic data of formation of Br_2Cl^- in aqueous solution are $\Delta_fG_{298} = -126.2$ kJ/mol, $\Delta_fH_{298} = -167.8$ kJ/mol, and $S_{298} = 190.8$ $J \cdot K^{-1} \cdot mol^{-1}$ [18] provided their reaction data are reliable. Δ_fG, Δ_fH, and S change at the most to -128, -179 kJ/mol, and 158 $J \cdot K^{-1} \cdot mol^{-1}$, respectively [18], when based on previously published reaction data [15, 33]. A compilation of thermodynamic data quotes $\Delta_fG° = -128.4$ kJ/mol, $\Delta_fH° = -170.3$ kJ/mol, and $S° = 188.7$ $J \cdot K^{-1} \cdot mol^{-1}$ at 298.15 K and 0.1 MPa [37] without giving a source.

Br_2Cl^- forms readily in solutions containing chloride and bromine, yet a number of side products were observed, e.g. Br_3^-, $BrCl_2^-$ [18, 29, 38], BrCl [18]. Acidification of the aqueous solution suppresses some side reactions; see for example [18]. Instead of the expected Br_2Cl^- anion, the monobromine complexes $R_3R'N^+Cl^- \cdot Br_2$ [39] and $R_3N^+(Cl^-)-R'-N^+(Cl^-)R_3 \cdot Br_2$ [40] were reported to form from respective mono- and diammonium chlorides where R' contains a C−C multiple bond. The assignment of Raman bands for Br_2-Cl^- mixtures in H_2O or CH_4CN [41] as well as for solutions of $(C_2H_5)_4N^+Cl^-$ or PCl_5 in Br_2 [42] probably needs to be reconsidered.

Molecular Properties and Spectra

There seems to be no experimental proof of the structure of Br_2Cl^-. Based on the structure of other trihalide anions (see, for example [1]) and by simple LCAO reasoning, Br_2Cl^- is expected to be linear [38, 43]; see also the discussion of the structure of the $BrCl_2^-$ trihalide anion on p. 225.

An IR band at 229 cm^{-1} was assigned to the antisymmetric stretching vibration of matrix-isolated $BrBrCl^-$ (Ar matrix, K^+ cations) [10] (data entered into a more recent compilation [44]). A tentative assignment to both stretching modes of $BrBrCl^-$ was based on the observation of two polarized Raman bands around 230 cm^{-1} and 192 cm^{-1} for mixtures of $(C_3H_7)_4N^+Cl^-$ and Br_2 in CH_3CN and for the solid product from the reaction of NH_4Cl with Br_2. IR bands were observed at about 227 and 189 cm^{-1} for the solid product from the $NH_4Cl + Br_2$ reaction and for the solution of $(C_4H_9)_4N^+Cl^-$ and Br_2 in C_6H_6. In a concurrent calculation, force constants with reasonable values ($k_{ClBr} = 1.0$, $k_{BrBr} = 0.9$, $k_{1/2} = 0.4$; all in mydn/Å) were used [38]. A probably erroneous assignment of the symmetric stretch was based on markedly different band positions around 280 cm^{-1} in the Raman spectra taken from solutions of $(C_2H_5)_4N^+Cl^-$ or PCl_5 in Br_2 or the otherwise unidentified product from PCl_5 and Br_2 dissolved in CH_4Cl_2 [42].

The color of alkali or ammonium salts of Br_2Cl^- was described to be orange [6, 7] or yellowish red [3]. The reflectance spectrum of Br_2Cl^- formed on the surface of NaCl has peaks at 228 and 217 nm [13]. Ar-matrix isolated Br_2Cl^- was reported to have a strong absorption at 248 nm [11]. Solution UV-VIS spectra are not precisely known, probably because of equilibria between the different absorbing species. Br_2Cl^- extinction coefficients were calculated with known equilibrium constants for Br_2Cl^- and $BrCl_2^-$ formation from the overall extinction of acidified aqueous solutions containing Br_2 and Cl^-. Data are as follows: $\varepsilon = 193$, 167, 184, 223, 325 $L \cdot mol^{-1} \cdot cm^{-1}$ at 320, 340, 360, 380, and 400 nm. A maximum above 400 nm was expected [29]. Based on these data, an earlier measurement of the UV-VIS spectrum of $Cs^+Br_2Cl^-$ dissolved in water ($\lambda_{max} = 265$ nm, log $\varepsilon = 2.4$ and

λ_{max}=395 nm, log ε=2.0) or in C_2H_5OH (λ_{max}=266 nm, log ε=3.7 and a broad shoulder between 360 and 440 nm, log ε between 2.7 and 2.5) [16] was questioned as being close to that of Br_3^- [29]. Spectrophotometric measurements on the Br_2+Cl^- equilibrium in acidified aqueous solution resulted in λ_{max}=237 nm, ε=2×10^4 $L \cdot mol^{-1} \cdot cm^{-1}$ [15], λ_{max}=236 nm [52]. Irradiation at wavelengths as long as 436 nm (ε=112 $L \cdot mol^{-1} \cdot cm^{-1}$ in H_2O, pH 2.6) was reported to cause Br_2Cl^- photodissociation [45]. UV-VIS spectra of acidified aqueous solutions containing Br_2 and Cl^- were depicted [19, 52].

Electrochemical Behavior

The first oxidation and reduction potentials of Br_2Cl^- were measured by cyclic voltammetry at 298 K. For the $(C_4H_9)_4N^+$ salt dissolved in tetrahydrofuran and benzonitrile, oxidation and reduction potentials of >1.2 and −0.07 V vs. SCE (in THF) and 1.15 and 0.10 V vs. SCE (in BN) were obtained [46]. Cyclic voltammograms of about equimolar Br_2 and Cl^- mixtures in a room temperature $AlCl_3$−imidazoliumchloride melt showed Br_2Cl^- reduction at 0.3 V. Br_2Cl^- was not stable under conditions of cathodic reduction via $Br_2Cl^- + 2\,e^- \rightleftharpoons 2\,Br^- + Cl^-$ [32].

Chemical Behavior

When Br_2Cl^- salts are dissolved in water, the anion dissociates to Br_2 and Cl^- to a considerable extent; see the equilibrium constants given above. Solid Br_2Cl^- salts are thermally unstable. The bromine pressure over $RbBr_2Cl$ reaches 30 Torr at 291 K and 760 Torr at 354 K [5], for example. The relative stability of trihalide anions in solid salts increases in the order Br_2Cl^-, $BrCl_2^-$, Br_3^- [5, 43].

There was some UV-VIS spectrophotometric evidence for the disproportionation reaction, $2\,Br_2Cl^- \rightleftharpoons Br_3^- + BrCl_2^-$, in aqueous solution; see a comment [29] concerning earlier spectrum [16]. Irradiation at 436 nm was thought to generate intermediate Cl atoms [45].

Br_2Cl^- salts were used as bromination agents in organic chemistry. Examples include bromination of alkenes [9, 47] (those of diethylfumarate and 2-chlorallylalcohol are relatively slow [47]), phenols [48], aromatic ethers [49], aromatic amines [50], and acetanilides [7]. Using benzyltrimethylammonium trihalides for bromination of aromatic ethers in CH_4Cl-CH_3OH mixtures, for example, the reactivity of the Br_2Cl^- salt is higher than the Br_3^- salt and lower than the $BrCl_2^-$ salt [49].

References:

[1] Popov, A. I. (in: Gutmann, V.; Halogen Chemistry, Vol. 1, Academic, London 1967, pp. 225/64).

[2] Berthelot, M. (Compt. Rend. **100** [1885] 761/7).

[3] Wells, H. L.; Penfield, S. L. (Z. Anorg. Allgem. Chem. **1** [1892] 85/103; Am. J. Sci. [3] **43** [1892] 17/32).

[4] Wells, H. L.; Wheeler, H. L. (Z. Anorg. Allgem. Chem. **1** [1892] 442/55; Am. J. Sci. [3] **43** [1892] 475/87).

[5] Ephraim, F. (Ber. Deut. Chem. Ges. **50** [1917] 1096/88).

[6] Mironenko, A. P.; Stepina, S. B.; Plyushchev, V. E.; Zotova, L. A. (Zh. Neorg. Khim. **13** [1968] 2838/42; Russ. J. Inorg. Chem. **13** [1968] 1460/2).

[7] Kajigaeshi, S.; Kakinami, T.; Yamasaki, H.; Fujisaki, S.; Okamoto, T. (Bull. Chem. Soc. Japan **61** [1988] 2681/3).

[8] Kanai, K.; Hashimoto, T.; Kitano, H.; Fukui, K. (Nippon Kagaku Zasshi **86** [1965] 534/9, A33).

[9] Appel, R.; Knoll, F.; Wihler, H.-D. (Angew. Chem. **89** [1977] 415/6).
[10] Ault, B. S.; Andrews, L. (J. Chem. Phys. **64** [1976] 4853/9).

[11] Andrews, L.; Prochaska, E. S.; Loewenschuss, A. (Inorg. Chem. **19** [1980] 463/5).
[12] Cremer, H. W.; Duncan, D. R. (J. Chem. Soc. **1933** 181/9).
[13] Kortüm, G.; Vögele, H. (Ber. Bunsenges. Physik. Chem. **72** [1968] 401/7).
[14] Bell, R. P.; Ramsden, E. N. (J. Chem. Soc. **1958** 161/7, 163).
[15] Daniele, G. (Gazz. Chim. Ital. **90** [1960] 1597/606).
[16] Gilbert, F. L.; Goldstein, R. R.; Lowry, T. M. (J. Chem. Soc. **1931** 1092/103).
[17] Job, P. (Ann. Chim. [Paris] [10] **9** [1928] 113/203, 146/8, 200/1).
[18] Kremer, V. A.; Onoprienko, T. A.; Zalkind, G. R.; Kashtanova, A. S. (Deposited Doc. VINITI 2363-76 [1976] 1/15).
[19] Mironenko, A. P.; Kulikova, L. N.; Alekseeva, I. I.; Stepina, S. B.; Plyushchev, V. E.; Pokrovskaya, L. I. (Zh. Neorg. Khim. **19** [1974] 3272/8; Russ. J. Inorg. Chem. **19** [1974] 1792/5).
[20] Tinkler, C. K. (J. Chem. Soc. **93** [1908] 1611/8).

[21] Dancaster, E. A. (J. Chem. Soc. **125** [1924] 2038/43).
[22] Jakowkin, A. A. (Z. Physik. Chem. **20** [1896] 19/39, 30).
[23] Mironenko, A. P.; Kulikova, L. N.; Alekseeva, I. I.; Stepina, S. B.; Plyushchev, V. E.; Pokrovskaya, L. I. (Zh. Neorg. Khim. **18** [1973] 1243/7; Russ. J. Inorg. Chem. **18** [1973] 656/8).
[24] Rây, P.; Sarkar, P. V. (J. Chem. Soc. **121** [1922] 1449/55, 1453).
[25] Korenman, I. M. (Zh. Obshch. Khim. **17** [1947] 1608/17; C.A. **1948** 4026).
[26] Lenarcik, B.; Kowalik, E. (Zes. Nauk. Mat. Fiz. Chem. Wyzsza Szk. Pedagog. Gdansku **4** [1964] 73/84).
[27] Mironenko, A. P.; Stepina, S. B. (Zh. Neorg. Khim. **27** [1982] 1043; Russ. J. Inorg. Chem. **27** [1982] 586/8).
[28] Oliveri-Mandalà, E. (Gazz. Chim. Ital. **50** II [1920] 89/98).
[29] Bell, R. P.; Pring, M. (J. Chem. Soc. A **1966** 1607/9).
[30] Benoit, R. L.; Guay, M. (Inorg. Nucl. Chem. Letters **4** [1968] 215/7).

[31] Salthouse, J. A.; Waddington, T. C. (J. Chem. Soc. A **1966** 1188/90).
[32] Dymek, C. J., Jr.; Reynolds, G. F.; Wilkes, J. S. (J. Electrochem. Soc. **134** [1987] 1658/63).
[33] Mironov, V. E.; Lastikova, N. P. (Zh. Fiz. Khim. **41** [1967] 1850/6; Russ. J. Phys. Chem. **41** [1967] 991/5).
[34] Artamonov, Yu. F.; Gergert, V. R. (Zh. Neorg. Khim. **22** [1977] 18/22; Russ. J. Inorg. Chem. **22** [1977] 8/11).
[35] Ray, S. K. (J. Indian Chem. Soc. **9** [1932] 259/69).
[36] Scott, R. L. (J. Am. Chem. Soc. **75** [1953] 1550/2).
[37] Wagman, D. D.; Evans, W. H.; Parker, V. B.; Schumm, R. H.; Halow, I.; Bailey, S. M.; Churney, K. L.; Nuttall, R. L. (J. Phys. Chem. Ref. Data Suppl. **11** No. 2 [1982] 2-51).
[38] Evans, J. C.; Lo, G. Y.-S. (J. Chem. Phys. **45** [1966] 1069/71).
[39] Gyul'nazaryan, A. Kh.; Khachatryan, N. G.; Saakyan, T. A.; Kinoyan, F. S.; Panosyan, G. A.; Babayan, A. T. (Zh. Org. Khim. **24** [1988] 504/9; Russ. J. Org. Chem. **24** [1988] 449/53).
[40] Khachatrian, N. G.; Gyulnazarian, A. Kh.; Churkina, N. P.; Sahakian, T. A.; Martirossian, N. R.; Babayan, A. T. (Arm. Khim. Zh. **39** [1986] 290/4).

[41] Delhaye, M.; Dhamelincourt, P.; Merlin, J.-C.; Wallart, F. (Compt. Rend. B **272** [1971] 1003/6).
[42] Finch, A.; Gates, P. N.; Muir, A. S. (Polyhedron **5** [1986] 1537/42).

[43] Wiebenga, E. H.; Havinga, E. E.; Boswiyk, K. H. (Advan. Inorg. Chem. Radiochem. **3** [1961] 133/69).
[44] Jacox, M. E. (J. Phys. Chem. Ref. Data **13** [1984] 945/1068).
[45] Rutenberg, A. C.; Taube, H. (J. Am. Chem. Soc. **73** [1951] 4426/31).
[46] Sakura, S.; Imai, H.; Anzai, H.; Moriya, T. (Bull. Chem. Soc. Japan **61** [1988] 3181/6).
[47] Bell, R. P.; Pring, M. (J. Chem. Soc. B **1966** 1119/26).
[48] Kajigaeshi, S.; Kakinami, T.; Moriwaki, M.; Fujisaki, S.; Okamoto, T. (Technol. Rept. Yamaguchi Univ. **4** [1987] 65/9).
[49] Kajigaeshi, S.; Kakinami, T.; Moriwaki, M.; Tanaka, T.; Fujisaki, S.; Okamoto, T. (Chem. Express **3** [1988] 219/22).
[50] Kajigaeshi, S.; Kakinami, T.; Shimizu, M.; Takahashi, M.; Fujisaki, S.; Okamoto, T. (Technol. Rept. Yamaguchi Univ. **4** [1988] 139/43).

[51] Huber, B. A.; Miller, T. M. (J. Appl. Phys. **48** [1977] 1708/11).
[52] Ben-Bassat, A. A. (Israel J. Chem. **11** [1973] 781/9).
[53] Ray, S. K.; Bhattacharya, R. R. (J. Indian Chem. Soc. **13** [1936] 456/63).

16.5 Dibromochlorate(1−), $BrClBr^-$

CAS Registry Number: *[15139-65-8]*

Vapors of heated KBr (800 K) and a 300:1 gas mixture of Ar and BrCl were codeposited on a CsI window held at 15 K. A weak IR band at 282 cm^{-1} was very tentatively assigned to the antisymmetric stretching vibration of the probably linear $BrClBr^-$ ion being formed in addition to the isomeric $BrBrCl^-$ ion [1] (data are compiled in [2]). Approximate wave numbers for the two stretching modes of a hypothetical $BrClBr^-$ ion and the degrees of polarization of IR and Raman bands have been estimated [3]. The synthesis of $BrClBr^-$ salts with $(C_6H_5)_3S^+$ and $(n\text{-}C_4H_9)_4N^+$ cations was reported [4]. However, the experimental conditions used favor the formation of $BrBrCl^-$ salts. Thus, experimental and theoretical UV absorption (λ_{max} = 267.6 and 259.2 nm, respectively) as well as the charge distribution (based on semiempirical MO calculations) [4] are tentative.

References:

[1] Ault, B. S.; Andrews, L. (J. Chem. Phys. **64** [1976] 4853/9).
[2] Jacox, M. E. (J. Phys. Chem. Ref. Data **13** [1984] 945/1068, 983).
[3] Evans, J. C.; Lo, G. Y.-S. (J. Chem. Phys. **45** [1966] 1069/71).
[4] Ohkubo, K.; Yoshinaga, K. (Bull. Japan Petrol. Inst. **19** [1977] 73/80), Ohkubo, K.; Aoji, T.; Yoshinaga, K. (Bull. Chem. Soc. Japan **50** [1977] 1883/4).

16.6 Bromine Monochloride, BrCl

CAS Registry Numbers: BrCl *[13863-41-7]*, ^{82}BrCl *[100551-50-6]*, ^{80}BrCl *[70786-53-7]*, ^{79}BrCl *[80500-88-5]*, ^{77}BrCl *[97716-68-2]*, $Br^{35}Cl$ *[80500-89-6]*, $^{81}Br^{37}Cl$ *[39705-22-1]*, $^{81}Br^{35}Cl$ *[29147-86-2]*, $^{79}Br^{37}Cl$ *[39705-21-0]*, $^{79}Br^{35}Cl$ *[29147-87-3]*

Bromine monochloride, which has been known since 1826 (see "Brom" 1931, pp. 339/42), is an orange-red liquid with a sharp penetrating odor. It is prepared by mixing the elements Br_2 and Cl_2 in the gas phase or in solution. Bromine monochloride exists in equilibrium with the elements; at room temperature, about 60% of the mixed halogens are present as BrCl. It is a strong oxidant and is used as a brominating agent in organic synthesis.

Some characteristic data are given below:

molecular weight	115.36
melting point	216.7 K
boiling point	274 K
dipole moment (g)	0.52 D
density (l)	2.34 g/cm^3 (293 K)
enthalpy of formation (g)	14.64 kJ/mol (298.15 K, 0.1 MPa)
solubility in water	8.5 g/100 g H_2O (293 K)

16.6.1 Preparation. Formation

BrCl forms from mixtures of Br_2 and Cl_2 via the equilibrium $Br_2+Cl_2\rightleftharpoons 2\,BrCl$ which has already been discussed in "Bromine" Suppl. Vol. A, 1985, pp. 374/7 and 467/8. The equilibrium constants listed there for the dissociation in the vapor phase and in CCl_4 solution correspond to about 40% dissociation at room temperature. In the liquid state, the dissociation seems to be significantly less (<20%) [1, 2]. The detailed description of the preparation given in [3] is based on results of [4 to 7]: Pure BrCl is prepared by UV irradiation of a solution containing Br_2 and excess Cl_2 in CCl_2F_2 at −79 °C. After equilibration, visible by a color change from black to orange-red, BrCl is precipitated after cooling to −120 to −100 °C. The solvent and the unreacted Cl_2 are pumped off from the product at the same temperature. Following sublimation at −79 °C, crystalline, orange BrCl is obtained which is stable in a closed vessel at this temperature [3].

A hydrochloric acid solution of BrCl is useful as titrant for analytic purposes. It is prepared by reacting stoichiometric quantities of $KBrO_3$ and KBr in a hydrochloric acid medium via

$$KBrO_3+2\,KBr+6\,HCl\rightarrow 3\,BrCl+3\,KCl+3\,H_2O$$

The BrCl solution is rather stable; after storing it for three months in the dark, less than 3 to 5% BrCl were lost [8 to 10].

When Cl^- ions are added to an acidic HBrO solution, BrCl is formed according to

$$HBrO+H^++Cl^-\rightleftharpoons BrCl+H_2O \text{ [11, 12]}$$

The formation constant is 3.16×10^4 L^2/mol^2 at 293 K [12].

$HgCl_2$ in aqueous solution reacts with Br_2 to give HgBrCl and BrCl which is separated by distillation [13].

The gas-phase reaction between HBr and Cl_2, which involves the formation and reaction of BrCl according to

$$Cl_2+HBr\rightarrow BrCl+HCl,\ BrCl+HBr\rightarrow Br_2+HCl,\ \text{and}\ Cl_2+Br_2\rightleftharpoons 2\,BrCl,$$

is discussed in "Bromine" Suppl. Vol. B 1, 1990, pp. 340 and 361.

The intermediate formation of BrCl in the oxidation of Br^- with Cl_2, e.g., in the commercial production of Br_2 by air desorption from natural brines or in eutectic melts of $ZnCl_2$-KCl or LiCl-KCl mixtures, is described in "Bromine" Suppl. Vol. A, 1985, pp. 9/12 and in "Bromine" Suppl. Vol. B1, 1990, pp. 481/2, respectively.

For the formation of BrCl from atomic or molecular bromine and Cl_2O, ClO_2, or ClNO in flow systems, the reader is referred to "Bromine" Suppl. Vol. A, 1985, pp. 414/5.

BrCl formation as a possible channel in the reaction of BrO with ClO was studied; implications for the stratospheric ozone depletion were discussed [14, 15].

For spectroscopic studies, BrCl is generally generated via the Br_2+Cl_2 or Br_2+ClO_2 reactions; see Section 16.6.4.

References:

[1] Mills, J. F.; Schneider, J. A. (Ind. Eng. Chem. Prod. Res. Develop. **12** [1973] 160/5).
[2] Mills, J. F. (in: Johnson, J. D., Disinfection: Water Wastewater, Ann Arbor Sci., Stoneham, Mass., 1975, pp. 113/43, 119).
[3] Huber, F.; Schmeisser, M. (in: Brauer, G., Handbuch der Präparativen Anorganischen Chemie, Vol. I, Enke, Stuttgart 1975, pp. 302/3).
[4] Lux, H. (Ber. Deut. Chem. Ges. **63** [1930] 1156/8).
[5] Popov, I.; Mannion, J. J. (J. Am. Chem. Soc. **74** [1952] 222/4).
[6] Schmeisser, M.; Taglinger, L. (Chem. Ber. **94** [1961] 1533/9).
[7] Schmeisser, M.; Tytko, K. H. (Z. Anorg. Allgem. Chem. **403** [1974] 231/42).
[8] Schulek, E.; Burger, K. (Talanta **1** [1958] 147/52).
[9] Schulek, E.; Burger, K. (Talanta **1** [1958] 219/23).
[10] Schulek, E.; Burger, K. (Magy. Tud. Akad. Kem. Tud. Oszt. Kozl. **12** [1959] 90/14).

[11] Derbyshire, D. H.; Waters, W. A. (J. Chem. Soc. **1950** 564/73).
[12] Voudrias, E. A.; Reinhard, M. (Environ. Sci. Technol. **22** [1988] 1049/56).
[13] Schulek, E.; Burger, K. (Talanta **7** [1960] 41/5).
[14] Toohey, D. W.; Anderson, J. G. (J. Phys. Chem. **92** [1988] 1705/8).
[15] Salawitch, R. J.; Wofsy, S. C.; McElroy, M. B. (Planet. Space Sci. **36** [1988] 213/24).

16.6.2 Toxicity. Handling. Uses

Liquid BrCl rapidly attacks the skin and other tissues producing irritation and burns; even at low concentrations, BrCl vapor is highly irritating and painful to the respiratory tract. The threshold limit value (TLV) for Br_2 of 0.1 ppm is the generally accepted limit of BrCl in air [1, 2].

BrCl is preferably stored in vessels containing a dip pipe, whereby the liquid can be removed under its own pressure (about 2 atm at 25 °C). Dry BrCl is less corrosive than Br_2 and is stored and shipped (classified as a corrosive liquid) in steel containers [1 to 3]. Highly resistant plastics, such as Kynar®, Teflon®, and Viton®, are preferred over PVC, rubber, or ABS plastics [1, 4].

BrCl is used in organic synthesis to brominate aromatics and to be added across double bonds; see e.g. [2]. Schulek and Burger [5 to 7] introduced standard solutions of BrCl into analysis. BrCl is environmentally less hazardous than Cl_2 which makes chlorobromination an attractive alternative to chlorination for wastewater disinfection [4, 8, 9]. The performance and safety of $Li/SOCl_2$ electrochemical cells is improved by using BrCl as an additive (so-called BCX cell); see e.g. [10].

References:

[1] Mills, J. F. (in: Johnson, J. D., Disinfection: Water Wastewater, Ann Arbor Sci., Stoneham, Mass., 1975, pp. 113/43, 119).
[2] Mills, J. F.; Schneider, J. A. (Ind. Eng. Chem. Prod. Res. Develop. **12** [1973] 160/5).
[3] Mills, J. F.; Oakes, B. D. (Chem. Eng. **80** [1973] 102/6).
[4] Mills, J. F. (in: Rubin, A. J., Chem. Wastewater Technol., Ann Arbor Sci., Stoneham, Mass., 1978, pp. 199/212, 203/212).
[5] Schulek, E.; Burger, K. (Talanta **1** [1958] 224/37).

[6] Burger, K.; Schulek, E. (Talanta **7** [1960] 46/50).
[7] Burger, K.; Schulek, E. (Ann. Univ. Sci. Budapest Rolando Eotvos Nominatae Sect. Chim. **2** [1960] 133/8).
[8] Mills, J. F. (Power **124** [1980] 127/9).
[9] Chiesa, R.; Geary, D. (Proc. Intern. Water Conf. Eng. Soc. West. Pa. **46** [1985] 414/26).
[10] Abraham, K. M.; Alamgir, M.; Perrotti, S. J. (J. Electrochem. Soc. **135** [1988] 2686/91).

16.6.3 The BrCl Molecule

16.6.3.1 Electron Configuration

16.6.3.1.1 Electronic Ground State

According to MO theory, the ground state X $^1\Sigma^+$ of the 52-electron molecule correlating with the ground-state atoms, $Br(^2P_{3/2})+Cl(^2P_{3/2})$, is represented by the electron configuration {KLM} $(10\sigma)^2$ $(11\sigma)^2$ $(12\sigma)^2$ $(5\pi)^4$ $(6\pi)^4$ [1] or, using Mulliken's notation [2, pp. 346/8], {KLM} $(z\sigma)^2$ $(y\sigma)^2$ $(x\sigma)^2$ $(w\pi)^4$ $(v\pi)^4$. {KLM} stands for the core orbitals 1σ to 9σ, 1π to 4π, and 1δ arising from the K, L, and M shells of the Br atom and from the K and L shells of the Cl atom. The valence orbitals are essentially the Br4s and Cl3s AO's, the Br4pσ-Cl3pσ bonding orbital, and the bonding and antibonding Br4pπ-Cl3pπ combinations, respectively. The energetical ordering of the highest three MO's is predicted to be $12\sigma<5\pi<6\pi$ by the more rigorous quantum chemical calculations. The photoelectron spectrum of BrCl, which could not decide whether $12\sigma<5\pi$ or $5\pi<12\sigma$, was interpreted by using an ab initio SCF MO calculation (see pp. 184/5).

The ground state X $^1\Sigma^+$ has been identified with the lower state of the absorption and emission systems B $^3\Pi(0^+)\leftrightarrow$X $^1\Sigma^+$ in the visible and near IR and with the lower state of a few UV absorption systems (see pp. 203/10), the analyses of which gave a number of ground-state properties.

A few quantum chemical ab initio calculations within the Hartree-Fock approximation [1, 3 to 5] or including electron correlation [6, 7], two pseudopotential calculations [8, 9], a relativistic Hartree-Fock-Slater calculation [10], and some 20 semiempirical calculations [11 to 36] deal with the electronic ground state giving total molecular and orbital energies, charge distribution and/or dipole moments, bond properties, spectroscopic and other molecular constants. The lowest value for the total molecular energy obtained with an extended STO basis set, $E_T=-3031.85918$ au at an optimized internuclear distance $r_e=2.1372$ Å, is the near Hartree-Fock limit [1], whereas a polarization configuration interaction calculation ("Full POL-CI") with an extended GTO basis set gave only $E_T=-3031.816883$ au at $r_e=2.2384$ Å [6].

Properties of the BrCl bond, especially its ionic character ($i\approx6$ to 10%), were derived also empirically and discussed in connection with the chlorine and bromine nuclear quadrupole coupling constants, electronegativity differences, and dipole moments [37 to 48]. A bond order $n=1.07$ resulted from the force constant [49].

References:

[1] Straub, P. A.; McLean, A. D. (Theor. Chim. Acta **32** [1974] 227/42).
[2] Herzberg, G. (Molecular Spectra and Molecular Structure, Vol. 1, Spectra of Diatomic Molecules, Van Nostrand, Princeton, N. J., 1961).
[3] Dunlavey, S. J.; Dyke, J. M.; Morris, A. (J. Electron Spectrosc. Relat. Phenom. **12** [1977] 259/63).
[4] Andzelm, J.; Klobukowski, M.; Radzio-Andzelm, E. (J. Computat. Chem. **5** [1984] 146/61).

[5] Dobbs, K. D.; Hehre, W. J. (J. Computat. Chem. **7** [1986] 359/78).
[6] Eades, R. A. (Diss. Univ. Minnesota 1983, pp. 1/214; Diss. Abstr. Intern. B **44** [1984] 3418).
[7] Kucharski, S. A.; Noga, J.; Bartlett, R. J. (J. Chem. Phys. **88** [1988] 1035/40).
[8] Ewig, C. S.; Van Wazer, J. R. (J. Chem. Phys. **63** [1975] 4035/41).
[9] Hyde, R. G.; Peel, J. B. (J. Chem. Soc. Faraday Trans. II **72** [1976] 571/8).
[10] Dyke, J. M.; Josland, G. D.; Snijders, J. G.; Boerrigter, P. M. (Chem. Phys. **91** [1984] 419/24).

[11] Grodzicki, M.; Lauer, S.; Trautwein, A. X.; Vera, A. (Advan. Chem. Ser. No. 194 [1981] 3/37).
[12] Grodzicki, M.; Männing, V.; Trautwein, A. X.; Friedt, J. M. (J. Phys. B **20** [1987] 5595/625).
[13] Bowmaker, G. A.; Boyd, P. D. W. (J. Mol. Struct. **150** [1987] 327/44).
[14] Cheesman, G. H.; Finney, A. J. T.; Snook, I. K. (Theor. Chim. Acta **16** [1970] 33/42).
[15] Bhattacharyya, S. P.; Chowdhury, M. (J. Phys. Chem. **81** [1977] 1602/4).
[16] Bhattacharyya, S. P. (Indian J. Chem. A **16** [1978] 4/6).
[17] Deb, B. M.; Coulson, C. A. (J. Chem. Soc. A **1971** 958/70).
[18] Dewar, M. J. S.; Healy, E. (J. Computat. Chem. **4** [1983] 542/51).
[19] Dewar, M. J. S.; Zoebisch, E. G. (J. Mol. Struct. **180** [1988] 1/21).
[20] Hase, H. L.; Schweig, A. (Theor. Chim. Acta **31** [1973] 215/20).

[21] Rhee, C. H.; Metzger, R. M.; Wiygul, F. M. (J. Chem. Phys. **77** [1982] 899/915).
[22] Scharfenberg, P. (Z. Chem. [Leipzig] **17** [1977] 388/9).
[23] Scharfenberg, P. (Theor. Chim. Acta **49** [1978] 115/22).
[24] Scharfenberg, P. (Theor. Chim. Acta **67** [1985] 235/43).
[25] Spurling, T. H.; Winkler, D. A. (Australian J. Chem. **39** [1986] 233/7).
[26] Sichel, J. M.; Whitehead, M. A. (Theor. Chim. Acta **11** [1968] 220/38).
[27] Sichel, J. M.; Whitehead, M. A. (Theor. Chim. Acta **11** [1968] 239/53).
[28] Sichel, J. M.; Whitehead, M. A. (Theor. Chim. Acta **11** [1968] 254/62).
[29] Sichel, J. M.; Whitehead, M. A. (Theor. Chim. Acta **11** [1968] 263/70).
[30] Boyd, R. J. (Diss. McGill Univ., Montreal, Canada, 1970 from Cornford, A. B.; Diss. Univ. Brit. Columbia, Canada, 1972, pp. 1/169, 71; Diss. Abstr. Intern. B **33** [1972] 2541).

[31] Wiebenga, E. H.; Kracht, D. (Inorg. Chem. **8** [1969] 738/46).
[32] Gázquez, J. L.; Ortiz, E. (J. Chem. Phys. **81** [1984] 2741/8).
[33] Kang, Y. K. (Bull. Korean Chem. Soc. **6** [1985] 107/11).
[34] Pohl, H. A.; Raff, L. M. (Intern. J. Quantum Chem. **1** [1967] 577/89).
[35] Lippincott, E. R. (J. Chem. Phys. **26** [1957] 1678/85).
[36] Iczkowski, R. (J. Am. Chem. Soc. **86** [1964] 2329/32).
[37] Gordy, W. (J. Chem. Phys. **19** [1951] 792/3).
[38] Gordy, W. (J. Chem. Phys. **22** [1954] 1470/1).
[39] Gordy, W. (Discussions Faraday Trans. No. 19 [1955] 14/29).
[40] Dailey, B. P. (J. Phys. Chem. **57** [1953] 490/6).

[41] Dailey, B. P.; Townes, C. H. (J. Chem. Phys. **23** [1955] 118/23).
[42] Wilmshurst, J. K. (J. Chem. Phys. **30** [1959] 561/5).
[43] Wilmshurst, J. K. (J. Chem. Phys. **33** [1960] 813/20).
[44] Bak, B.; Hansen-Nygaard, L. (Z. Elektrochem. **61** [1957] 895/900).
[45] Whitehead, M. A. (J. Chem. Phys. **34** [1961] 2204).
[46] Whitehead, M. A.; Jaffé, H. H. (Trans. Faraday Soc. **57** [1961] 1854/62).
[47] Whitehead, M. A.; Jaffé, H. H. (Theor. Chim. Acta **1** [1962/63] 209/21).
[48] Narayana, K. L.; Santhamma, C. (Indian J. Phys. **37** [1963] 261/74).
[49] Siebert, H. (Z. Anorg. Allgem. Chem. **274** [1953] 34/6).

16.6.3.1.2 Excited States

Valence States. Excitations from the highest occupied into the lowest unoccupied MO give rise to the states ... $(12\sigma)^2$ $(5\pi)^4$ $(6\pi)^3$ $(13\sigma)^1$ $^3\Pi_i$ (2, 1, 0^+) and $^1\Pi(1)$ (notation for (Λ, S) and, in parentheses, (Ω, ω) coupling) which correlate with the neutral atoms, $Br(^2P_{3/2,\,1/2})+Cl(^2P_{3/2,1/2})$. The triplet states have been labeled A′ $^3\Pi(2)$, A $^3\Pi(1)$, and $B^3\Pi(0^+)$.

The bound B $^3\Pi(0^+)$ state has been experimentally observed and well characterized by detailed studies of the B $^3\Pi(0^+)\leftrightarrow X$ $^1\Sigma^+$ absorption, excitation, and emission spectra and by lifetime measurements for B-state rovibrational levels (see pp. 203/8 and 196/9). These experiments show that the B state which is expected to correlate with a ground-state $Br(^2P_{3/2})$ and a spin-orbit excited $Cl(^2P_{1/2})$ atom predissociates into two ground-state atoms, $Br(^2P_{3/2})+Cl(^2P_{3/2})$, in the v′=5 to 7 levels; predissociation was assumed to result from curve crossing with a repulsive 0^+ state that correlates with the ground-state atoms. By allowing for this perturbation and applying a non-linear least-squares technique, Coxon [1] derived the term values $T_e=16879.91\pm0.09$ and 16881.19 ± 3.56 cm^{-1} for $^{79}Br^{35}Cl$ and $^{81}Br^{35}Cl$, respectively, from the B←X absorption spectrum. Using Coxon's [1] method, McFeeters et al. [2] obtained similar T_e values from the B→X emission spectrum. Analyses without consideration of the B $^3\Pi(0^+)$-$0^+(Br(^2P_{3/2})+Cl(^2P_{3/2}))$ interaction resulted in somewhat lower term values [3 to 5].

The A $^3\Pi(1)$ state has been tentatively identified with the lower state of a weak ultraviolet emission system β $^3\Pi(1)\rightarrow$A $^3\Pi(1)$ appearing in the stronger D′ $^3\Pi(2)\rightarrow$A′ $^3\Pi$ system (see below); a term value of $T_e=15030(100)$ cm^{-1} (standard deviation in parentheses) for $^{79}Br^{35}Cl$ has been derived [6]. Several bands at the long-wavelength end of the B←X absorption spectrum have been tentatively assigned to the A←X transition but not analyzed [1].

The A′ $^3\Pi(2)$ state has been identified with the lower state of the ultraviolet emission system D′ $^3\Pi(2)\rightarrow$A′ $^3\Pi(2)$ (see p. 210), from which the term value $T_e=14690(100)$ cm^{-1} (standard deviation in parentheses) for $^{79}Br^{35}Cl$ followed [7].

For the $^3\Pi_i$ state without spin-orbit splitting, a configuration interaction (Full POL-CI) calculation has been carried out giving its potential energy function and spectroscopic constants [8].

The $^1\Pi(1)$ state was predicted to be repulsive by the Full POL-CI calculation [8].

Full POL-CI calculations for a number of further excited valence states, 2 $^1\Pi$, 2 $^3\Pi$ and 2 $^1\Sigma^+$, 1 $^3\Sigma^+$, 2 $^3\Sigma^+$, 1 $^1\Sigma^-$, 1 $^3\Sigma^-$, 1 $^1\Delta$, 1 $^3\Delta$ (presumably arising from 13σ←5π and 13σ←12σ and/or 7π←6π and/or $(13\sigma)^2\leftarrow(6\pi)^2$ excitations, see [9, pp. 335/7]) showed these all to be repulsive and to dissociate into $Br(^2P)+Cl(^2P)$ (spin-orbit coupling neglected) [8].

Ion-Pair States. By comparison with other diatomic interhalogens the lowest group of ion-pair states correlating with $Br^+(^3P_2)+Cl^-(^1S_0)$ is expected to consist of the three states D′ $^3\Pi(2)$, β $^3\Pi(1)$, and E $^3\Pi(0^+)$. The D′ and β states have been identified with the upper states of two ultraviolet emission systems, D′ $^3\Pi(2)\rightarrow$A′ $^3\Pi(2)$ and β $^3\Pi(1)\rightarrow$A $^3\Pi(1)$, appearing around 314 nm, the E state has been identified with the upper state of the nearby E $^3\Pi(0^+)\rightarrow$B $^3\Pi(0^+)$ system (see p. 211) [6, 7]; term values (standard deviation in parentheses) $T_e(D')=48410(100)$ cm^{-1}, $T_e(E)=48759(4)$ cm^{-1} [7], and $T_e(\beta)=48750(100)$ cm^{-1} [6] were derived for $^{79}Br^{35}Cl$, $T_0(E)=48854.29(5)$ cm^{-1} (zero level of E relative to the minimum of the ground state) for $^{81}Br^{37}Cl$ [10]. The zero level of the β state was predicted to lie 24(3) cm^{-1} below that of the E state [10].

The E(0^+) state and (higher by 3840 cm^{-1}) the first member of the $Br^+(^3P_0)+Cl^-(^1S_0)$ ion-pair cluster, f(0^+), were identified with the upper states of transitions observed in the vacuum-UV absorption and excitation spectra and in the vacuum-UV and UV fluorescence spectra by Hopkirk et al. [11].

Rydberg States. The first members of the $\{BrCl^+\}$ ns Rydberg series, that are the states $\{BrCl^+, X\ ^2\Pi_{3/2}\}$ 5sσ $^{1,3}\Pi$ and $\{BrCl^+, X\ ^2\Pi_{1/2}\}$ 5sσ $^{1,3}\Pi$, were identified with the upper states of two absorption systems, $a_5\leftarrow X$ and $b_5\leftarrow X$, in the vacuum UV [11]. These are identical with the much earlier observed C←X and D←X band systems [12] which had given term values $T_e(C)=59325\ cm^{-1}$ and $T_e(D)=61570\ cm^{-1}$ [13]. The first members of the np series, $\{BrCl^+, X\ ^2\Pi_{3/2}\}$ 5pσ $^{1,3}\Pi$ and $\{BrCl^+, X\ ^2\Pi_{1/2}\}$ 5pσ $^{1,3}\Pi$, were calculated to lie at 69248 and 71318 cm^{-1} [11].

References:

[1] Coxon, J. A. (J. Mol. Spectrosc. **50** [1974] 142/65).

[2] McFeeters, B. D.; Perram, G. P.; Crannage, R. P.; Dorko, E. A. (Chem. Phys. **139** [1989] 347/57).

[3] Hadley, S. G.; Bina, M. J.; Brabson, G. D. (J. Phys. Chem. **78** [1974] 1833/6).

[4] Clyne, M. A. A.; Coxon, J. A. (J. Chem. Soc. Chem. Commun. **1966** 285/6).

[5] Clyne, M. A. A.; Coxon, J. A. (Proc. Roy. Soc. [London] A **298** [1967] 424/52).

[6] Chakraborty, D. K. (Diss. Vanderbilt Univ. 1987, pp. 1/189, 1/12, 76/84; Diss. Abstr. Intern. B **48** [1987] 772).

[7] Chakraborty, D. K.; Tellinghuisen, P. C.; Tellinghuisen, J. (Chem. Phys. Letters **14** [1987] 36/40).

[8] Eades, R. A. (Diss. Univ. Minnesota 1983, pp. 1/214, 73/5, 79/81, 90, 104/5, 132/5; Diss. Abstr. Intern. B **44** [1984] 3418).

[9] Herzberg, G. (Molecular Spectra and Molecular Structure, Vol. 1, Spectra of Diatomic Molecules, Van Nostrand, Princeton, N. J., 1961).

[10] Brown, S. W.; Dowd, C. J., Jr.; Tellinghuisen, J. (J. Mol. Spectrosc. **132** [1988] 178/92).

[11] Hopkirk, A.; Shaw, D.; Donovan, R. J.; Lawley, K. P.; Yencha A. J. (J. Phys. Chem. **93** [1989] 7338/42).

[12] Cordes, H.; Sponer, H. (Z. Physik **79** [1932] 170/85).

[13] Huber, K. P.; Herzberg, G. (Molecular Spectra and Molecular Structure, Vol. 4, Constants of Diatomic Molecules, Van Nostrand Reinhold, New York 1979, pp. 108/9).

16.6.3.2 Ionization Potentials

An upper limit for the first vertical ionization potential, $E_i(vert)=11.1\pm0.2$ eV, has been estimated from the appearance potential of $BrCl^+$ measured by electron impact mass spectrometry [1].

In the He I photoelectron (PE) spectrum of BrCl ($Br_2+Cl_2\leftrightarrow2\ BrCl$ equilibrium) three bands were observed and assigned to the ionizations of the three uppermost orbitals on the basis of ab initio SCF MO calculations (Koopmans' theorem, $E_i=-\varepsilon_i$) as given in Table 22, p. 185 [2] (results from a relativistic Hartree-Fock-Slater calculation [4], which is able to predict the spin-orbit splitting of the ionic $^2\Pi$ states, are included in the table).

The first PE band shows the spin-orbit splitting of the ionic ground state, $\Delta E=2070\pm30$ cm^{-1}; each component shows vibrational structure corresponding to ionic vibrations which is consistent with the removal of an antibonding 6π electron [2]. By comparison with other diatomic halogens and interhalogens and using atomic E_i's, the vertical ionization potentials

Table 22
BrCl, Ionization Potentials.

MO	ionic state	E_i(vert) in eV observed [2]	calc. [2][a)]	calc. [3][b)]	calc. [4][c)]
6π	$^2\Pi_{3/2}$	11.012±0.005	9.75	11.30	10.12
	$^2\Pi_{1/2}$	11.271±0.005			10.35
5π	$^2\Pi_{3/2}$	13.70	12.98	14.48	12.71
	$^2\Pi_{1/2}$				12.81
12σ	$^2\Sigma^+$	15.27	13.19	15.18	15.07
11σ	$^2\Sigma^+$	–	23.82	26.35	22.93

a) Minimum STO basis. – b) Extended STO basis; near Hartree-Fock limit. – c) Relativistic HFS calculation.

E(6π) = 10.9 and 11.3 eV (mean value 11.1 ± 0.2 eV, ΔE = 2275 cm^{-1}), E_i(5π) = 13.75 ± 0.4 eV, and E_i(12σ) = 15.5 ± 0.4 eV were predicted [5].

Further theoretical E_i (= $-\varepsilon_i$) values for the valence electrons have been derived by two pseudopotential and a number of semiempirical calculations; see references [8, 9] and [12, 15 to 18, 20, 27, 30] of Section 16.6.3.1.1.

The near Hartree-Fock calculation of [3] also gives E_i values for all core electrons.

References:

[1] Irsa, A. P.; Friedman, L. (J. Inorg. Nucl. Chem. **6** [1958] 77/90).
[2] Dunlavey, S. J.; Dyke, J. M.; Morris, A. (J. Electron Spectrosc. Relat. Phenom. **12** [1977] 259/63).
[3] Straub, P. A.; McLean, A. D. (Theor. Chim. Acta **32** [1974] 227/42).
[4] Dyke, J. M.; Josland, G. D.; Snijders, J. G.; Boerrigter, P. M. (Chem. Phys. **91** [1984] 419/24).
[5] Cornford, A. B. (Diss. Univ. Brit. Columbia, Canada, 1972, pp. 1/169, 68/9, 71, 88; Diss. Abstr. Intern. B **33** [1972] 2541).

16.6.3.3 Dipole Moment. Quadrupole Moment

The **dipole moment** |μ| = 0.519(4) D (two standard deviations in parentheses) for the vibronic ground state was derived from Stark effect measurements on nine hyperfine components of the J = 1←0, v = 0 transition in the microwave spectrum of $^{79}Br^{35}Cl$ and $^{81}Br^{35}Cl$ [1, 2]. The older microwave value, |μ| = 0.57 ± 0.02 D [3, 4], was estimated from J = 1←0, v = 0 Stark effect data neglecting quadrupole interactions. Studies of molecular-beam electric resonance on J = 1, F_1 (= J + I_{Br}) = 5/2, 3/2, 1/2 levels of BrCl (equilibrated Br_2-Cl_2 mixture in Ar carrier gas) resulted in |μ| = 0.5237(4) D (standard deviation) [5]. The polarization Br^+Cl^- is expected according to quantum chemical calculations (see below) and electronegativity arguments.

A great number of the quantum chemical studies on BrCl include the calculation of μ. In reasonable agreement with the experimental value are the results of a many-body perturbation calculation, μ = 0.564 D [6], of the most rigorous ab initio SCF MO (near Hartree-

Fock limit), $\mu = 0.588$ D [7], and of another ab initio SCF MO calculation with an extended GTO basis set, $\mu = 0.524$ D [8]. Further ab initio calculations [8], a pseudopotential calculation [9], and various semiempirical calculations (see references [11 to 14, 17, 18, 20 to 25, 28, 33] of Section 16.6.3.1.1) result in values between 0.03 and 1.24 D.

For the **molecular quadrupole moment** experimental data are not available. The following different values were obtained from quantum chemical calulations: Θ (in 10^{-26} esu·cm^2) = 0.3520 (ab initio SCF MO calculation, near Hartree-Fock limit) [7], 0.799 (pseudopotential calculation) [9], and −0.197 (SCF MS Xα calculation) [10].

References:

[1] Nair, K. P. R.; Hoeft, J.; Tiemann, E. (Chem. Phys. Letters **58** [1978] 153/6).
[2] Nair, K. P. R. (Kem. Kozlem. **52** [1979] 431/50).
[3] Smith, D. F.; Tidwell, M.; Williams, D. V. P. (Phys. Rev. [2] **79** [1950] 1007/8).
[4] Lovas, F. J.; Tiemann, E. (J. Phys. Chem. Ref. Data **3** [1974] 609/769).
[5] Sherrow, S. A. (LBL-16545 [1983] 1/96, 10/25; C.A. **100** [1984] No. 129056).
[6] Kucharski, S. A.; Noga, J.; Bartlett, R. J. (J. Chem. Phys. **88** [1988] 1035/40).
[7] Straub, P. A.; McLean, A. D. (Theor. Chim. Acta **32** [1974] 227/42).
[8] Andzelm, J.; Klobukowski, M.; Radzio-Andzelm, E. (J. Computat. Chem. **5** [1984] 146/61).
[9] Hyde, R. G.; Peel, J. B. (J. Chem. Soc. Faraday Trans. II **72** [1976] 571/8).
[10] Bowmaker, G. A.; Boyd, P. D. W. (J. Mol. Struct. **150** [1987] 327/44).

16.6.3.4 Polarizability

The average molecular polarizability $\bar{\alpha} = (\alpha_{\parallel} + 2\alpha_{\perp})/3 = 5.743$ Å^3, the parallel and perpendicular components $\alpha_{\parallel} = 9.299$ Å^3, $\alpha_{\perp} = 3.965$ Å^3, and the anisotropy $k = 0.310$ with $k^2 = [(\alpha_{\parallel} - \bar{\alpha})^2 + 2(\alpha_{\perp} - \bar{\alpha})^2]/6\bar{\alpha}^2$ result from a perturbation calculation within the CNDO approximation [1].

Using IR intensity data of the fundamental absorption band and its first overtone [2], the atomic polarization $P_A = 0.018$ cm^3 (0.018 cm^3) [3, 4] and the atomic polarizability $\alpha_A = 0.00746$ Å^3 (0.0073 Å^3) at 300 K and 1 atm [4] have been calculated in the anharmonic oscillator-rigid rotor (harmonic oscillator-rigid rotor) approximation.

References:

[1] Rhee, C. H.; Metzger, R. M.; Wiygul, F. M. (J. Chem. Phys. **77** [1982] 899/915).
[2] Brooks, W. V. F.; Crawford, B., Jr. (J. Chem. Phys. **23** [1955] 363/5).
[3] Illinger, K. H.; Smith, C. P. (J. Chem. Phys. **32** [1960] 787/98).
[4] Illinger, K. H. (J. Chem. Phys. **35** [1961] 409/20).

16.6.3.5 Nuclear Quadrupole Coupling Constants eqQ

The hyperfine splitting of the $J = 1 \leftarrow 0$, $v = 0$ microwave transitions of all four isotopic species of BrCl gave the following values [1] (quoted also in the microwave spectral tables of [2]):

$$eqQ(^{79}Br) = +876.8 \pm 0.9 \text{ MHz} \qquad eqQ(^{35}Cl) = -103.6 \pm 0.1_5 \text{ MHz}$$

$$eqQ(^{81}Br) = +732.9 \pm 0.5 \text{ MHz} \qquad eqQ(^{37}Cl) = -\ 81.1_4 \pm 0.1_5 \text{ MHz}$$

Refits of the microwave data for $^{79}Br^{35}Cl$ reported by Smith et al. [1] and Nair et al. [3] resulted in somewhat lower values for both eqQ(Br) and eqQ(Cl) obtained by six different fitting procedures, the most reliable of which gives [4]:

$$eqQ(^{79}Br) = +875.008 \pm 0.131\ \text{MHz} \qquad eqQ(^{35}Cl) = -102.269 \pm 0.110\ \text{MHz}$$

Preliminary studies of molecular-beam electric resonance on J=1, F_1 $(=J+I_{Br})=5/2$, 3/2, 1/2 levels of BrCl (equilibrated Br_2-Cl_2 mixture in Ar carrier gas) resulted in $eqQ(^{35}Cl) = -102.294 \pm 0.099$ MHz; transitions with $\Delta F \neq 0$ $(F=J+I_{Br}+I_{Cl})$ but $\Delta F_1=0$ have only been analyzed which did not allow to derive the bromine quadrupole coupling constant [4].

A few theoretical values for the electric field gradients q at the Br and Cl nuclei and/or the quadrupole coupling constants result from ab initio SCF MO and semiempirical calculations; see references [1] and [11, 13, 15 to 17, 29] of Section 16.6.3.1.1.

References:

[1] Smith, D. F.; Tidwell, M.; Williams, D. V. P. (Phys. Rev. [2] **79** [1950] 1007/8).
[2] Lovas, F. J.; Tiemann, E. (J. Phys. Chem. Ref. Data **3** [1974] 609/769).
[3] Nair, K. P. R.; Hoeft, J.; Tiemann, E. (Chem. Phys. Letters **58** [1978] 153/6).
[4] Sherrow, S. A. (LBL-16545 [1983] 1/96, 10/25; C.A. **100** [1984] No. 129056).

16.6.3.6 Rotational and Vibrational Constants. Internuclear Distances. Mean Amplitudes of Vibration

Electronic Ground State. Analyses of the microwave absorption spectra in the region of the J=1←0 and J=14←13 to 34←33 transitions, of the high-resolution IR spectrum in the region of the fundamental band, and analyses of the B $^3\Pi(0^+)\leftrightarrow X\ ^1\Sigma^+$ absorption and emission spectra in the visible and near-IR region gave for each isotopic species, $^{79}Br^{35}Cl$, $^{81}Br^{35}Cl$, $^{79}Br^{37}Cl$, and $^{81}Br^{37}Cl$, the rotational constant B_e and centrifugal stretching constants D_e, H_e, L_e, the corresponding rotation-vibration interaction constants α_e, γ_e, β_e, δ_e, and the vibrational constants ω_e, $\omega_e x_e$, $\omega_e y_e$. The equilibrium internuclear distance r_e was obtained by converting the rotational constant. The results derived from the microwave spectra are given in Table 23, p. 188, those from the IR and near-IR-visible spectra in Table 24, p. 189. A few explanatory remarks and additional references are given below the tables.

Using the atomic weights of Br and Cl and the fundamental frequency $\omega_e = 440\ \text{cm}^{-1}$, the mean amplitudes of vibration u have been calculated for T=0 to 1000 K at 100 K intervals: u=0.0395, 0.0446, 0.0528, and 0.0604 Å at T=0, 300, 500, and 1000 K, respectively [13]; using $\omega_e = 430\ \text{cm}^{-1}$, the amplitudes u=0.0458 and 0.0538 Å at T=298.16 and 500 K were obtained [14].

A few quantum chemical calculations deal with the derivation of ground-state spectroscopic constants. Results from polarization configuration interaction calculations (Full POL-CI) by use of extended GTO basis sets for Br and Cl are not very satisfactory [15]. Ab initio SCF MO and pseudopotential calculations of r_e, ω_e and semiempirical calculations of r_e are also available, see references [1, 4, 5, 8] and [17, 18, 20, 22 to 25, 34] of Section 16.6.3.1.1.

Table 23
BrCl, Ground State X $^1\Sigma^+$. Rotational and Vibrational Constants and Internuclear Distance from Microwave Spectra.

	$^{79}Br^{35}Cl$	$^{81}Br^{35}Cl$	$^{79}Br^{37}Cl$	$^{81}Br^{37}Cl$	$^{79}Br^{35}Cl$	$^{81}Br^{35}Cl$	$^{79}Br^{37}Cl$	$^{81}Br^{37}Cl$
B_e in MHz	4571.0295(26)	4536.3793(25)	4399.9082(25)	4365.2527(25)	4570.92(4)	4536.14(4)	[4499.84(4)]	4365.01(4)
α_e in MHz	23.143(3)	22.880(3)	21.855(3)	21.598(3)	23.22(3)	22.95(3)	21.94(3)	21.67(3)
γ_e in kHz	−61.9(18)	−61.0(18)	−57.4(17)	−56.5(16)	−	−	−	−
D_e in kHz	2.1527(7)	2.1202(7)	1.9945(7)	1.9632(7)	−	−	−	−
β_e in Hz	−10.7(8)	−10.5(8)	−9.7(8)	−9.5(8)	−	−	−	−
ω_e in cm^{-1}	444.322(62)	442.634(62)	435.926(61)	434.206(61)	−	−	−	−
$\omega_e x_e$ in cm^{-1}	1.856(20)	1.849(20)	1.787(19)	1.771(20)	−	−	−	−
r_e in Å	2.136124(5)	2.136124(5)	2.136124(5)	2.136124(5)	2.138(10)	2.138(10)	2.138(10)	2.138(10)
remark	a)	a)	a)	a)	b)	b)	b)	b)
Ref.	[2, 3]	[2, 3]	[2, 3]	[2, 3]	[4]	[4]	[4]	[4]

a) Analysis of the $J=13\leftarrow12$ to $34\leftarrow33$, $v=0$ to 3 transitions using Dunham's [1] energy level equation $E(v, J)=\Sigma_{ik} Y_{ik}(v+\frac{1}{2})^i\cdot J^k(J+1)^k$; the values given for α_e, γ_e, D_e, and β_e are the Dunham coefficients $-Y_{11}$, Y_{21}, $-Y_{02}$, and Y_{12}, respectively (the signs of the interaction constants follow the definitions given in "Bromine" Suppl. Vol. A, 1985, p. 73); further constants derived are: $Y_{31}=-0.9(3)$, $-0.9(3)$, $-0.8(3)$, $-0.8(3)$ kHz, $10^4\cdot Y_{03}$ ($\approx 10^4\cdot H_e$) $=-4.656(3)$, $-4.551(3)$, $-4.152(3)$, $-4.055(3)$ Hz, $10^5\cdot Y_{13}$ ($\approx -10^5\cdot\delta_e$) $=-3.0(2)$, $-2.9(2)$, $-2.6(2)$, $-2.6(2)$ Hz, $10^{10}\cdot Y_{04}$ ($\approx 10^{10}\cdot L_e$) $=9.3(8)$, 9.0(8), 8.0(8), 7.7(7) Hz, and $Y_{00}=11.3(7)$, 11.2(7), 10.9(7), 10.8(7) GHz for $^{79}Br^{35}Cl$, $^{81}Br^{35}Cl$, $^{79}Br^{37}Cl$, and $^{81}Br^{37}Cl$, respectively; one standard deviations are given in parentheses, Willis and Clark [2, 3].

b) Analysis of the $J=1\leftarrow0$, $v=0$, 1 transitions by Smith et al. [4] and quotation by Lovas and Tiemann [5]. In view, however, of the good agreement of the B_e and α_e values with their own results, Willis and Clark [2] suspected a misprint in the B_e value for $^{79}Br^{37}Cl$ of [4]. Mean deviations (presumably) are given in parentheses.

Table 24
BrCl, Ground State X $^1\Sigma^+$. Rotational and Vibrational Constants from IR and Near-IR-Visible Spectra.

		$^{79}Br^{35}Cl$	$^{81}Br^{35}Cl$	$^{79}Br^{37}Cl$	$^{81}Br^{37}Cl$	$^{79}Br^{35}Cl$	$^{81}Br^{35}Cl$
B_e	in cm^{-1}	0.152468(2)	0.151309(3)	0.146755(4)	0.145617(5)	0.152469(1)	0.151309(1)
$10^4 \cdot \alpha_e$	in cm^{-1}	7.720(3)	7.622(3)	7.280(5)	7.210(4)	7.697(24)	7.640(26)
$10^6 \cdot \gamma_e$	in cm^{-1}	−2.22(5)	−2.41(5)	−2.09(10)	−2.02(9)	−2.56(38)	−1.87(41)
$10^8 \cdot D_e$	in cm^{-1}	7.14(3)	7.00(5)	6.49(7)	6.66(10)	7.183	7.074
$10^{10} \cdot \beta_e$	in cm^{-1}	−3.5(3)	−3.5(4)	−5.5(7)	−2.1(7)	−3.487	−3.336
ω_e	in cm^{-1}	444.2622(8)	442.5736(10)	435.8670(10)	434.1468(11)	444.276(14)	442.589(14)
$\omega_e x_e$	in cm^{-1}	1.8413(5)	1.8269(5)	1.7727(6)	1.7580(6)	1.843(3)	1.829(3)
$10^3 \cdot \omega_e y_e$	in cm^{-1}	−3.97(8)	−3.96(9)	−3.65(10)	−3.78(10)	−4.00(25)	−3.94(24)
remark		a)	a)	a)	a)	b)	b)
Ref.		[6]	[6]	[6]	[6]	[7]	[7]

a) Diode laser IR spectrum in the region of the v=1←0, 2←1, and 3←2 bands analyzed by Nakagawa et al. [6]; fit of the wave numbers to the eight Dunham coefficients Y_{01}, $-Y_{11}$, Y_{21}, $-Y_{02}$, Y_{12}, Y_{10}, $-Y_{20}$, Y_{30}, which are set equal to the constants given above (compare remark a) of Table 23).

b) Analysis of the high-resolution B $^3\Pi(0^+)$←X $^1\Sigma^+$ absorption spectrum in the region of the v′=2 to 8, v″=1 to 7 transitions by Coxon [7]. The values for $^{79}Br^{35}Cl$ have been adopted by Huber and Herzberg [8] who derived the equilibrium internuclear distance $r_e=2.13606_5$ Å for $^{79}Br^{35}Cl$. Rotational analysis of the B→X, v=0→12, 1→12, 0→11, and 2→10 emission bands gave very precise B_v''(v″=10 to 12) values for $^{79}Br^{35}Cl$ and $^{81}Br^{35}Cl$ [9]. Similar but less precise vibrational constants than those given above were derived for $^{79}Br^{35}Cl$ and $^{81}Br^{35}Cl$ from the B→X emission spectrum in the region of the v′=0 to 6, v″=5 to 19 transitions [10] and for $^{79,81}Br^{35}Cl$ from the B→X emission spectra in the regions of the v′=0 to 8, v″=2 to 19 [11] and v′=2 to 8, v″=1 to 14 [12] transitions.

Excited State B $^3\Pi(0^+)$. Constants for the B state result from rotational-vibrational analyses of the B $^3\Pi(0^+)\leftrightarrow$X $^1\Sigma^+$ absorption and emission spectra. The derivation of upper-state equilibrium values, however, is complicated by the predissociation of the B state which is assumed to be caused by an avoided curve crossing with a repulsive 0^+ state (cf. p. 192). A perturbational, non-linear least-squares fitting procedure has been suggested and applied by Coxon [7] and adopted by McFeeters et al. [10] for the analysis of their high-resolution B$\leftrightarrow$X absorption and emission spectra, respectively; the derived constants are compiled in Table 25. Vibrational constants obtained from B$\leftrightarrow$X spectra with lower resolution and without including the B-state perturbation [11, 12, 17, 18] are therefore improved.

Table 25
BrCl, B $^3\Pi(0^+)$ State. Rotational and Vibrational Constants and Internuclear Distance.

		$^{79}Br^{35}Cl$	$^{81}Br^{35}Cl$	$^{79}Br^{35}Cl$	$^{81}Br^{35}Cl$
B_e	in cm^{-1}	0.107704	0.106911	0.10874	0.10845
$10^3 \cdot \alpha_e$	in cm^{-1}	–	–	−1.169	−1.084
$10^7 \cdot D_e$	in cm^{-1}	[1.01]	[0.993]	–	–
ω_e	in cm^{-1}	222.68(2)	221.87(58)	(226.78)	222.18
$\omega_e x_e$	in cm^{-1}	2.884(1)	2.848(1)	(3.144)	2.82
$\omega_e y_e$	in cm^{-1}	−0.0673(4)	−0.0653(100)	–	−0.07
r_e	in Å	2.5415	2.5415	–	–
remark		a)	a)	b)	b)
Ref.		[7]	[7]	[10]	[10]

a) Analysis of the high-resolution B $^3\Pi(0^+)\leftarrow$X $^1\Sigma^+$ absorption spectrum in the region of the v′=2 to 8, v″=1 to 7 transitions by non-linear least-squares fits up to the second order (i.e., terms up to $\omega_e x_e(v+\frac{1}{2})^2$) and third order (i.e., terms up to $\omega_e y_e(v+\frac{1}{2})^3$); results from the third-order fit are given above; $D_e=4B_e^3/\omega_e^2$; standard deviation in parentheses, Coxon [7]. (A preliminary rotational analysis of the v=8, 7, 6←3 and 6←2 bands and B_v' and D_v' values for v′=6, 7, 8 were reported [16].) The results for $^{79}Br^{35}Cl$ have been adopted by Huber and Herzberg [8].

b) Analysis of the high-resolution laser induced B→X fluorescence spectrum (v′=0 to 6, v″=5 to 19) by McFeeters et al. [10] applying Coxon's [7] fitting procedures; results from third-order and, in parentheses, second-order fits. For precise B_v' (v′=0, 1, 2) values for $^{79}Br^{35}Cl$ and $^{81}Br^{35}Cl$ from rotational analyses of the B→X, v=0→12, 1→12, 0→11, and 2→10 emission bands see also [9].

A polarization configuration interaction calculation (Full POL-CI) gave the constants B_e, α_e, ω_e, $\omega_e x_e$, and r_e for the lowest bound $^3\Pi(0^+, 1, 2)$ state, however, without including spin-orbit interaction [15].

Other Excited States. Vibrational analysis of three ultraviolet emission systems assigned to transitions between ion-pair states and the three lowest excited valence states D′ $^3\Pi(2)\rightarrow$A′ $^3\Pi(2)$, β $^3\Pi(1)\rightarrow$A $^3\Pi(1)$, and E $^3\Pi(0^+)\rightarrow$B $^3\Pi(0^+)$ [19, 20] and rotational analyses of the D′→A′, 0→8, 9, 10 bands [19, 20] and the E→B, 0→6, 7, 8 and 1→6 bands [21] resulted in the following spectroscopic constants for $^{79}Br^{35}Cl$ if not otherwise indicated (rotational and vibrational constants in cm^{-1}, r_e in Å):

constant	A′ $^3\Pi(2)$	A $^3\Pi(1)$	D′ $^3\Pi(2)$	β $^3\Pi(1)$	E $^3\Pi(0^+)$
B_e [B_0]	0.1066(5)	–	0.07177(7)	–	[0.07167(7)][a)]
$10^4 \cdot \alpha_e$	9.41(104)	–	9.7	–	–
$10^5 \cdot \gamma_e$	−4.7(6)	–	–	–	–
ω_e	232.15(88)	227.00(3.48)	197.88(16)[b)]	192.31(60)	196.4(8)
$\omega_e x_e$	3.91(7)	3.301(287)	0.71(3)	0.7346(1518)	1.7(3)
$\omega_e y_e$	−0.007(2)	−0.0292(77)	–	–	–
r_e	2.555(6)	(2.54)[c)]	3.113(2)[b)]	3.14(2)	3.042(2)[a)]
Ref.	[20]	[19]	[20]	[19]	[20, 21]

a) $^{81}Br^{37}Cl$. – b) Approximation of the D′ state potential function by a Rittner potential (cf. p. 194) resulted in $\omega_e = 243\ cm^{-1}$ and $r_e = 2.778$ Å [22]. – c) Value for the B $^3\Pi(0^+)$ state.

In the rotational analysis of the E→B system, a perturbation was incorporated which was assumed to be due to the interaction of the E state with the β state; therefrom, $B_0 = 0.06980(15)\ cm^{-1}$ and $r_e = 3.082(3)$ Å were predicted for the β state [21].

For the two $\{BrCl^+, X\ ^2\Pi_{3/2,1/2}\}$ 5sσ Rydberg states, labeled C and D [8, 23] or a_5 and b_5 [24], vibrational constants, $\omega_e = 519\ cm^{-1}$ and $\omega_0 = 504\ cm^{-1}$ [8, 23], or average vibrational separations (v′ = 0 to 5) of 507 ± 3 and 508 ± 4 cm^{-1} [23] were derived from the absorption spectra in the vacuum UV.

References:

[1] Dunham, J. L. (Phys. Rev. [2] **41** [1932] 721/31).
[2] Willis, R. E., Jr.; Clark, W. W., III (J. Chem. Phys. **72** [1980] 4946/50).
[3] Willis, R. E., Jr. (Diss. Duke Univ. 1979, pp. 1/158, 98, 107/11; Diss. Abstr. Intern. B **40** [1980] 4880).
[4] Smith, D. F.; Tidwell, M.; Williams, D. V. P. (Phys. Rev. [2] **79** [1950] 1007/8).
[5] Lovas, F. J.; Tiemann, E. (J. Phys. Chem. Ref. Data **3** [1974] 609/769).
[6] Nakagawa, K.; Horiai, K.; Konno, T.; Uehara, H. (J. Mol. Spectrosc. **131** [1988] 233/40).
[7] Coxon, J. A. (J. Mol. Spectrosc. **50** [1974] 142/65).
[8] Huber, K. P.; Herzberg, G. (Molecular Spectra and Molecular Structure, Vol. 4, Constants of Diatomic Molecules, Van Nostrand Reinhold, New York 1979, pp. 108/9).
[9] Clyne, M. A. A.; Toby, S. (J. Photochem. **11** [1979] 87/100).
[10] McFeeters, B. D.; Perram, G. P.; Crannage, R. P.; Dorko, E. A. (Chem. Phys. **139** [1989] 347/57).

[11] Hadley, S. G.; Bina, M. J.; Brabson, G. D. (J. Phys. Chem. **78** [1974] 1833/6).
[12] Clyne, M. A. A.; Coxon, J. A. (Proc. Roy. Soc. [London] A **298** [1967] 424/52).
[13] Baran, E. J. (Z. Physik. Chem. [Leipzig] **255** [1974] 1022/6).
[14] Pandey, A. N.; Mithal, A. K.; Shukla, M. M.; Singh, G. C. (Indian J. Pure Appl. Phys. **11** [1973] 69/71).
[15] Eades, R. A. (Diss. Univ. Minnesota 1983, pp. 1/214; Diss. Abstr. Intern. B **44** [1984] 3418).
[16] Clyne, M. A. A.; Coxon, J. A. (J. Phys. B **3** [1970] L9/L11).
[17] Clyne, M. A. A.; Coxon, J. A. (J. Chem. Soc. Chem. Commun. **1966** 285/6).
[18] Clyne, M. A. A.; Coxon, J. A. (Nature **217** [1968] 448/9).
[19] Chakraborty, D. K.; Tellinghuisen, P. C.; Tellinghuisen, J. (Chem. Phys. Letters **141** [1987] 36/40).

[20] Chakraborty, D. K. (Diss. Vanderbilt Univ. 1987, pp. 1/189, 1/12, 76/100; Diss. Abstr. Intern. B **48** [1987] 772).

[21] Brown, S. W.; Dowd, C. J., Jr.; Tellinghuisen, J. (J. Mol. Spectrosc. **132** [1988] 178/92).
[22] Diegelmann, M.; Hohla, K.; Rebentrost, F.; Kompa, K. L. (J. Chem. Phys. **76** [1982] 1233/47).
[23] Cordes, H.; Sponer, H. (Z. Physik **79** [1932] 170/85).
[24] Hopkirk, A.; Shaw, D.; Donovan, R. J.; Lawley, K. P.; Yencha A. J. (J. Phys. Chem. **93** [1989] 7338/42).

16.6.3.7 Potential Energy Functions

Ground State. High-resolution absorption and emission data for the B $^3\Pi(0^+)\leftrightarrow$X $^1\Sigma^+$ system were used to evaluate Rydberg-Klein-Rees (RKR) potential functions. Coxon [1] evaluated the energy levels $E(v'')=G(v'')+Y_{00}$ (Dunham correction Y_{00} neglected) up to $\sim$3230 cm^{-1} ($v''=0$ to 7) and the classical turning points, r_{min} and r_{max}, for $^{79}Br^{35}Cl$ and $^{81}Br^{35}Cl$. On the basis of their data on laser-induced fluorescence of BrCl, McFeeters et al. [2] extended these RKR potential functions up to $\sim$8300 cm^{-1} corresponding to $v''=20$. In the region of the vibrational levels $v''=0$ to 7, a Morse potential curve for X $^1\Sigma^+$ of $^{79,81}Br^{35}Cl$ [3], which is reproduced in **Fig. 7a**, p. 193, agrees well with the RKR curves [1]. The Morse potential depicted in [4] is based on a wrong vibrational assignment (cf. p. 205).

The coefficients a_i of Dunham's power series expansion of the potential energy around the equilibrium internuclear distance r_e, $V(x)=a_0x^2(1+a_1x+a_2x^2+a_3x^3+...)$ with $x=(r-r_e)/r_e$, were derived from the Dunham coefficients Y_{ik} (cf. p. 188) which resulted from the analysis of the microwave spectrum: $a_0=3.2369(9)\times10^5$ cm^{-1}, $a_1=-3.4590(5)$, $a_2=6.85(8)$, and $a_3=-11.2(7)$ [5, 6].

A few empirical potential functions have been compared and/or tested for their ability to reproduce a certain molecular parameter for the equilibrium position or a certain part of the potential curve [7 to 10].

Quantum chemical ab initio calculations for the ground state and several excited states (see below) using GVB(pp) (perfect pairing generalized valence bond), POL-CI (polarization configuration interaction), and Full POL-CI wave functions resulted in numerical energy values for r=3.2000 to 25.0000 au (1.69 to 13.2 Å); a graphical representation of the Full POL-CI potential curve at r=1.6 to 3.7 Å is given [11]. Three semiempirical calculations also gave ground-state potential energy functions, see references [14, 34, 35] of Section 16.6.3.1.1.

Excited State B $^3\Pi(0^+)$. Predissociation. As in the case of BrF (cf. p. 17), the evaluation of B-state potential functions is complicated by predissociation of rovibrational levels with $v'\geq6$. On the basis of B $^3\Pi(0^+)\leftrightarrow$X $^1\Sigma^+$ spectroscopic data [1, 3, 4, 12] and lifetime measurements on B-state rovibrational levels [13, 14] (cf. pp. 196/8), the predissociation was suggested to arise from an avoided curve crossing between the B-state potential curve and that of a repulsive $0^+(Br(^2P_{3/2})+Cl(^2P_{3/2}))$ state (Herzberg's case I (c) predissociation [15]) which causes the former to display a potential barrier; as a consequence, its upper rovibrational levels are resonances bound only by this barrier and predissociation occurs via tunneling through the potential barrier.

Rydberg-Klein-Rees (RKR) potential curves for $^{79}Br^{35}Cl$ and $^{81}Br^{35}Cl$, i.e., the energy levels E(v′) up to $\sim$1540 cm^{-1} (with respect to the B-state potential minimum) for $v'=0$

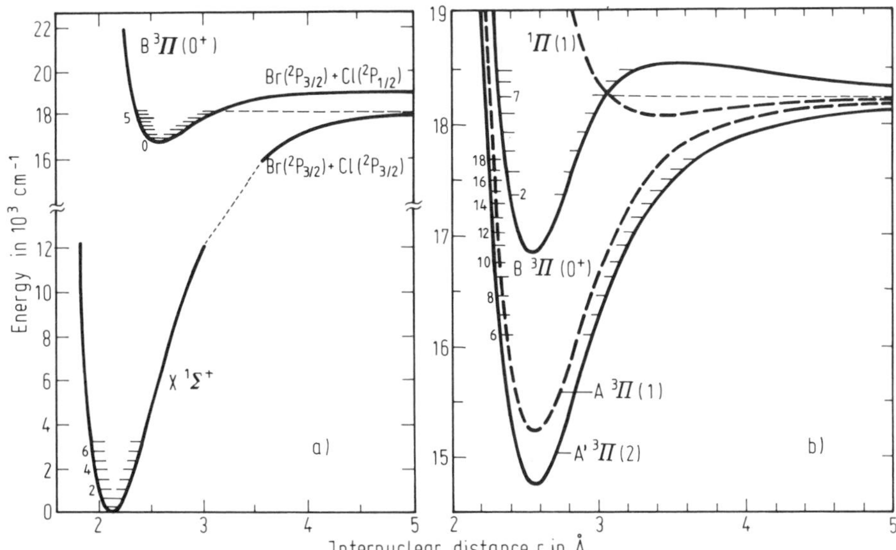

Fig. 7. Potential energy curves for BrCl. Vertical bars and integral numbers indicate observed vibrational levels, the dotted horizontal lines indicate the first dissociation limit. a) Morse potential energy functions [3]. b) RKR potentials (presumably) for A′ and B, estimated potentials for A and $^1\Pi(1)$ [18].

to 8 and the corresponding r_{min} and r_{max} values, were derived by Coxon [1] on the basis of his own high-resolution B←X absorption data and applying a perturbational method (cf. p. 183). A Morse potential reproduces the RKR curve well in that v′ region [3], see **Fig. 7a**. A modified Morse potential arising from the interaction of a Morse potential V_M with a simple repulsive potential V_R was fitted to Coxon's [1] spectroscopic data which resulted in (V in cm^{-1}, r in Å) $V(r)=(V_M+V_R-T)/2$ with $V_M(r)=17478.2+7662\ \{1-\exp[-1.671\ (r-2.52382)]\}^2$, $V_R(r)=18162+1.82\times10^5/r^4$, and $T=((V_R-V_M)^2+4\times1900^2)^{1/2}$ (1900 cm^{-1} is the interaction energy) [16]; this potential represents well the RKR points and reproduces the experimental G(v′) and B'_v values of [1]; it shows a potential barrier of 18503.5 cm^{-1} (above the X-state minimum) at r = 3.33 Å [16, figure 4]. Analysis of the ultraviolet E $^3\Pi(0^+)\rightarrow$B $^3\Pi(0^+)$, v = 0→8, 7, 6, and 1→6 emission spectrum located the barrier at 18550(20) cm^{-1} above the ground-state minimum and near r = 3.4 Å [16].

On the basis of earlier emission data (v′ = 2 to 8) [4] and continuous absorption data [17], Child and Bernstein [12] calculated an RKR potential for the bound region and an exponential approximation for the repulsive part of the potential curve; the latter has the form $V(r)=-1400+5345\cdot\exp[-6.51\ (r-2.1380)]$ with V in cm^{-1} relative to the ground-state dissociation limit and r in Å; for the perturbing repulsive 0^+ state crossing at r ≈ 3.2 Å, the potential $V(r)=82000\cdot\exp[-5.6\ (r-2.138)]$ was estimated [12].

More recently, rotational-electronic coupling with a weakly bound $^1\Pi(1)$ state, which intersects the attractive branch of the B potential near the first dissociation limit as in the cases of I_2 and Br_2 (see "Bromine" Suppl. Vol. A, 1985, pp. 186, 199), was found to better account for the lifetime measurements; for the intersecting $^1\Pi(1)$ state the potential $V(r)=2.8\times10^8/r^{11}-8.1\times10^6/r^8-3.1\times10^5/r^6+4.6\times10^4/r^5$ (V in cm^{-1}, r in Å) was derived;

coupling with the A $^3\Pi(1)$ state via Franck-Condon overlap of the A- and B-state repulsive branches as in Cl_2 was ruled out [18]; **Fig. 7b**, p. 193, shows the relevant excited states near the first dissociation limit [18] (the B-state vibrational levels, v=2 to 9, result from B←X [1] and E→B [19] spectra).

Ab initio calculations using the Full POL-CI wave function resulted in numerical energy values for r=2.8000 to 25.0000 au (1.48 to 13.2 Å) and in a graphical representation at r=2.0 to 4.0 Å for the bound 1 $^3\Pi(0^+$, 1, 2) state (the calculations do not include spin-orbit effects) [11].

Other Excited States. For the lowest excited valence state, A′ $^3\Pi(2)$, a potential curve (presumably RKR) has been constructed [18] using D′ $^3\Pi(2)$→A′ $^3\Pi(2)$ emission data [19]; that for the next higher valence state, A $^3\Pi(1)$, has been guessed from data for the ICl molecule [18], see **Fig. 7b**, p. 193.

For the ion-pair excited state D′ $^3\Pi(2)$ ($Br^+(^3P_2)+Cl^-(^1S_0)$) the potential curve was approximated by a Rittner potential [20] and by using literature values for the relevant ionic parameters [21, 22]. That of the ion-pair state E $^3\Pi(0^+)$ was approximated by a Morse potential using E→B emission data [16].

Excited states, which possibly cause B-state predissociation, are dealt with above.

For several low-lying (i.e., arising from ground-state Br(^{2}P) and Cl(^{2}P) atoms) repulsive singlet (1 $^1\Pi$, 2 $^1\Pi$, 1 $^1\Delta$, 2 $^1\Sigma^+$, 1 $^1\Sigma^-$) and triplet (2 $^3\Pi$, 1 $^3\Sigma^+$, 1 $^3\Sigma^-$, 1 $^3\Delta$, 2 $^3\Sigma^+$) valence states numerical energy values for r=3.7000 to 25.0000 au (1.96 to 13.2 Å) and potential curves at r=2.1 to 4.0 Å (singlets) and 2.0 to 4.0 Å (triplets) are presented which result from ab initio Full POL-CI (cf. above) calculations [11].

References:

[1] Coxon, J. A. (J. Mol. Spectrosc. **50** [1974] 142/65).
[2] McFeeters, B. D.; Perram, G. P.; Crannage, R. P.; Dorko, E. A. (Chem. Phys. **139** [1989] 347/57).
[3] Clyne, M. A. A.; McDermid, I. S. (J. Chem. Soc. Faraday Trans. II **74** [1978] 798/806).
[4] Clyne, M. A. A.; Coxon, J. A. (Proc. Roy. Soc. [London] A **298** [1967] 424/52).
[5] Willis, R. E., Jr.; Clark, W. W., III (J. Chem. Phys. **72** [1980] 4946/50).
[6] Willis, R. E., Jr. (Diss. Duke Univ. 1979, pp. 1/158, 101; Diss. Abstr. Intern. B **40** [1980] 4880).
[7] Lippincott, E. R.; Schroeder, R. (J. Chem. Phys. **23** [1955] 1131/41).
[8] Lippincott, E. R.; Steele, D.; Caldwell, P. (J. Chem. Phys. **35** [1961] 123/41).
[9] Borkman, R. F.; Simons, G.; Parr, R. G. (J. Chem. Phys. **50** [1969] 58/65).
[10] Varshni, Y. P. (Can. J. Chem. **66** [1988] 763/6).
[11] Eades, R. A. (Diss. Univ. Minnesota 1983, pp. 1/214, 104/5, 132/5; Diss. Abstr. Intern. B **44** [1984] 3418).
[12] Child, M. S.; Bernstein, R. B. (J. Chem. Phys. **59** [1973] 5916/25).
[13] Clyne, M. A. A.; McDermid, I. S. (J. Chem. Soc. Faraday Trans. II **74** [1978] 807/20).
[14] Clyne, M. A. A.; McDermid, I. S. (Faraday Discussions Chem. Soc. No. 67 [1979] 316/28).
[15] Herzberg, G. (Molecular Spectra and Molecular Structure, Vol. 1, Spectra of Diatomic Molecules, Van Nostrand, Princeton, N.J., 1961, pp. 413/4, 420/4).
[16] Brown, S. W.; Dowd, C. J., Jr.; Tellinghuisen, J. (J. Mol. Spectrosc. **132** [1988] 178/92).
[17] Seery, D. J.; Britton, D. (J. Phys. Chem. **68** [1964] 2263/6).
[18] Tellinghuisen, J. (J. Chem. Phys. **89** [1988] 6150/6).

[19] Chakraborty, D. K.; Tellinghuisen, P. C.; Tellinghuisen, J. (Chem. Phys. Letters **141** [1987] 36/40).
[20] Rittner, E. S. (J. Chem. Phys. **19** [1951] 1030/5).
[21] Diegelmann, M.; Hohla, K.; Rebentrost, F.; Kompa, K. L. (J. Chem. Phys. **76** [1982] 1233/47).
[22] Diegelmann, M. (Max-Planck-Ges. Foerd. Wiss. Projektgruppe Laserforsch. Ber. PLF-33 [1980] 1/124).

16.6.3.8 Dissociation Energy

The thermochemical value for the ground-state dissociation energy,

$$D_0^\circ = 18010 \pm 100\ \text{cm}^{-1} = 2.23_3\ \text{eV} = 215.4\ \text{kJ/mol}\ [1, 2]$$

is obtained from an average value $\Delta_r H^\circ_{298.15} = 1.699$ kJ/mol for the reaction $Br_2(g) + Cl_2(g) \rightarrow 2\ BrCl(g)$ as recommended in the JANAF Tables [3] and from reliable values for the dissociation energies of Br_2 and Cl_2.

Spectroscopic values of the ground-state dissociation energy are based on the B-state predissociation studied in high-resolution B $^3\Pi(0^+) \leftarrow X\ ^1\Sigma^+$ absorption and excitation spectra and by lifetime measurements of B-state rovibrational levels. An upper limit, $D_0^\circ \le 18035\ \text{cm}^{-1}$, provided by the energy of the B, v′=7, J′=0 level (all v′=7 rotational levels are predissociated) [1] is refined to $D_0^\circ = 17934 \pm 26\ \text{cm}^{-1}$ for $^{81}Br^{35}Cl$ by the observation of the highest known stable level, v′=5, J′=70, and from the sharp onset of predissociation at v′=6, J′=42 [4]. A computational study and reinterpretation of the spontaneous predissociation in BrCl (cf. pp. 193/4) by Tellinghuisen [5] lead to a revised value $D_e = 18244(5)\ \text{cm}^{-1}$ (preliminary result $D_e = 18245(15)\ \text{cm}^{-1}$ [6]) or, using $(1/2\omega_e - 1/4\ \omega_e x_e) = 221\ \text{cm}^{-1}$ for $^{79,81}Br^{35}Cl$ (cf. pp. 188/9), to

$$D_0^\circ = 18023\ \text{cm}^{-1} = 2.235\ \text{eV} = 215.6\ \text{kJ/mol}$$

A few theoretical values of D_e are available. A polarization configuration interaction calculation (Full POL-CI) results in only $D_e = 168.6$ kJ/mol [7]. SCF MO calculations with basis sets of various sizes give too low D_e values or even are not able to predict a stable molecular ground state ($D_e < 0$); semiempirical methods and simple models of bonding have also been applied; see references [4, 26, 33, 34] of Section 16.6.3.1.1.

For the dissociation of BrCl into $Br^+ + Cl^-$ or $Br^- + Cl^+$, D = 1125 and 1017 kJ/mol, respectively, were estimated from dipole moment and polarizability data [8].

References:

[1] Clyne, M. A. A.; McDermid, I. S. (J. Chem. Soc. Faraday Trans. II **74** [1978] 798/806).
[2] Huber, K. P.; Herzberg, G. (Molecular Spectra and Molecular Structure, Vol. 4, Constants of Diatomic Molecules, Van Nostrand Reinhold, New York 1979, pp. 108/9).
[3] Chase, M. W., Jr.; Davies, C. A.; Downey, J. R., Jr.; Frurip, D. J.; McDonald, R. A.; Syverud, A. N. (JANAF Thermochemical Tables, 3rd Ed., J. Phys. Chem. Ref. Data **14** Suppl. No. 1 [1985] 425), Stull, D. R.; Prophet, H. (JANAF Thermochemical Tables, 2nd Ed., NSRDS-NBS-37 [1971]).
[4] Clyne, M. A. A.; McDermid, I. S. (Faraday Discussions Chem. Soc. No. 67 [1979] 316/28).
[5] Tellinghuisen, J. (J. Chem. Phys. **89** [1988] 6150/6).
[6] Brown, S. W.; Dowd, C. J., Jr.; Tellinghuisen, J. (J. Mol. Spectrosc. **132** [1988] 178/92).

[7] Eades, R. A. (Diss. Univ. Minnesota 1983, pp. 1/214, 90; Diss. Abstr. Intern. B **44** [1984] 3418).
[8] Cantacuzène, J. (J. Chim. Phys. Physicochim. Biol. **65** [1968] 502/15).

16.6.3.9 Relaxation Processes

16.6.3.9.1 Deactivation of X $^1\Sigma^+$ (v″, J″) Levels

Vibrational and rotational deexcitation by collisions (V-T and R-T energy transfer) was studied by Farthing et al. [1] in a supersonic BrCl beam seeded in Cl_2 (12.3% BrCl, 87.6% Cl_2, 0.1% Br_2) and expanded through a nozzle heated to ~355 K; the vibrational and rotational cooling was measured by analysis of B→X band intensities in the laser-induced fluorescence spectrum which resulted in $T_{vib}=195\pm5$ K and $T_{rot}=55\pm5$ K, and the final translational temperature $T_{trans}=49\pm2$ K was obtained from time-of-flight measurements; these gave the collision numbers for vibrational and rotational relaxation, $Z_v=28$ and $Z_r\approx1$, or collision probabilities, $P_v=0.04$ and $P_r\approx1$. Thus, the rate for rotational relaxation is close to gas kinetic. Vibrational relaxation presumably is enhanced at the low temperature of the beam (Z_v of BrCl is about an order of magnitude smaller than, for example, that for a supersonic Br_2 beam) by the formation of metastable orbiting BrCl···Cl_2 complexes (resonances), in which the vibrational energy is redistributed into translational and rotational energies of both collision partners, rather than by the formation of van der Waals complexes, which lose energy by vibrational predissociation [1] (for vibrational relaxation enhancement at low temperatures due to complex formation, see for example [2]).

References:

[1] Farthing, J. W.; Fletcher, I. W.; Whitehead, J. C. (Mol. Phys. **48** [1983] 1067/73).
[2] Ewing, G. (Chem. Phys. **29** [1978] 253/70, 259).

16.6.3.9.2 Deactivation of B $^3\Pi(0^+)$ (v′, J′) Levels

General. As in the case of BrF (cf. pp. 20/4) the dynamics of the B $^3\Pi(0^+)$ state were studied by laser-induced, time-resolved fluorescence in the B↔X system, which for BrCl, however, is complicated by the low transition probability of the B↔X system, by the occurrence of four different BrCl isotopes, and by the presence of Br_2 in the $Br_2+Cl_2\rightleftharpoons2$ BrCl equilibrium. The latter difficulty, suppression of the Br_2 fluorescence, was overcome by allowing mixtures of Br_2 with excess Cl_2 to equilibrate. Time-resolved, laser-induced fluorescence (LIF) at room temperature and total pressures of 0.1 mTorr to 11 Torr was measured covering the range of collision-free to thermalized conditions. The observations can be summarized as follows:

B-state rovibrational levels with v′=1 to 4, v′=5 up to at least J′=71, and v′=6 up to J′=41 are stable with radiative lifetimes τ_{rad} of about 40 µs. Predissociation sharply commences at v′=6, J′=42 where τ_{rad} suddenly falls below 10 µs and decreases monotonically with increasing J′; all v′=7 levels are predissociated with $\tau_{rad}\leq1$ µs.

At "high" total pressures, 0.5 to 11 Torr, the fluorescence decay was found to be independent of the excitation wavelength and thus of the quantum numbers v′ and J′, which shows that the initially excited B-state rovibrational level or rovibrational manifold rapidly relaxes (e.g., within 2 µs at p=5 Torr) to a Boltzmann or thermalized rovibrational distribution (~25% of v′=0 and ~75% of v′=1 at T≈300 K). The fluorescence is then depleted by slow collisional electronic quenching, i.e., self-quenching by the Cl_2-BrCl-Br_2 mixture, where Cl_2 is the most abundant collision partner, or quenching by added foreign gases. Electronic quenching of BrCl(B) is very inefficient for all collision partners studied ($k_Q \approx$ (1 to 4) $\times 10^{-13}$ $cm^3 \cdot molecule^{-1} \cdot s^{-1}$) except for BrCl itself (estimated k_Q value) and O_2 with which resonant electronic energy transfer is possible.

At "low" total pressures, $p \leq 200$ mTorr, approaching single-collision (i.e., one radiative lifetime between collisions) or collision-free conditions, a pressure and v′-dependent rapid fluorescence decay which becomes more rapid with increasing v′ (= 1 to 6) was observed in the pure equilibrated Cl_2-BrCl-Br_2 mixture and with added foreign-gas collision partners. This was attributed to upward vibrational transfer as a primary deactivation mechanism, i.e., rovibrational ladder climbing into predissociated levels, which is much more efficient than electronic quenching for the B state of BrCl.

However, after the first studies of BrCl(B) state dynamics by Wright et al. [1, 2] at pressures of 2 to 50 Torr and using broad band excitation, Clyne and co-workers [3 to 6] reported on systematic measurements of fluorescence lifetimes and fluorescence decay rate constants at total pressures of 0.1 to 1.0 mTorr, 30 to 200 mTorr, and 0.5 to 11 Torr and employing (v′, J′) state-selected excitation by use of pulsed, narrow-band tunable dye lasers. Summaries of these experiments and results are given by Clyne and McDermid [7] and by Heaven [8], and a compilation of lifetimes, quenching rate constants, and vibrational energy-transfer rate constants is presented by Steinfeld [9].

More recently, during the search for visible chemical lasers, Wolf, Perram, and Davis [10, 11] began detailed studies of the B $^3\Pi(0^+)$ state collisional dynamics in interhalogens, including BrCl, which are still in progress [12]. These authors do not accept all of Clyne's [4 to 6] interpretations, but could not yet give a complete description of the quenching and energy transfer processes responsible for the B→X fluorescence depletion.

Collision-Free Lifetimes τ_0. Radiative Lifetimes τ_{rad}. The lifetime data were analyzed by the Stern-Volmer formulation, $1/\tau = 1/\tau_0 + k_X$ [X], where k_X is the rate constant for collision-induced deactivation (electronic quenching, k_Q, and/or rovibrational energy transfer, $k_{v'}$, see below) and [X] is the concentration of the quenching species. The collision-free lifetime τ_0 is independent of v′ and equal to the B-state radiative lifetime τ_{rad} for the stable (not predissociated) rovibrational levels with v′ = 1 to 6. The results from the low-pressure measurements of Clyne et al. [4 to 6] and Perram and Davis [11] are in good agreement, as shown in the following table:

v′	τ_0 in µs [5] p = 0.25 to 1.0 mTorr	τ_0 in µs [6] p = 2 to 100 mTorr	τ_0 in µs [11] p = 0.1 to 200 mTorr
0	–	–	40.9 ± 2.0
1	–	41.5 ± 0.5 (J′ = 25)	38.8 ± 2.6
2	–	–	38.0 ± 1.5
3	42.0 ± 2.7 (J′ = 15 to 35)	–	–
4	40.0 ± 2.0 (J′ = 4 to 45)	–	38.7 ± 1.4
5	39.8 ± 2.0 (J′ = 6 to 60)	–	36.8 ± 1.2
6	40.0 ± 0.7 (J′ = 10 to 41)	–	40.1 ± 1.3
average	40.2 ± 1.8*)	–	38.7 ± 1.7

*) Single τ_0 values for the v′ = 5, J′ = 6, 10, 15, 20, ..., 50, 55 levels are given in [5]; data for the various J′ levels of v′ = 3, 4, and 6 are available from the authors [5].

The results obtained under multi-collision conditions (p=30 to 200 mTorr), $\tau_0(v'=3)=29^{+10}_{-8}$ μs, $\tau_0(v'=4)=40^{+12}_{-9}$ μs, τ_0(average)$=35^{+11}_{-9}$ μs [4], and the high-pressure results for vibrationally thermalized BrCl, $\tau_0(v'=0, 1)=13^{+15}_{-5}$ μs [4] and $\tau_0(v'=0, 1)=18.5\pm3$ μs [1, 2], can be considered only as rough estimates. The preliminary result for $\tau_0(v'=5, J'=35)$ [3] has been refuted [5].

Predissociation appreciably shortens the collision-free lifetimes of the higher v'=6 and all observed v'=7 levels. Above the last stable level, v'=6, J'=41 with $\tau_0=40.6$ μs, a monotonical decrease of τ_0 from 8.8 μs for J'=42 to 4.4 μs for J'=50 was measured at pressures of 0.1 to 1.0 mTorr, and a linear dependence of $1/\tau_0$ upon J'(J'+1) was found [5]. The lifetimes of the v'=7 rotational levels also show a systematic decrease with increasing rotational energy. Values of τ_0=0.91, 0.78, 0.45, 0.35, and 0.24 μs for J'=7, 10, 13, 16, and 19 were obtained at 70 mTorr, respectively [4]. At pressures below 1 mTorr, the fluorescence intensity from v'=7 was too low for systematic measurements; however, it could be confirmed at least that $\tau_0<1$ μs for rovibrational levels with v'=7 [5].

Rate Constants. In high-pressure Cl_2-BrCl-Br_2 mixtures, where rovibrational thermalization of the B state of BrCl (to the v'=0 and 1 levels) was accomplished within a few microseconds, slow collisional electronic quenching takes place, i.e., self-quenching by BrCl and quenching by Cl_2. Quenching by added inert gases or N_2 is even slower, whereas O_2 turned out to be a very efficient quencher, which may be due to resonant electronic energy transfer between BrCl(B $^3\Pi(0^+)\rightarrow$X $^1\Sigma^+$) and O_2(a $^1\Delta_g\leftarrow$X $^3\Sigma_g$). Rate constants k_Q(X) at room temperature obtained from the slopes of Stern-Volmer plots are as follows:

collision partner X	quenching rate constant k_Q(X) in 10^{-13} $cm^3\cdot molecule^{-1}\cdot s^{-1}$		
	k_Q(X) [1]	k_Q(X) [4, 6]	k_Q(X) [11]
M [a)]	3.4 [b)]	3.9±1.2 [4] 2.2 [6]	4.3±0.7
He	1.18 [c)]	–	1.5±0.2
Ne	–	–	0.78±0.19
Ar	–	–	0.81±0.30
Kr	–	–	0.98±0.14
N_2	–	–	1.6±0.3
O_2	–	–	65±6
air	16.5 [c)]	–	–

[a)] Cl_2-BrCl-Br_2 mixture with ~15% BrCl, ~85% Cl_2, and negligible Br_2 amount. – [b)] Derived by [4, 5] from the data of [1]. The separated values, $k_Q(Cl_2)=4\times10^3$ $s^{-1}\cdot Torr^{-1}=1.24\times10^{-13}$ $cm^3\cdot molecule^{-1}\cdot s^{-1}$ and $k_Q(BrCl)=33\times10^3$ $s^{-1}\cdot Torr^{-1}=10.2\times10^{-13}$ $cm^3\cdot molecule^{-1}\cdot s^{-1}$, respectively, were obtained by varying the Cl_2 pressure and extrapolating to $p(Cl_2)=0$ [1]. – [c)] Converted from $k_Q(He)=3.8\times10^3$ $s^{-1}\cdot Torr^{-1}$, $k_Q(air)=53\times10^3$ $s^{-1}\cdot Torr^{-1}$ in [1].

At pressures approaching single collision conditions, a rapid and strongly v'-dependent fluorescence decay was observed. Rapid upward vibrational energy transfer into predissociated rovibrational levels (v'>6), which becomes more efficient with increasing v', most probably superimposes the inefficient electronic quenching. The following total rate constants $k_{Q,v'}$(X) were measured in pure Cl_2-BrCl-Br_2 mixtures and with added foreign gases; the rates quoted for v'=3 to 6 were obtained at truly single-collision conditions, whereas those for v'=0 to 3 are slightly affected by multiple collision effects:

v′	total rate constant $k_{Q,v'}(X)$ in 10^{-11} $cm^3 \cdot$ molecule$^{-1} \cdot s^{-1}$ $k_{Q,v'}(M)$ [11] [a)]	$k_{Q,v'}(M)$ [4 to 6] [a), b)]	$k_{Q,v'}(He)$ [11]
0	0.043 ± 0.007	(0.039 ± 0.012)	0.015 ± 0.003
1	0.180 ± 0.070	0.30 ± 0.02	0.10 ± 0.02
2	0.270 ± 0.060	–	0.30 ± 0.04
3	0.96 ± 0.53	0.40 ± 0.3	0.42 ± 0.05
4	4.7 ± 1.4	6.2 ± 2.3	1.03 ± 0.06
5	6.2 ± 1.7	9.6 ± 2.8	2.65 ± 1.0
6	14.0 ± 3.0	20.7 ± 4.1	10.4 ± 3.8
	$k_{Q,v'}(Ar)$ [6]	$k_{Q,v'}(Cl_2)$ [6]	$k_{Q,v'}(O_2)$ [6]
1	0.13 ± 0.02	0.27 ± 0.01	0.58 ± 0.02

a) M = Cl_2-BrCl-Br_2 mixture with ~15% BrCl, ~85% Cl_2, and negligible amounts of Br_2. — b) Result for v′ = 0 for quenching from a thermalized BrCl(B, v′) distribution (see above) [4]; values for v′ = 1 and 3 from [6] and values for v′ = 4, 5, 6 from [5]. Preliminary $k_{v'}$ (v′ = 3 to 6) values obtained under multiple-collision conditions [4] are considerably discrepant from those given above.

Preliminary studies on rotational energy transfer in collisions with Cl_2, He, Ar, and Xe indicated a high efficiency with nearly gas kinetic rate constants: k_r(in 10^{-10} $cm^3 \cdot$ molecule$^{-1} \cdot s^{-1}$) = 1.6 ± 0.3, 1.9 ± 0.3, 2.4 ± 0.4, and 2.2 ± 0.3, respectively, for the total rotational removal from J′ = 37 in the v′ = 6 level [10]. Detailed studies are in progress [12].

References:

[1] Wright, J. J.; Spates, W. S.; Davis, S. J. (J. Chem. Phys. **66** [1977] 1566/70).
[2] Wright, J. J.; Spates, W. S.; Davis, S. J. (Electron. Transition Lasers 2 Proc. 3rd Summer Colloq. 6, Snowmass Village, Colo., 1976 [1977], pp. 288/92).
[3] McDermid, I. S.; Clyne, M. A. A. (Lasers Chem. Proc. Conf., London 1977, pp. 122/7).
[4] Clyne, M. A. A.; McDermid, I. S. (J. Chem. Soc. Faraday Trans. II **74** [1978] 807/20).
[5] Clyne, M. A. A.; McDermid, I. S. (Faraday Discussions Chem. Soc. No. 67 [1979] 316/28).
[6] Clyne, M. A. A.; Zai, L. C. (J. Chem. Soc. Faraday Trans. II **78** [1982] 1221/9).
[7] Clyne, M. A. A.; McDermid, I. S. (Advan. Chem. Phys. **50** [1982] 1/104, 48/52).
[8] Heaven, M. C. (Chem. Soc. Rev. **15** [1986] 405/48).
[9] Steinfeld, J. I. (J. Phys. Chem. Ref. Data **13** [1984] 445/553, 460/2).
[10] Wolf, P. J.; Perram, G. P.; Davis, S. J. (Springer Proc. Phys. **15** [1987] 556/61).
[11] Perram, G. P.; Davis, S. J. (J. Chem. Phys. **93** [1990] 1720/31).
[12] Perram, G. P.; Davis, S. J. (to be published from [10]).

16.6.4 Spectra

16.6.4.1 Microwave Absorption Spectrum

The MW spectrum of BrCl was measured in gaseous Br_2-Cl_2 mixtures in the 3.5 to 3.2 cm and 0.25 to 0.01 cm regions (9 and 122 to 309 GHz) which pertain to the J = 1←0 and J = 14←13 to 34←33 rotational transitions.

For the J = 1←0 transition Smith et al. [1] observed some 29 lines near 9 GHz at room temperature; they identified them with the hyperfine (hf) transitions of all four isotopic species, $^{79}Br^{35}Cl$, $^{79}Br^{37}Cl$, $^{81}Br^{35}Cl$, and $^{81}Br^{37}Cl$, in their vibrational ground and first excited levels (v=0 and 1) arising from quadrupole coupling with the Br and Cl nuclei: $J+I_{Br}=F_1$, $F_1+I_{Cl}=F$ with $I_{Br}=I_{Cl}=3/2$. The measurements were not precise enough to include also spin-rotation interaction. Uncertainties in the wave numbers of 0.50 MHz were estimated [2]. More precise results for the J=1←0 transition were published only partially; these include the nine hf lines (with uncertainties of 0.025 MHz) of $^{79}Br^{35}Cl$ and $^{81}Br^{35}Cl$, for which Stark effect measurements had been carried out [3, 4]. Measured hf components with two decimals from [1, 2], with three decimals from [3, 4], and assignments are as follows (v=0 level if not indicated otherwise; ν in MHz):

F′←F″	F′ ←F″₁	$^{79}Br^{35}Cl$	$^{81}Br^{35}Cl$	$^{79}Br^{37}Cl$	$^{81}Br^{37}Cl$
2←3	3/2 ←3/2	9307.96	9209.57	8964.19	–
3←3	3/2 ←3/2	9291.61	9193.26	8951.38	8852.93
0←1	3/2 ←3/2	9274.709	9176.444	–	–
1←0	5/2 ←3/2	9088.61	9026.17	8745.17	8683.06
4←3	5/2 ←3/2	9080.73	9018.71	8738.47	8676.3
	5/2 ←3/2	9080.840	9018.794	–	–
2←3	5/2 ←3/2	9074.91	9012.97	8733.84	8671.87
2←1	5/2 ←3/2	9074.981	9013.175	–	–
3←2	5/2 ←3/2	9064.035	9001.864	–	–
3←3	5/2 ←3/2	9063.77	9001.44	8725.49	8663.4
	5/2 ←3/2	9063.724	–	–	–
4←3	5/2 ←3/2*)	9034.14	8972.41	–	–
2←3	1/2 ←3/2	8899.50	8865.66	8559.58	8525.53

*) v=1 level.

Stark effect measurements for the nine J=1←0, v=0 hf transitions (given above with three decimal places) in electric fields up to 1000 V/cm gave Stark coefficients of 2.17 to 2.84 $Hz\cdot cm^2\cdot V^{-2}$ and the dipole moment of BrCl (see p. 185) [3, 4] (earlier result [1]).

Measurements with Br_2-Cl_2 mixtures at 20 to 200 °C and 122 to 309 GHz enabled a number of higher rotational transitions from J=14←13 to J=34←33 of the four isotopic species in their lowest vibrational levels, v=0 to 3, to be observed. The hf components with $\Delta F=+1$ dominate in strength and generally appear as doubly degenerate doublets corresponding to the transitions $F_1''+1\leftarrow F_1''$ with $F_1''=J-I_{Br}$ and $J-I_{Br}+1$ and with $F_1''=J+I_{Br}-1$ and $J+I_{Br}$; the weaker quadrupole coupling due to the ^{35}Cl and ^{37}Cl nuclei caused line broadenings only. The frequencies of the unsplit rotational lines were determined from the measured hf splitting as well as by using the more accurate eqQ values from [1], and the average of both results was chosen to yield the final values for the reduced rotational frequencies [5, 6]. The measured and reduced rotational frequencies are listed in [5, pp. 125/31], the latter are compiled in the following table (ν in MHz):

J′←J″	v	$^{79}Br^{35}Cl$	$^{81}Br^{35}Cl$	$^{79}Br^{37}Cl$	$^{81}Br^{37}Cl$
14←13	0	–	–	122867.50	121901.16
15←14	0	136752.32	–	–	–
	1	136054.06	–	–	–
16←15	0	145864.37	–	–	–
	1	145120.00	–	–	–
17←16	0	154976.57	153803.42	149183.63	148010.41
	1	154185.23	153021.07	148436.39	147272.01
	2	153389.30	152234.27	–	–
19←18	0	173197.29	171886.28	–	–
	1	172312.75	171011.84	–	–
20←19	0	182306.21	–	175492.42	174112.39
	2	180438.74	–	–	–
	3	179496.82	178148.84	–	–
23←22	0	209626.51	208040.08	–	–
24←23	0	–	217075.70	–	–
	3	–	213742.24	–	–
25←24	2	–	–	217115.84	–
30←29	0	273329.76*)	–	–	–
34←33	0	309698.50*)	–	–	–

*) Quadrupole structure not resolved.

References:

[1] Smith, D. F.; Tidwell, M.; Williams, D. V. P. (Phys. Rev. [2] **79** [1950] 1007/8).

[2] Lovas, F. J.; Tiemann, E. (J. Phys. Chem. Ref. Data **3** [1974] 609/769).

[3] Nair, K. P. R.; Hoeft, J.; Tiemann, E. (Chem. Phys. Letters **58** [1978] 153/6).

[4] Nair, K. P. R. (Kem. Kozlem. **52** [1979] 431/50).

[5] Willis, R. E., Jr. (Diss. Duke Univ. 1979, pp. 1/158, 125/31; Diss. Abstr. Intern. B **40** [1980] 4880).

[6] Willis, R. E., Jr.; Clark,W. W., III (J. Chem. Phys. **72** [1980] 4946/50).

16.6.4.2 IR Absorption Spectrum

The IR spectrum of gaseous Br_2-Cl_2 mixtures, recorded at room temperature and total pressures of about 200 Torr in the region of the fundamental vibration of BrCl, exhibited a double band corresponding to the P- and R-branch envelopes of the $v=1 \leftarrow 0$ band; the minimum between the maxima was at $439.5 \pm 0.5\ cm^{-1}$, their separation $15 \pm 1\ cm^{-1}$ [1]. The IR absorption and localization of the intensity minimum at $440 \pm 2\ cm^{-1}$ was reinvestigated by [2]. In mixtures of Br_2 and Cl_2 ($p < 130$ Torr) with N_2 added to make a total pressure between 1.6 and 8 atm, the integrated intensity of the fundamental band ($439.5 \pm 0.5\ cm^{-1}$) was measured to be $A(1 \leftarrow 0) = 105 \pm 14$ cm/mmol, that of the first overtone ($876 \pm 1\ cm^{-1}$) was estimated to be $A(2 \leftarrow 0) < 2.5$ cm/mmol [3].

A high-resolution diode laser spectrum of the fundamental region was observed in equimolar Br_2-Cl_2 mixtures at room temperature and pressures of about 0.1 to 0.5 and 3 to 5 Torr. Out of the 388 lines recorded between 409 and 460 cm^{-1}, 320 (including 23 overlapping) lines could be assigned to the 1←0, 2←1, and 3←2 bands (P and R branches with J up to 94, 76, and 63, respectively) of the four isotopic species $^{79}Br^{35}Cl$, $^{79}Br^{37}Cl$, $^{81}Br^{35}Cl$, and $^{81}Br^{37}Cl$. The estimated wave number uncertainties are between 0.001 and 0.010 cm^{-1}, in a few cases 0.015 cm^{-1}. Least-squares fits to a Dunham expansion with eight parameters resulted in the Dunham coefficients Y_{ik} (i=0 to 3, k=0 to 2) for each of the isotopic species (cf. p. 189) [4].

References:

[1] Mattraw, H. C.; Pachucki, C. F.; Hawkins, N. J. (J. Chem. Phys. **22** [1954] 1117/9).
[2] Clyne, M. A. A.; Coxon, J. A. (Proc. Roy. Soc. [London] A **298** [1967] 424/52, 426/7).
[3] Brooks, W. V. F.; Crawford, B., Jr. (J. Chem. Phys. **23** [1955] 363/5).
[4] Nakagawa, K.; Horiai, K.; Konno, T.; Uehara, H. (J. Mol. Spectrosc. **131** [1988] 233/40).

16.6.4.3 Raman Effect

According to the UV-visible absorption continuum of BrCl, resonance Raman scattering (RRS) is observed upon excitation with the Ar^+ ion laser lines at 514.5 to 457.9 nm in gaseous [1, 2] and matrix-isolated [3] BrCl. The fundamental band at $\nu=440\pm1$ and 432 ± 1 cm^{-1} and the first overtone at $\nu=874\pm2$ and 856 ± 2 cm^{-1} (chlorine isotopic splitting) were observed in a Br_2-Cl_2 mixture (with an excess of Cl_2) upon excitation at 488 nm; a depolarization ratio of $\varrho_S=0.22$ for the fundamental band in the Stokes region and a ratio of 1.00: ~0.15 for the Raman scattering cross sections of the fundamental band and the overtone were found [1]. RRS upon excitation by the 514.5, 501.7, 496.5, 488.0, 476.5, and 457.9 nm laser lines has been extensively studied in Br_2-Cl_2 mixtures (t=25 °C, partial pressure of BrCl ~250 Torr) and overtones up to the fifth order have been observed; the resonance phenomenon was weak with 514.5 nm excitation (two overtones) and became stronger with decreasing wavelength of the exciting line (three, three, four, five, and five overtones, respectively, for the above given wavelengths). High-resolution measurements in the fundamental ($\Delta v=1$) and overtone ($\Delta v=2$, 3, and 4) regions revealed isotopic splitting due to the four species $^{79}Br^{35}Cl$, $^{81}Br^{35}Cl$, $^{79}Br^{37}Cl$, and $^{81}Br^{37}Cl$ and progressions of hot bands with $\Delta v=1$ and v′ up to 6, $\Delta v=2$ and v′ up to 9, $\Delta v=3$ and v′ up to 5; thus, using the molecular constants of [4], the 31 observed Raman peaks between 440.2 and 1285.0 cm^{-1} have been identified with 53 partially overlapping Q-branch heads in the various vibrational bands of the four isotopic species. For the resonance enhancement by hot bands, collisional excitation of BrCl by excited $Br_2(B\ ^3\Pi(0^+))$ molecules or $Br(^2P_{1/2})$ atoms present in the irradiated Br_2-Cl_2 mixture was proposed to be responsible [2]. RRS was also observed upon irradiation of matrix-isolated BrCl (cocondensation of an equilibrated Br_2-Cl_2 mixture with excess Ar or Kr at 12 K) with Ar^+ and Kr^+ ion laser lines; for $\lambda_{exc}=488$ nm, for example, the fundamental band was observed at 430 ± 2 cm^{-1}, the first overtone at 861 ± 3 cm^{-1} [3].

The Raman spectrum of liquid BrCl at 300 K irradiated by the Kr^+ ion laser line at 647.1 nm revealed four isotopic Stokes lines in the fundamental region at 433.8 ($^{79}Br^{35}Cl$), 432.4 ($^{81}Br^{35}Cl$), 425.6 ($^{79}Br^{37}Cl$), 424.2 ($^{81}Br^{37}Cl$) cm^{-1} and the corresponding hot bands ($v=2-1$) shifted by about 3.5 cm^{-1} to lower wave numbers [5]. Earlier, the fundamental band of BrCl was observed upon excitation by He radiation at 587.56 and 667.82 nm in a 0.5 M solution of Br_2 in liquid Cl_2 to be 434.1 ± 0.3 ($^{79,81}Br^{35}Cl$) and 425.5 ± 0.3 ($^{79,81}Br^{37}Cl$) cm^{-1}, in mixtures of 1 M Br_2+ ~1.8 M Cl_2 in CCl_4 to be 431 ± 1 ($^{79,81}Br^{35}Cl$) and 423 ± 1 ($^{79,81}Br^{37}Cl$) cm^{-1} (the diffuseness of the spectra prevented the observation of the ^{79}Br-^{81}Br isotopic splitting), and in a 0.75 M solution of Br_2 in C_6H_6 saturated in Cl_2 to be 418 ± 2 ($^{79,81}Br^{35}Cl$) cm^{-1} [6]; a preliminary result for Br_2-Cl_2 in CCl_4 was 428 ± 2 cm^{-1} [7].

References:

[1] Holzer, W.; Murphy, W.; Bernstein, H. J. (J. Chem. Phys. **52** [1970] 399/407).
[2] Chang, H.; Lin, H. M. (Proc. Natl. Sci. Council Repub. China B **5** [1981] 263/72).

[3] Wight, C. A.; Ault, B. S.; Andrews, L. (J. Mol. Spectrosc. **56** [1975] 239/50).
[4] Willis, R. E., Jr.; Clark, W. W., III (J. Chem. Phys. **72** [1980] 4946/50).
[5] Wallart, F. (Can. J. Spectrosc. **17** [1972] 128/31).
[6] Stammreich, H.; Forneris, R.; Tavares, Y. (Spectrochim. Acta **17** [1961] 1173/84).
[7] Stammreich, H.; Forneris, R. (J. Chem. Phys. **21** [1953] 944/5).

16.6.4.4 Near-IR, Visible, and UV Spectra

16.6.4.4.1 Near-IR and Visible Region. The B $^3\Pi(0^+)\leftrightarrow$X $^1\Sigma^+$ System

Absorption Spectrum. Conventional absorption measurements and laser-induced fluorescence (LIF) measurements were carried out for the study of the absorption spectrum of BrCl. Equilibrium mixtures of gaseous Br_2, Cl_2, and BrCl (with an excess of Cl_2 to minimize the bromine absorption) were used for the absorption measurements [1 to 3]. LIF measurements were also made with equilibrium mixtures of gaseous Br_2, Cl_2, and BrCl [4 to 6] and with a supersonic beam of BrCl seeded in Cl_2 [7].

A system of absorption bands of BrCl in the 550 to 720 nm region and between 294 and 690 K was observed for the first time by Clyne and Coxon [1], who compared it with the already known emission spectrum of BrCl (see below) and with the spectra of other interhalogens and assigned 31 band heads at 550.0 to 677.1 nm to short vibrational progressions originating from ground-state vibrational levels v″ = 0 to 6. The first rotational analysis of the 6←2, 6←3, 7←3, and 8←3 bands at 588, 603, 598, and 593 nm, which revealed pairs of single P and R branches of $^{79}Br^{35}Cl$ and $^{81}Br^{35}Cl$, confirmed the B $^3\Pi(0^+)\leftarrow$X $^1\Sigma^+$ assignment with the B state belonging to Hund's case (c) [2]. A systematic high-resolution study (inverse dispersion 0.55 to 0.85 Å/mm) of the B←X system in the 560 to 720 nm region between 300 and 700 K and partial pressures of BrCl in the Br_2-Cl_2 mixtures of about 10 to 30 Torr has been reported by Coxon [3]: the spectrum comprises 24 bands of the v″ = 1 to 7 progressions, each exhibiting simple PR structure and isotopic lines with relative intensities of about 3:3:1:1 corresponding to the natural abundancies of the four species $^{79}Br^{35}Cl$, $^{81}Br^{35}Cl$, $^{79}Br^{37}Cl$, and $^{81}Br^{37}Cl$; only the dominating lines due to the ^{35}Cl species have been analyzed. The band origins λ_0, vibrational assignments, and examined J″ ranges are as follows:

	$^{79}Br^{35}Cl$		$^{81}Br^{35}Cl$	
v′←v″	J″	λ_0 in nm	J″	λ_0 in nm
8←1	8 to 38	563.9810	7 to 40	564.0252
7←1	11 to 40	568.2108	11 to 40	568.2619
8←2	7 to 36	578.2312	7 to 36	578.2232
7←2	27 to 53	582.6783	27 to 54	582.6766
6←2	11 to 53	587.8629	12 to 53	587.8564
8←3	15 to 49	593.0888	17 to 48	593.0240
7←3	9 to 55	597.7683	10 to 56	597.7092
6←3	12 to 55	603.2261	13 to 55	603.1609
8←4	14 to 45	608.5901	16 to 44	608.4633
7←4	9 to 53	613.5184	10 to 54	613.3968
6←4	9 to 71	619.2690	13 to 71	619.1397
5←4	11 to 64	625.7053	10 to 66	625.5607

$v'\leftarrow v''$	$^{79}Br^{35}Cl$ J″	λ_0 in nm	$^{81}Br^{35}Cl$ J″	λ_0 in nm
7←5	6 to 56	629.9703	8 to 56	629.7809
4←4	22 to 52	632.7565	24 to 51	632.5897
6←5	5 to 70	636.0350	7 to 72	635.8362
5←5	6 to 76	642.8265	7 to 75	642.6100
4←5	9 to 71	650.2711	5 to 72	650.0297
6←6	10 to 55	653.5706	22 to 55	653.2960
5←6	7 to 76	660.7438	8 to 76	660.4490
4←6	11 to 76	668.6118	14 to 81	668.2888
3←6	10 to 67	677.1377	6 to 67	676.7805
2←6	13 to 60	686.3074	21 to 63	685.9089
4←7	11 to 62	687.8357	8 to 60	687.4236
3←7	12 to 66	696.8623	16 to 65	696.4117

A list of wave numbers for more than two thousand identified rotational lines of $^{79}Br^{35}Cl$ and $^{81}Br^{35}Cl$ is available on request from the Editor of the Journal of Molecular Spectroscopy or from the author [3].

At higher wavelengths, a number of bands were recorded which showed extensive rotational structure. Their band origins (approximate estimates), however, were inconsistent with the identification of these bands as part of the B←X system; they were tentatively assigned to the A $^3\Pi(1)\leftarrow X\ ^1\Sigma^+$ system by analogy with the spectra of other interhalogens; no details on the A←X system are reported [3].

High-resolution excitation spectra (LIF) of the B←X system have been recorded at room temperature by Clyne and co-workers [4 to 6] using gaseous Br_2-Cl_2 mixtures (with Cl_2 in excess) and tunable dye lasers in the 550 to 570, 570 to 595, and 630 to 670 nm regions. This spectroscopic method enables the observation of the shortest wavelength bands which are obscured in the absorption spectrum by the strong absorption continuum.

Studies with Br_2-Cl_2 (1:15) mixtures at total pressures of about 5 Torr allowed rotational structure of several vibrational bands and isotopic splitting due to the $^{79}Br^{35}Cl$ and $^{81}Br^{35}Cl$ species to be analyzed (the weak branches due to $^{79}Br^{37}Cl$ and $^{81}Br^{37}Cl$ were observed but not assigned). The 550 to 570 nm region, which is overlapped by strong Br_2 absorption lines, contains the $v''=0$ progressions with $v'=3$ to 8; the origins of the 8←0 and 4←0 bands were observed near 550.3 and 570.0 nm, respectively; a few rotational lines of the 6←0 and 5←0 bands could be assigned. In the 570 to 595 nm region, the $v''=1$ ($v'=3$ to 6) and $v''=2$ ($v'=5$ to 7) progressions with rotational structure up to $J'=61$ have been analyzed. The fact, that the 7←2 band is anomalously weak (i.e., weaker by a factor of about 5 than predicted from Franck-Condon factors and thermal populations) and that the 8←1 and 8←2 bands are completely absent from the LIF spectrum, supplements the lifetime measurements (cf. pp. 196/8) which demonstrated rotationally dependent predissociation of the B-state $v'=7$ level [4]. At wavelengths above 630 nm and with Br_2-Cl_2 mixtures of 0.1 to 0.2 Torr (~15% BrCl, ~85% Cl_2), rotational and isotopic ($^{79}Br^{35}Cl$-$^{81}Br^{35}Cl$) structures of five bands with $v'=0$ to 3 have been analyzed [5]. Band origins in the 570 to 595 and 630 to 670 nm regions are as follows [4, 5]:

$^{79}Br^{35}Cl$ v′←v″	λ_0 in nm	$^{79}Br^{35}Cl$ v′←v″	λ_0 in nm	$^{79,81}Br^{35}Cl$ v′←v″	λ_0 in nm
6←1	573.140	7←2	582.678	2←3	631.181
5←1	578.649	6←2	587.863	1←4	657.521
4←1	584.674	5←2	593.660	3←5	658.515
3←1	591.184			0←4	666.844
				2←5	667.181

Studies under approximately collision-free conditions (total pressures near 5 mTorr and partial pressures of BrCl near 0.4 mTorr) in the 570 to 590 nm region enabled rotational assignment in all four isotopic P and R branches of the 4←1 and 5←1 bands as well as an improved rotational assignment in the 6←1 band of $^{81}Br^{35}Cl$ [6].

With a supersonic beam of BrCl seeded in Cl_2 and a dye laser tunable in the 580 to 600 nm region, LIF revealed the 7←2, 4←1, 6←2, 3←1, and 5←2 bands [7].

The electronic transition dipole moment for the B←X transition, $|R_e|^2 = 0.096 \pm 0.010$ D^2 (without any significant variation with v′), was calculated from improved radiative lifetimes measured for the vibrational manifolds v′ = 3 to 6 (cf. p. 197) [6]. A previous result based on the lifetimes for the v′ = 3 and 4 manifolds was $|R_e|^2 = 0.11 \pm 0.04$ D^2; using this value, the derivation of the absorption coefficient for rovibrational lines of the B←X system was illustrated for the very intense R(20) line of the 6←1 band of $^{79}Br^{35}Cl$ [8].

Emission Spectrum. The B→X emission spectrum has been observed as the deep red chemiluminescence arising from the reaction of bromine atoms or molecules with chlorine dioxide or from the recombination of ground-state bromine and chlorine atoms in the presence of a third body ("halogen afterglow") [9 to 13] and as resonance fluorescence series excited by laser irradiation of gaseous [14 to 18] or matrix-isolated [19] BrCl molecules. Observations and analyses of the B→X spectrum cover the visible and near-IR regions from 565 to 1107 nm.

The B→X chemiluminescence of **gaseous** BrCl was first observed by Clyne and Coxon [9, 10] in a fast flow system at total pressures of 1 to 4 Torr when Br atoms or Br_2 molecules were mixed with ClO_2. Using Br atoms, the rapid reaction $Br + ClO_2 \rightarrow BrCl(B\ ^3\Pi(0^+)) + O_2$ occurred directly; using Br_2 molecules, the reaction proceeded only at comparatively high reagent partial pressures (e.g., $p(Br_2) = 0.3$, $p(ClO_2) = 0.4$, $p(Ar) = 1.5$ Torr) and when the mixture was slightly heated or sparked with a Tesla coil. The spectrum was recorded in the 565 to 880 nm region at a resolution, which allowed vibrational analysis of the red-degraded forty-nine $^{79,81}Br^{35}Cl$ bands (v′ = 2 to 8, v″ = 1 to 14) and sixteen $^{79,81}Br^{37}Cl$ bands (v′ = 2 and 3, v″ = 6 to 13). The identification of the BrCl emitter was confirmed by recording the emission from a mixture of approximately equal concentrations of Br and Cl atoms [10]. A reinvestigation of the spectrum ($Br_2 + ClO_2$ reaction) and extension into the IR to 1105 nm by Hadley et al. [11] confirmed the vibrational assignment for $^{79,81}Br^{35}Cl$ in the 570 to 750 nm region (v′ = 2 to 8, v″ = 2 to 9); in the long-wavelength region, however, the v′ = 0, v″ = 9 to 19 and v′ = 1, v″ = 8 to 16 progressions were identified and the earlier assignments revised: bands assigned to the 2→10 to 2→14 and 3→9 to 3→14 progressions by [10] turned out to belong to the 0→9 to 0→13 and 1→8 to 1→13 progressions, respectively [11].

The modifications of Hadley et al. [11] have been confirmed by Clyne and Smith [12] who presented the first quantitative study of the BrCl afterglow (640 to 900 nm; 28 bands with v′=0 to 4, v″=6 to 14) resulting from the radiative recombination of $Br(^2P_{3/2})$ and $Cl(^2P_{3/2})$ atoms. Recording the Br_2+ClO_2 chemiluminescence at high resolution (0.5 Å), Clyne and Toby [13] succeeded in the first rotational analysis for the strong 0→12, 1→12, 0→11, and 2→10 bands: P- and R-branch lines of $^{79}Br^{35}Cl$ and $^{81}Br^{35}Cl$ could be assigned for up to 94 J values, and the following band origins ν_0 (in cm^{-1}) were obtained:

v′→v″	$\nu_0(^{79}Br^{35}Cl)$	$\nu_0(^{81}Br^{35}Cl)$	number of lines
0→12	11699.98±0.04	11718.22±0.03	88 and 94
1→12	11912.33±0.03	11929.97±0.04	83 and 91
0→11	12097.26±0.08	12114.77±0.06	73 and 85
2→10	12917.61±0.06	12931.78±0.11	70 and 60

Laser-induced resonance fluorescence of gaseous BrCl was excited in equilibrated Br_2-Cl_2 mixtures. First, irradiation with a dye laser tuned to the B←X, v=6←2 transition at 587.8 nm and at a total pressure of 275.5 Torr caused emission of only two short resonance fluorescence series, the v′=0, v″=8 to 12 and the v′=1, v″=9 to 12 progression, in the 700 to 860 nm region (the emission from the v′=0 and 1 levels only indicates a thermalized vibrational distribution attained to by collisional V-T transfer, cf. p. 198) [14, 15]. More recently, use of a dye laser of very narrow line width, total pressures of 10 to 12 Torr and in some cases 140 mTorr, and an appropriate fluorescence detection system enabled excitation of single rovibrational and isotopic lines and the observation of transitions from all B-state v′=0 to 6 levels into high vibrational levels of the ground state in the 650 to 1100 nm region. By pumping to the R(44) line of the 6←1 transition of $^{79}Br^{35}Cl$ at 17347.00 cm^{-1} and to the P(37) line of the 6←1 transition of $^{81}Br^{35}Cl$ at 17353.31 cm^{-1}, McFeeters et al. [16] obtained the following band heads, λ_H (in nm):

v′→v″	$\lambda_H(^{81}Br^{35}Cl)$	$\lambda_H(^{79}Br^{35}Cl)$	v′→v″	$\lambda_H(^{81}Br^{35}Cl)$	$\lambda_H(^{79}Br^{35}Cl)$
0→8	–	750.75	2→5	666.66	–
9	773.82	774.54	6	685.89	686.34
10	798.76	799.65	7	706.02	706.57
11	825.14	826.34	8	727.26	727.85
12	853.11	854.40	9	749.50	750.21
13	882.74	884.23	10	773. 12	773.78
14	914.00	915.86	11	797.91	798.57
15	947.30	949.64			
16	982.70	985.10	3→5	657.94	–
17	1020.5	1022.58	6	676.79	–
18	1060.8	1063.81	7	696.36	696.89
19	1104.5	1107.41	8	716.96	717.55
			10	–	762.47
1→5	675.95	–	12	–	811.83
6	695.58	696.15	14	865.50	866.84
7	716.36	716.93			
8	738.24	–	4→5	649.93	–
9	761.12	762.04	7	687.50	687.88
10	785.47	786.21			
11	810.96	812.13	5→5	642.51	–

v′→v″	$\lambda_H(^{81}Br^{35}Cl)$	$\lambda_H(^{79}Br^{35}Cl)$	v′→v″	$\lambda_H(^{81}Br^{35}Cl)$	$\lambda_H(^{79}Br^{35}Cl)$
1→12	837.94	839.14	5→7	679.07	–
13	866.45	867.95	9	–	719.89
			11	763.36	–
			12	787.42	–
			6→4	619.02	–
			5	635.75	–
			11	754.20	–

For the spectrally resolved LIF emission spectra in the 700 to 900 nm region recorded at 50 mTorr and 1 Torr after excitation of the v′=6, J′=37 level, see [18].

Laser action of BrCl has not been observed in the near-IR-visible region until now. By considering the radiative lifetimes for several rovibrational levels [8] (cf. p. 198) Davis [19] demonstrated the B→X, v=0→15 transition to be an attractive candidate for lasing in the near IR and predicted the optical gain that would be available from such a BrCl laser.

In **matrix-isolated** BrCl (generated by codeposition of an equilibrated Br_2-Cl_2 mixture with Ar or Kr gas at 12 K) resonance fluorescence was also excited in the 660 to 900 nm region by the 457.9, 476.5, 488.0, and 514.5 nm Ar^+ and the 568.1 and 530.9 nm Kr^+ ion laser lines; a series consisting of pairs of doublets with spacings of approximately 420 cm^{-1} was attributed to the B→X, v=0→5 to 0→12 progression. Band positions ν (in cm^{-1}) with error limits of ±2 cm^{-1} for BrCl in Ar excited at 514.5 nm are as follows [20]:

v′→v″	$\nu(^{79}Br^{35}Cl)$	$\nu(^{81}Br^{35}Cl)$	$\nu(^{79}Br^{37}Cl)$	$\nu(^{81}Br^{37}Cl)$
0→ 4(?)	~14800*)	–	–	–
0→ 5	14391	–	–	–
0→ 6	13970	–	14018	–
0→ 7	13555	13566	13612	13623
0→ 8	13147	13159	13211	13222
0→ 9	12741	12754	12810	12826
0→10	12341	12355	12418	12433
0→11	11942	11957	12025	12042
0→12	11547	11564	11638	–

*) Read from [20, figure 1].

In the Kr matrix the resonance series is shifted by 76 cm^{-1} to lower wave numbers [20].

Franck-Condon Factors, r-Centroids. The Franck-Condon (FC) factors computed on the basis of various spectroscopic data show the low transition probability for the B↔X system. Using the RKR potentials for the X and B states of $^{79}Br^{35}Cl$ and $^{81}Br^{35}Cl$ based on his own absorption data, Coxon [3] calculated FC factors and r-centroids for all B↔X transitions between vibrational levels with v″=0 to 7 and v′=0 to 8 in $^{79}Br^{35}Cl$ and $^{81}Br^{35}Cl$; they are listed in [3, table 15]. Using the RKR potentials for the X and B states of $^{79}Br^{35}Cl$ and $^{81}Br^{35}Cl$ based on their own emission data, McFeeters et al. [16] extended the v″

and v′ range and calculated FC factors for transitions with v″=0 to 20 and v′=0 to 8, see [16, tables 5 and 6]; agreement with Coxon's [3] reported values is excellent. Applying Morse potentials for the X and B states of $^{79}Br^{35}Cl$ based on their own emission data, Hadley et al. [11] derived FC factors (values have not been published) presumably for the v″, v′ range covered by their emission measurements, v″=2 to 19, v′=0 to 8, and compared predicted and observed intensity maxima in the v′=0 and 1 progressions.

References:

[1] Clyne, M. A. A.; Coxon, J. A. (Nature **217** [1968] 448/9).
[2] Clyne, M. A. A.; Coxon, J. A. (J. Phys. B **3** [1970] L9/L11).
[3] Coxon, J. A. (J. Mol. Spectrosc. **50** [1974] 142/65).
[4] Clyne, M. A. A.; McDermid, I. S. (J. Chem. Soc. Faraday Trans. II **74** [1978] 798/806).
[5] Clyne, M. A. A.; Zai, L. C. (J. Chem. Soc. Faraday Trans. II **78** [1982] 1221/9).
[6] Clyne, M. A. A.; McDermid, I. S. (Faraday Discussions Chem. Soc. No. 67 [1979] 316/28).
[7] Farthing, J. W.; Fletcher, I. W.; Whitehead, J. C. (Mol. Phys. **48** [1983] 1067/73).
[8] Clyne, M. A. A.; McDermid, I. S. (J. Chem. Soc. Faraday Trans. II **74** [1978] 807/20).
[9] Clyne, M. A. A.; Coxon, J. A. (J. Chem. Soc. Chem. Commun. **1966** 285/6).
[10] Clyne, M. A. A.; Coxon, J. A. (Proc. Roy. Soc. [London] A **298** [1967] 424/52).
[11] Hadley, S. G.; Bina, M. J.; Brabson, G. D. (J. Phys. Chem. **78** [1974] 1833/6).
[12] Clyne, M. A. A.; Smith, D. (J. Chem. Soc. Faraday Trans. II **75** [1979] 704/24).
[13] Clyne, M. A. A.; Toby, S. (J. Photochem. **11** [1979] 87/100).
[14] Wright, J. J.; Spates, W. S.; Davis, S. J. (J. Chem. Phys. **66** [1977] 1566/70).
[15] Wright, J. J.; Spates, W. S.; Davis, S. J. (Electron. Transition Lasers 2 Proc. 3rd Summer Colloq. 6, Snowmass Village, Colo., 1976 [1977], pp. 288/92).
[16] McFeeters, B. D.; Perram, G. P.; Crannage, R. P.; Dorko, E. A. (Chem. Phys. **139** [1989] 347/57).
[17] Wolf, P. J.; Perram, G. P.; Davis, S. J. (Springer Proc. Phys. **15** [1987] 556/61).
[18] Perram, G. P.; Davis, S. J. (J. Chem. Phys. **93** [1990] 1720/31).
[19] Davis, S. J. (Laser Interact. Relat. Plasma Phenom. **6** [1984] 33/45).
[20] Wight, C. A.; Ault, B. S.; Andrews, L. (J. Mol. Spectrosc. **56** [1975] 239/50).

16.6.4.4.2 Absorption Continuum in the Visible and Near UV

Using gaseous Br_2–Cl_2 (2:1, 1:1, 1:2) mixtures at 25 °C, Seery and Britton [1] measured the absorption spectrum in the 600 to 220 nm region; the molar absorption coefficients ε(BrCl) were derived for the 510 to 220 nm region from the total optical density using the extinction coefficients of pure Br_2 and Cl_2 and the equilibrium constant of BrCl from [2]: a broad absorption band between 510 and 300 nm with $\lambda_{max}=370$ nm and $\varepsilon_{max}=107\ L\cdot mol^{-1}\cdot cm^{-1}$ is followed by a second with $\lambda_{max}\approx 230$ nm and $\varepsilon_{max}=18.8\ L\cdot mol^{-1}\cdot cm^{-1}$. By fitting the experimental data to theoretical equations for ε as a function of temperature and wavelength, as suggested by Sulzer and Wieland [3], the absorption continuum has been resolved into three overlapping Gaussian absorption curves characterized by three parameters at T=0 K, the molar absorption coefficient ε^{o}_{max} and the wave number ν^{o}_{max} at the absorption maximum, and the half-width $\Delta\nu^{o}$, as follows [1]:

ε^{o}_{max} in $L\cdot mol^{-1}\cdot cm^{-1}$	ν^{o}_{max} in cm^{-1}	$\Delta\nu^{o}$ in cm^{-1}
25.1 ± 1.4	21952 ± 59	1945 ± 64
121.2 ± 0.4	26768 ± 17	2537 ± 19
21.3 ± 0.2	43898 ± 64	5023 ± 75

According to Clyne and McDermid [4], the absorption continuum is due mainly to transitions from the X $^1\Sigma^+$,v″=0 level to unbound levels above the dissociation limit of the B $^3\Pi(0^+)$ state. No assignment for the three absorption curves was given by [1]; compare, however, "Bromine" Suppl. Vol. A, 1985, p. 250.

The absorption continuum of BrCl in the 250 to 500 nm region was studied also in various solutions; these were prepared by mixing various quantities of Br_2 and Cl_2 with the solvent, and the BrCl absorption had to be extracted from the total absorption (cf. above). The absorption maximum was found at 370 nm for a 0.005 M solution of BrCl in CCl_4 [5], at ~350 nm for a 0.005 M solution in H_2O, and at 343 nm for 0.005 M solutions in 1.0 N H_2SO_4, 0.2 N and 0.5 N NaCl, 0.5 N and 1.5 N HCl [6, 7], at 343 nm for BrCl in 2 N $HClO_4$ [8, 9], at 385 nm for a 0.006 M solution in $SOCl_2$ (Br_2-Cl_2 1:1 mixture), and at 373 nm for "pure" BrCl in $SOCl_2$ (Br_2-Cl_2 1:10 mixture) [10]. In the aqueous solutions, the absorption coefficient at the absorption maximum is independent of the H_2SO_4 concentration (1.0 to 5.0 N), but increases with the concentration of the chloride ion in the solutions containing NaCl or HCl [6, 7].

References:

[1] Seery, D. J.; Britton, D. (J. Phys. Chem. **68** [1964] 2263/6).
[2] Evans, W. H.; Munson, T. R.; Wagman, D. D. (J. Res. Natl. Bur. Std. **55** [1955] 147/64).
[3] Sulzer, P.; Wieland, K. (Helv. Phys. Acta **25** [1952] 653/76).
[4] Clyne, M. A. A.; McDermid, I. S. (J. Chem. Soc. Faraday Trans. II **74** [1978] 807/20).
[5] Popov, A. I.; Mannion, J. J. (J. Am. Chem. Soc. **74** [1952] 222/4).
[6] Pungor, E.; Burger, K.; Schulek, E. (J. Inorg. Nucl. Chem. **11** [1959] 56/61).
[7] Pungor, E.; Burger, K.; Schulek, E. (Magy. Kem. Foly. **65** [1959] 301/5).
[8] Grove, J. R.; Raphael, L. (J. Inorg. Nucl. Chem. **25** [1963] 130/2).
[9] Raphael, L. (Bromine Compounds Chem. Appl. 1st Intern. Conf. Chem. Appl. Bromine Its Compounds, Salford, U.K., 1986 [1988], pp. 369/84).
[10] Abraham, K. M.; Alamgir, M. (J. Electrochem. Soc. **134** [1987] 2112/80).

16.6.4.4.3 Ultraviolet Absorption and Emission Systems

Absorption. Using gaseous Br_2-Cl_2 mixtures with Br_2 in excess, Cordes and Sponer [1] recorded the absorption spectrum of BrCl in the vacuum UV between 172 and 156 nm; they arranged the heads of 21 violet-degraded bands into two systems with v″=0, v′=0 to 5; v″=1, v′=0, 1, 5, 6; v″=2, v′=0, 6 (172 to 161 nm) and v″=0, v′=0 to 5; v″=1 and 2, v′=0 (165 to 156 m), but could not identify the electronic states involved. These band systems also appeared in the spectrum resulting from flash photolysis of BrCl [2]. Recently, Hopkirk et al. [3] reported on vacuum-UV absorption, fluorescence excitation, and dispersed fluorescence (see below) spectra of BrCl using Br_2-Cl_2 mixtures with Cl_2 in excess and synchrotron radiation at 180 to 137 nm; the early vibrational assignment of [1] was confirmed, and the absorption features in the 170 to 163 and 164 to 155 nm regions were identified as transitions from the ground state X $^1\Sigma^+$ to the first members of the nsσ Rydberg series, $\{BrCl^+, X\ ^2\Pi_{3/2}\}$ 5sσ $^{1,3}\Pi$ and $\{BrCl^+, X\ ^2\Pi_{1/2}\}$ 5sσ $^{1,3}\Pi$ (not decided whether $^1\Pi$ or $^3\Pi$), labeled a_5 and b_5, respectively. The absorption maxima were assigned as follows (wave numbers ν in cm^{-1}; Table see p. 210).

The average upper-state vibrational separations are 507±3 and 508±8 cm^{-1} in the a_5 and b_5 band system, respectively. The Rydberg bands corresponding to the $\{BrCl^+, X\ ^2\Pi_{3/2}\}$ 5pσ and $\{BrCl^+, X\ ^2\Pi_{1/2}\}$ 5pσ states were estimated to lie at 144.5 and 140.2 nm, which is slightly longer than the wavelengths of a group of Rydberg-like bands assigned to these Rydberg transitions [3].

$v' \leftarrow v''$	$\nu(a_5 \leftarrow X)$	$\nu(b_5 \leftarrow X)$
$0 \leftarrow 1$	58952	61169
$0 \leftarrow 0$	59403	61614
$1 \leftarrow 0$	59916	62131
$2 \leftarrow 0$	60419	62637
$3 \leftarrow 0$	60920	63135
$4 \leftarrow 0$	61429	63637
$5 \leftarrow 0$	61939	64152

The absorption between 155 and 144 nm is predominantly due to transitions from the ground state X $^1\Sigma^+$ to ion-pair states; these are the E(0^+) state correlating with $Br^+(^3P_2) + Cl^-(^1S_0)$ and the f(0^+) state lying 3840 cm^{-1} higher and correlating with $Br^+(^3P_0) + Cl^-(^1S_0)$ [3].

The corresponding fluorescence excitation spectrum in the 180 to 137 nm region (310 to 450 nm detection) indicates resonant interactions between the ion-pair and the Rydberg states. The onset of fluorescence excitation attributed to a quasi-continuous ion-pair$\leftarrow$X absorption is at ~167 nm followed by an increase in intensity to a maximum at 147 nm; a number of "dip resonances" in the 155 to148 nm region show the reduction of the ion-pair fluorescence by predissociated Rydberg vibrational levels [3] (for more detailed explanations of this resonance phenomenon see [4]).

Emission. In the near UV between 330 and 297 nm, two or three emission systems have been observed which arise from transitions between ion-pair excited upper states belonging to the ion-pair cluster $Br^+(^3P_2) + Cl^-(^1S_0)$ and excited valence states; these are the dominating D′ $^3\Pi(2) \rightarrow$ A′ $^3\Pi(2)$ transition occurring in the 320 to 297 nm region and the less intense E $^3\Pi(0^+) \rightarrow$ B $^3\Pi(0^+)$ transition occurring in the 329 to 323 nm region; the third transition, β $^3\Pi(1) \rightarrow$ A $^3\Pi(1)$, was observed as a weak band system in the D′$\rightarrow$A′ spectrum between 318 and 308 nm.

The D′$\rightarrow$A′ emission was first observed and identified by Diegelmann et al. [5, 6] when mixtures of CF_3Br and Cl_2 in helium gas were excited by short pulses of high-energy electrons. The maximum of emission at 314.2 nm (the $0 \rightarrow 10$ band, see below) has been suggested to be an appropriate candidate for lasing in the UV [7]. Vibrational and partial rotational analyses of the D′$\rightarrow$A′ system have been presented by Tellinghuisen and coworkers [8, 9]: the emission was excited by Tesla discharges in mixtures of appropriate isotopes of Br_2 and Cl_2 (1:5) with Ar and the vibrational assignments included band heads for all four isotopic species of BrCl; a list of all assigned band heads is available on request from the authors; the band heads and assignments for $^{79}Br^{35}Cl$ and $^{79}Br^{37}Cl$ ($v'=0$ to 4, $v''=5$ to 18) are listed in [9]. High-resolution spectra of the most prominent bands $0 \rightarrow 8$, $0 \rightarrow 9$, and $0 \rightarrow 10$ of $^{79}Br^{35}Cl$ and $^{81}Br^{37}Cl$ allowed rotational analysis. The band origins ν_0 are as follows (wave numbers ν_0 in cm^{-1}; standard deviation in parentheses) [8, 9]:

$v' \rightarrow v''$	$\nu_0(^{79}Br^{35}Cl)$	$\nu_0(^{81}Br^{37}Cl)$
$0 \rightarrow 8$	32136.66(1)	32166.24(1)
$0 \rightarrow 9$	31976.58(1)	32008.31(2)
$0 \rightarrow 10$	31824.47(1)	31857.86(1)

Lists of all assigned P- and R-branch lines of $^{79}Br^{35}Cl$ and $^{81}Br^{37}Cl$ (J=8 to 90) are given; the Q branches were too weak for rotational analysis [9].

Franck-Condon factors for the D′→A′, v′=0 to 5, v″=3 to 18, transitions in $^{79}Br^{35}Cl$ and $^{79}Br^{37}Cl$ were calculated by assuming a Morse potential for the D′ state and a Morse-RKR potential for the A′ state [9].

The E→B spectrum, a system of red-degraded bands, appears to the red of the D′→A′ system and is about a factor 10 less intense. Aided by the known B-state vibrational energies [10], the bands have been assigned to transitions of the v′=0, 1, and 2 progressions [8, 9]. Rotational analyses were performed for the 0→8, 7, 6 and 1→6 bands of $^{81}Br^{37}Cl$ which resulted in about 370 assigned P- and R-branch lines with J=16 to 71, 9 to 86, 13 to 85, and 17 to 81, respectively; spectral congestion near the band origins hindered the assignment of low J values [11]. The following band heads ν_H [8] and band origins ν_0 [11] were obtained (wave numbers ν in cm^{-1}; standard deviation in parentheses):

v′→v″	$\nu_0(^{81}Br^{37}Cl)$	$\nu_H(^{81}Br^{37}Cl)$	$\nu_H(^{79}Br^{37}Cl)$	$\nu_H(^{81}Br^{35}Cl)$	$\nu_H(^{79}Br^{35}Cl)$
0→9	–	30382.3	30379.2	–	30365.7
0→8	30489.514(17)	30490.8	30476.8	30472.6	30469.0
0→7	30622.449(17)	30623.9	30619.0	30604.4	30600.8
0→6	30772.938(15)	30773.5	30769.8	–	30753.4
1→9	–	30571.2	30568.2	30560.0	39557.8
1→7	–	30811.8	30809.2	–	30793.6
1→6	30961.509(22)	30962.5	30959.5	–	30946.7
2→9	–	30755.3	–	30749.5	30749.5

Franck-Condon factors and r-centroids for the E→B (v′=0 to 5, v″=0 to 9) system were calculated using a Morse potential for the E state and a modified Morse potential (interaction with a repulsive state) for the B state [11].

A few red-degraded bands in the D′→A′ region possibly belong to the expected β→A transition [8, 9]; wave numbers and assignments for $^{79}Br^{35}Cl$ and $^{79}Br^{37}Cl$ (v′=0 to 3, v″=7 to 17) are listed in [9]. Mixing of the β state into the E state was thought to be the reason for anomalous rotational behavior of the E state [11].

Franck-Condon factors for the β→A, v′=0 to 5, v″=3 to 18, transitions in $^{79}Br^{35}Cl$ and $^{79}Br^{37}Cl$ were calculated by assuming a Morse potential for the β state and a Morse-RKR potential for the A state [9].

Six continuous emission systems in the near UV at 360 to 352, 339 to 330, 319 to 307, 295 to 282.5, 275 to 271, and 260.5 to 255 nm arise from condensed discharges and from high-frequency discharges in Br_2-Cl_2 mixtures and were reported earlier without assignment [12]. The lack of any discrete structure, for example for the D′→A′ region at 319 to 307 nm, probably is due to the high effective temperature in the discharge sources employed [8].

Dispersed fluorescence spectra in the 140 to 400 nm region were recorded following irradiation of BrCl with 147 and 153 nm synchrotron radiation which causes excitation of the ion-pair state $E(0^+)$ and to a lesser extent of the ion-pair state $f(0^+)$ (cf. above). An intense fluorescence peak labeled j, close to the excitation wavelength, was assigned to E→X, v″=0, 1, 2, ... transitions. Several irregularly spaced peaks between 370 and 230 nm, labeled a to i, most probably are due to transitions from the ion-pair E and f states into four of the five shallow-bound $\Omega=0^+$ valence states that correlate with $Br(^2P_{3/2}$ or $^2P_{1/2})+Cl(^2P_{3/2}$ or $^2P_{1/2})$ [3].

In contrast to BrF (see p. 32) laser action in the UV has not yet been observed with BrCl. However, four laser transitions in the UV were expected by Parks [13] to originate from the lowest ion-pair states and to terminate in the lowest excited valence states $^3\Pi(2)$ or $^3\Pi(1)$. The upper state of the transitions at 340 and 370 nm correlates with $Br^+(^3P_{1,0})+Cl^-(^1S_0)$ and $Br^+(^3P_2)+Cl^-(^1S_0)$, and of the transitions at 260 and 270 nm the upper state correlates with $Cl^+(^3P_{1,0})+Br^-(^1S_0)$ and $Cl^+(^3P_2)+Br^-(^1S_0)$.

References:

[1] Cordes, H.; Sponer, H. (Z. Physik **79** [1932] 170/85).
[2] Donovan, R. J.; Husain, D. (Trans. Faraday Soc. **64** [1968] 2325/31).
[3] Hopkirk, A.; Shaw, D.; Donovan, R. J.; Lawley, K. P.; Yencha A. J. (J. Phys. Chem. **93** [1989] 7338/42).
[4] Yencha, A. J.; Donovan, R. J.; Hopkirk, A.; Shaw, D. (J. Phys. Chem. **92** [1988] 5523/9).
[5] Diegelmann, M. (Max-Planck-Ges. Foerd. Wiss. Projektgruppe Laserforsch. Ber. PLF-33 [1980] 1/124, 42/5).
[6] Diegelmann, M.; Hohla, K.; Rebentrost, F.; Kompa, K. L. (J. Chem. Phys. **76** [1982] 1233/47).
[7] Diegelmann, M.; Grieneisen, H. P. (Ger. Offen. 3031954 [1980/82] 1/14).
[8] Chakraborty, D. K.; Tellinghuisen, P. C.; Tellinghuisen, J. (Chem. Phys. Letters **141** [1987] 36/40).
[9] Chakraborty, D. K. (Diss. Vanderbilt Univ. 1987, pp. 1/189, 1/12, 76/100, 126/35; Diss. Abstr. Intern. B **48** [1987] 772).
[10] Coxon, J. A. (J. Mol. Spectrosc. **50** [1974] 142/65).
[11] Brown, S. W.; Dowd, C. J., Jr.; Tellinghuisen, J. (J. Mol. Spectrosc. **132** [1988] 178/92).
[12] Haranath, P. B. V.; Rao, P. T. (Indian J. Phys. **31** [1957] 368/75).
[13] Parks, J. H. (AD-A085520 [1979] 1/142, 119/31; C.A. **93** [1980] No. 212997).

16.6.5 Heat of Formation. Free Energy of Formation. Thermodynamic Functions

The NBS tables of chemical thermodynamic properties recommend the following values for the standard heat of formation for $1/2\ Br_2(l)+1/2\ Cl_2(g)\rightarrow BrCl(g)$ at 298.15 and 0 K and 0.1 MPa [1]:

$$\Delta_fH^\circ_{298.15}=14.64\ \text{kJ/mol and}\ \Delta_fH^\circ_0=22.09\ \text{kJ/mol}$$

The corresponding free energy of formation is $\Delta_fG^\circ_{298.15}=-0.98$ kJ/mol. These values are given without references and obviously are based on the same or similar data as those used by the authors of the JANAF Tables [2, 3]: Using an average value $\Delta_rH^\circ_{298.15}=1.699$ kJ/mol for the reaction $Br_2(g)+Cl_2(g)\rightleftharpoons 2\ BrCl(g)$, derived from spectroscopic [4, 5], T-p [6], and mass spectrometric [7] equilibrium measurements, and $\Delta_fH^\circ_{298.15}(1/2\ Br_2(g))=15.455$ kJ/mol [2], they derived [3]

$$\Delta_fH^\circ_{298.15}=14.644\pm1.25\ \text{kJ/mol and}\ \Delta_fH^\circ_0=22.09\pm1.25\ \text{kJ/mol},$$

$$\Delta_fG^\circ_{298.15}=-0.967\ \text{kJ/mol}$$

These are the earlier JANAF data [2] converted from a standard state pressure of 1 atm to that of 0.1 MPa. Also tabulated are Δ_fH° and Δ_fG° values for T=0 to 6000 K at 100 K intervals [2, 3].

The heat of formation of BrCl from the gaseous elements, $1/2\ Br_2(g)+1/2\ Cl_2(g)\rightarrow BrCl(g)$, was derived from the equilibrium measurements of [4 to 8] to give $\Delta_fH^\circ_0(=-1/2\ \Delta_rH^\circ_0)=-0.84$ kJ/mol and $\Delta_fH^\circ_{298.15}=-0.89$ kJ/mol [9].

Two semiempirical SCF MO calculations gave too high values for $-\Delta_f H_0^\circ$; see references [18, 19] of Section 16.6.3.1.1.

The NBS tables [1] recommend a standard heat of formation of $\Delta_f H_{298.15}^\circ = -6.473$ kJ/mol for a dilute solution of BrCl in CCl_4.

The heat capacity C_p° and the thermodynamic functions S°, $-(G^\circ - H_{298}^\circ)/T$, $H^\circ - H_{298}^\circ$ for BrCl as an ideal gas and the logarithm of the equilibrium constant K_f for the formation of BrCl from the elements have been calculated for a standard state pressure of 1 atm and tabulated for T=298.15 K and between T=0 and 6000 K at 100 K intervals in the old JANAF Tables [2] (rotational [10] and vibrational [7, 11] constants and the ground-state configuration $^1\Sigma^+$ were used to establish the partition function). These were converted to a standard state pressure of 0.1 MPa and tabulated in the more recent JANAF Tables [3]; selected values are as follows:

T in K	C_p°	S°	$-(G^\circ - H_{298}^\circ)/T$	$H^\circ - H_{298}^\circ$ in kJ/mol	log K_f
	in $J\cdot mol^{-1}\cdot K^{-1}$				
0	0.00	0.00	∞	−9.402	∞
100	29.709	204.809	269.641	−6.483	−7.216
200	32.987	226.411	243.147	−3.347	−1.509
298.15	34.990	240.001	240.001	0.000	0.169
400	36.029	250.445	241.389	3.622	0.407
600	36.927	255.253	247.032	10.932	0.372
800	37.323	275.937	252.984	18.362	0.354
1000	37.557	284.292	258.441	25.851	0.344
1500	37.918	299.594	269.776	44.727	0.329
2000	38.180	310.540	278.663	63.754	0.321
3000	38.637	326.108	292.053	102.166	0.309
4000	39.068	337.283	302.028	141.020	0.295
5000	39.492	346.046	309.986	180.301	0.281
6000	39.914	353.284	316.617	220.004	0.270

Tabulations of C_p°, S°, $-(G^\circ - H_0^\circ)/T$, $-(H^\circ - H_0^\circ)/T$, $(H^\circ - H_0^\circ)$, and log K_f for T=0, 250, 273.16, 298.16, and 300 to 1500 K at 100 K intervals based on the same spectroscopic constants are given in [12], tabulations of C_p°, S°, $-(G^\circ - H_0^\circ)/T$, and $(H^\circ - H_0^\circ)/T$ for T=298.16 K, T=300 to 1000 K at 100 K intervals, and T=1200 to 2000 K at 200 K intervals based on estimated and interpolated molecular constants are given in [9].

More recent spectroscopic data for ground-state and electronically excited BrCl [13] were used to give polynomial expansions of the partition function and the equilibrium constant (Br+Cl$\rightleftharpoons$BrCl) for the temperature range T=1000 to 9000 K (astrophysical interest) [14].

References:

[1] Wagman, D. D.; Evans, W. H.; Parker, V. B.; Schumm, R. H.; Halow, I.; Bailey, S. M.; Churney, K. L.; Nuttall, R. L. (J. Phys. Chem. Ref. Data **11** Suppl. No. 2 [1982] 2-1/2-392, 2-51).

[2] Stull, D. R.; Prophet, H. (JANAF Thermochemical Tables, 2nd Ed., NSRDS-NBS-37 [1971]).

[3] Chase, M. W., Jr.; Davies, C. A.; Downey, J. R., Jr.; Frurip, D. J.; McDonald, R. A.; Syverud, A. N. (JANAF Thermochemical Tables, 3rd Ed., J. Phys. Chem. Ref. Data **14** Suppl. No. 1 [1985] 425).
[4] Vesper, H. G.; Rollefson, G. K. (J. Am. Chem. Soc. **56** [1934] 620/5).
[5] Brauer, G.; Victor, E. (Z. Elektrochem. **41** [1935] 508/9).
[6] Beeson, C. M.; Yost, D. M. (J. Am. Chem. Soc. **61** [1939] 1432/6).
[7] Mattraw, H. C.; Pachucki, C. F.; Hawkins, N. J. (J. Chem. Phys. **22** [1954] 117/9).
[8] Gray, L. T. M.; Style, D. W. G. (Proc. Roy. Soc. [London] A **126** [1939] 603/12).
[9] Cole, L. G.; Elverum, G. W., Jr. (J. Chem. Phys. **20** [1952] 1543/51).
[10] Smith, D. F.; Tidwell, M.; Williams, D. V. P. (Phys. Rev. [2] **79** [1950] 1007/8).

[11] Stammreich, H.; Forneris, R. (J. Chem. Phys. **21** [1953] 944/5).
[12] Evans, W. H.; Munson, T. R.; Wagman, D. D. (J. Res. Natl. Bur. Std. **55** [1955] 147/64).
[13] Huber, K. P.; Herzberg, G. (Molecular Spectra and Molecular Structure, Vol. 4, Constants of Diatomic Molecules, Van Nostrand Reinhold, New York 1979, pp. 108/9).
[14] Sauval, A. J.; Tatum, J. B. (Astrophys. J. Suppl. Ser. **56** [1984] 193/209).

16.6.6 Transition Points. Heats of Transition. Vapor Pressure. Density

A sharp and constant melting point, $T_m = 216.7$ K, was observed for BrCl crystals [1]. This fairly confirms an earlier result, $T_m = 219$ K, for an ochre-yellow solid obtained by low-pressure distillation of a Br_2-Cl_2 (1:1) mixture at 203 to 183 K and claimed to be solid BrCl [2] (cf. "Brom" 1931, p. 340). The melting and boiling points were estimated to be at $T_m = 207$ K and $T_b = 278$ K [3] from the earliest measurements of the freezing-point curves and the boiling- and melting-point curves for the Br_2-Cl_2 system at the beginning of the century, when the existence of BrCl was still doubted (cf. "Brom" 1931, p. 340). These values, repeatedly quoted in numerous reviews on interhalogen compounds, are in fact not those of the pure substance, but rather are the equilibrium temperatures for the phase changes, i.e., they indicate approximately the temperatures between which the dissociating BrCl exists as a liquid phase [4].

Vapor-pressure measurements on solid and liquid BrCl [2] yield the approximate heats of sublimation, vaporization, and fusion, 45, 35, and 10 kJ/mol, respectively, and extrapolation gives a boiling point of 274 K and a triple-point pressure of 14 Torr [4] (the values reported for the heats of vaporization and fusion by [5] with reference to [4] differ appreciably from the original data).

Vapor-pressure measurements between ~263 and ~423 K have been reported by [5]; from a log p vs. log T plot (with p in psi and T in °F) the following values were estimated:

T in K	263	273	293	373	423
p in MPa	0.1	0.1	0.3	0.9	1.9

Also with reference to [4], densities are reported to be 2.352, 2.339, 2.324, and 2.310 g/cm^3 for the liquid at 288, 293, 298, and 303 K, respectively, and 5.153 g/L for the vapor at 273 K and 1 atm [5].

References:

[1] Schmeisser, M.; Tytko, K.-H. (Z. Anorg. Allgem. Chem. **403** [1974] 231/42).
[2] Lux, H. (Ber. Deut. Chem. Ges. **63** [1930] 1156/8).
[3] Sidgwick, N. V. (The Chemical Elements and Their Compounds, Vol. 2, Clarendon Press, Oxford 1950, pp.1148/51).

[4] Greenwood, N. N. (Mellor's Comprehensive Treatise on Inorganic and Theoretical Chemistry, Suppl. II, Pt. I, Longmans Green, London 1956, pp. 476/513, 476/82).
[5] Mills, J. F.; Schneider, J. A. (Ind. Eng. Chem. Prod. Res. Develop. **12** [1973] 160/5).

16.6.7 Chemical Behavior

Reactions with Elements and Inorganic Compounds

H, D. The reactions $H+BrCl \rightarrow HCl(v \leq 7)+Br$, $\Delta_rH^\circ_0=-212.1$ kJ/mol and

$$H+BrCl \rightarrow HBr(v \leq 4)+Cl,\ \Delta_rH^\circ_0=-122.2 \text{ kJ/mol}$$

were studied by the infrared-chemiluminescence method to obtain the product energy distribution over vibration, rotation, and translation. The average fractions of available energies entering vibration, rotation, and translation are 0.55, 0.09, and 0.36 in HCl and 0.58, 0.12, and 0.30 in HBr. The branching ratio Γ(HCl/HBr) was 0.40 ± 0.05 [1]. Classical trajectory calculations [5] yielded product energy distributions over vibration and rotation in satisfactory agreement with the experiment [1] as well as the temperature dependence of the rate constants for both reactions. Potential energy surfaces for both reaction channels were calculated using the semiempirical DIM-3C method. The potential barrier height and the activation energy are 6.3 and 7.1 kJ/mol for $H+BrCl \rightarrow HCl+Br$ and 7.1 and 7.5 kJ/mol for $H+BrCl \rightarrow HBr+Cl$, respectively, in the linear arrangement [2, 3]. Similar values for the activation energies were obtained with the semiempirical "bond energy – bond length" method [4].

The classical trajectory method to calculate product energy distributions and rate constants was also applied to the equivalent reactions with deuterium [6].

H_2. Activation energies of 213.4 [7] and 165.7 kJ/mol [8] were calculated for the four-center reaction $H_2+BrCl \rightarrow HBr+HCl$ using two different empirical models.

O. The reaction $O+BrCl \rightarrow BrO+Cl$, $\Delta_rH^\circ_{298}=-16$ kJ/mol, was studied in a discharge flow system using atomic resonance fluorescence for following the decay of O atoms. The rate constant of $(2.1 \pm 0.7) \times 10^{-11}$ $cm^3 \cdot molecule^{-1} \cdot s^{-1}$ at 298 K is nearly independent of the temperature up to 619 K, indicating a low or zero activation energy [9].

Cl. The reaction $Cl+BrCl \rightarrow Cl_2+Br$, $\Delta_rH^\circ_{298}=-24$ kJ/mol, was studied in the same manner as the O + BrCl reaction. A rate constant of $k=(1.45 \pm 0.20) \times 10^{-11}$ $cm^3 \cdot molecule^{-1} \cdot s^{-1}$ was measured at 298 K [10]. An equation of $k=(3.3 \pm 1.7) \times 10^{-12} \exp[-(550 \pm 200)/T]$ $cm^3 \cdot molecule^{-1} \cdot s^{-1}$ for T between 293 and 333 K was determined in an earlier study on the photochemical reaction between bromine and chlorine [11].

OH. The gas-phase reaction of the OH radical with BrCl can proceed via four pathways:

(1a) $OH+BrCl \rightarrow HOCl+Br$ $\Delta H=-16.7$ kJ/mol

(1b) $OH+BrCl \rightarrow HOBr+Cl$ $\Delta H=-12.1$ kJ/mol

(1c) $OH+BrCl \rightarrow HCl+BrO$ $\Delta H=-20.9$ kJ/mol

(1d) $OH+BrCl \rightarrow HBr+ClO$ $\Delta H=-10.9$ kJ/mol

The reactions (1a) and (1b) likely proceed through a linear transition state and (1c) and (1d) through a cyclic four-center transition state. The total rate constant $k=(1.49 \pm 0.40) \times 10^{-12}$ $cm^3 \cdot molecule^{-1} \cdot s^{-1}$ was derived from measurements, which were carried out in He at 298 K using a discharge-flow system, and by considering that BrCl exists in equilibrium with Br_2 and Cl_2 [12]. An apparent activation energy for the OH + BrCl reaction was estimated to be 8.8 [12] and 13.0 kJ/mol [13].

H_2O. BrCl is quickly hydrolyzed to HBrO. The hydrolysis constant was found to be 5.5 $\times 10^{-6}$ mol/L at 290.0 K and 2.6×10^{-5} mol/L at 299.5 K [14, 15], and 2.94×10^{-5} mol/L at 273.2 K [16].

NH_3. BrCl reacts with ammonia to give monobromoamine and dibromoamine. The disinfectant properties of these products and of HBrO (from BrCl hydrolysis) make BrCl attractive to many applications in water treatment [17, 18].

$ClONO_2$. The reaction of BrCl with $ClONO_2$ gives $BrONO_2$ at low yields [19 to 21].

HBr. The reaction of BrCl with HBr is described in "Bromine" Suppl. Vol. B1, 1990, p. 361.

Miscellaneous. The strong oxidizing capability of BrCl makes it useful for oxidimetric titration. For example, NH_2OH [22, 23], N_2H_4 [24, 25], SO_3^{2-} [26], $S_3O_6^{2-}$ [24], CNS^- [24, 27 to 29], $H_2PO_2^-$ [25, 30], PO_3S^{3-} [24], As^{3+}, and Sn^{2+} [26] can be determined with BrCl.

Reactions with Organic Compounds

BrCl serves as a brominating and bromochlorinating agent in organic syntheses. In brominations, it has several advantages over elemental bromine, such as higher reactivity, greater product yields, and the fact that HCl is the main by-product instead of HBr. Reactions with organic chemicals are reviewed in [31].

The most important reactions of BrCl are electrophilic brominations of aromatic compounds [31, 44]. The high reactivity and selectivity of these reactions with phenolic compounds can be used for the analytic determination of phenols, cresols, nitrophenols, and salicylic acids [32, 33]. BrCl in solution or in the vapor phase is rapidly added across olefinic bonds producing bromochloro compounds [31]. Primary and secondary alcohols are completely oxidized by aqueous BrCl [34]. Radical reactions are also known [31]. BrCl forms complexes with organic molecules in which BrCl acts as an electron acceptor. Frequently studied electron donors are, for example, pyridine and its derivatives [35 to 43].

References:

[1] Polanyi, J. C.; Skrlac, W. J. (Chem. Phys. **23** [1977] 167/94).
[2] Last, I. (Chem. Phys. **69** [1982] 193/203).
[3] Last, I.; Baer, M. (J. Phys. Chem. **80** [1984] 3246/52).
[4] Karachevtsev, G. V.; Savkin, V. V. (Khim. Fiz. **1983** No. 9, pp. 1286/8; C.A. **99** [1983] No. 164682).
[5] Konoplev, N. A.; Stepanov, A. A.; Shcheglov, V. A. (Khim. Fiz. **1984** No. 3, pp. 828/32; C.A. **101** [1984] No. 60938).
[6] Konoplev, N. A.; Stepanov, A. A.; Shcheglov, V. A. (Kratk. Soobshch. Fiz. **1985** 17/21; C.A. **103** [1985] No. 221749).
[7] Benson, S. W.; Haugen, G. R. (J. Am. Chem. Soc. **87** [1965] 4036/44).
[8] Noyes, R. M. (J. Am. Chem. Soc. **88** [1966] 4318/25).
[9] Clyne, M. A. A.; Monkhouse, P. B.; Townsend, L. W. (Intern. J. Chem. Kinet. **8** [1976] 425/49).
[10] Clyne, M. A. A.; Cruse, H. W. (J. Chem. Soc. Faraday Trans. II **68** [1972] 1377/87).
[11] Christie, M. I.; Roy, R. S.; Thrush, B. A. (Trans. Faraday Soc. **55** [1959] 1139/48).
[12] Loewenstein, L. M.; Anderson, J. G. (J. Phys. Chem. **88** [1984] 6277/86).
[13] Loewenstein, L. M.; Anderson, J. G. (J. Phys. Chem. **91** [1987] 2993/7).
[14] Schulek, E.; Pungor, E. (Anal. Chim. Acta **7** [1952] 402/7).

[15] Schulek, E.; Pungor, E.; Burger, K. (Chem. Zvesti **13** [1959] 669/79).
[16] Kanyaev, N. P.; Shilov, E. A. (Tr. Ivanov. Khim. Tekhnol. Inst. No. 3 [1940] 69/73 from C.A. **1941** 2775).
[17] Mills, J. F. (Chem. Wastewater Technol. **1978** 199/212).
[18] Mills, J. F. (in: Johnson, J. D.; Disinfection: Water Wastewater, Ann Arbor Sci., Stoneham, Mass., 1975, pp. 113/43).
[19] Schmeisser, M.; Taglinger, L. (Chem. Ber. **94** [1961] 1533/9).
[20] Spencer, J. E.; Rowland, F. S. (J. Phys. Chem. **82** [1978] 7/10).

[21] Wilson, W. W.; Christe, K. O. (Inorg. Chem. **26** [1987] 1573/80).
[22] Burger, K.; Gaizer, F.; Schulek, E. (Talanta **5** [1960] 97/101).
[23] Burger, K.; Gaizer, F.; Schulek, E. (Magy. Kem. Foly. **67** [1961] 173/5 from C.A. **1961** 21968).
[24] Nair, C. G. R.; Lalitha Kumari, R. (Indian J. Chem. A **14** [1976] 115/7).
[25] Burger, K.; Schulek, E. (Ann. Univ. Sci. Budapest Rolando Eotvos Nominatae Sect. Chim. **2** [1960] 133/8).
[26] Korneva, L. E.; Gengrinovich, A. I.; Murtazaev, A. M. (Dokl. Akad. Nauk Uzbek. SSR **22** [1965] 28/31; C.A. **64** [1966] 5724).
[27] Schulek, E.; Pungor, E. (Talanta **2** [1959] 280/2).
[28] Mathur, P. K. (Rev. Roumaine Chim. **23** [1978] 1607/9).
[29] Sivasankara Pillai, V. N. (Indian J. Chem. A **19** [1980] 1031).
[30] Burger, K.; Landanyi, L. (Acta Pharm. Hung. **30** [1960] 80/3 from C.A. **1960** 21643).

[31] Mills, J. F.; Schneider, J. A. (Ind. Eng. Chem. Prod. Res. Develop. **12** [1973] 160/5).
[32] Schulek, E.; Burger, K. (Talanta **1** [1958] 147/52).
[33] Schulek, E.; Burger, K. (Ann. Univ. Sci. Budapest Rolando Eotvos Nominatae Sect. Chim. **2** [1960] 139/43).
[34] Konishi, K.; Mori, Y.; Inoue, H.; Nozoe, M. (Anal. Chem. **40** [1968] 2198/200).
[35] Ginn, S. G. W.; Haque, I.; Wood, J. L. (Spectrochim. Acta A **24** [1968] 1531/42).
[36] Surles, T.; Popov, A. I. (Inorg. Chem. **8** [1969] 2049/52).
[37] Heasley, V. L.; Griffith, C. N.; Heasley, G. E. (J. Org. Chem. **40** [1975] 1358/60).
[38] Heasley, G. E.; McCall Bundy, J.; Heasley, V. L.; Arnold, S.; Gipe, A.; McKee, D.; Orr, R.; Rodgers, S. L.; Shellhamer, D. F. (J. Org. Chem. **43** [1978] 2793/9).
[39] Rubenacker, G. V.; Brown, T. L. (Inorg. Chem. **19** [1980] 392/8).
[40] Rubenacker, G. V.; Brown, T. L. (Inorg. Chem. **19** [1980] 398/401).

[41] Williams, D. M. (J. Chem. Soc. **1931** 2783/7).
[42] Barooah, S. K.; Haque, I. (J. Indian Chem. Soc. **63** [1986] 490/4).
[43] Zingaro, R. A.; Witmer, W. B. (J. Phys. Chem. **64** [1960] 1705/11).
[44] Voudrias, E. A.; Reinhard, M. (Environ. Sci. Technol. **22** [1988] 1049/56).

16.7 Bromine Chloride Hydrate, $BrCl \cdot n\,H_2O$, n = 1 to 10

CAS Registry Numbers: $BrCl \cdot n\,H_2O$ *[25131-46-8]*, $^{81}Br^{37}Cl \cdot n\,H_2O$ *[63129-50-0]*, $^{79}Br^{37}Cl \cdot n\,H_2O$ *[63129-49-7]*, $^{81}Br^{35}Cl \cdot n\,H_2O$ *[63129-48-6]*, $^{79}Br^{35}Cl \cdot n\,H_2O$ *[63129-47-5]*, $BrCl \cdot 4\,H_2O$ *[68349-66-6]*

Pure, orange-red crystals of bromine chloride hydrate form when liquid BrCl is dissolved in cold H_2O ($\sim$5 °C) which is acidified by H_2SO_4 to suppress hydrolysis [1]. The product obtained by bubbling Cl_2 through a layer of Br_2 kept under ice-cold water [2] is contaminated with Br_2 [3]. Early reports on the formation of the hydrate are quoted in "Brom" 1931, pp. 339/40.

The most probable composition of the hydrate is $BrCl\cdot(7.34\pm0.60)\ H_2O$ as obtained by two independent methods (chemical analysis and a thermodynamic approach). This is close to what one would expect for a cubic, type-I gas hydrate, a clathrate with filled medium-size cavities [1]. The lattice parameter of the clathrate is a = 12.02 Å [4]. The structure of the BrCl hydrate is probably similar to that of the better characterized hydrates of Br_2 and Cl_2 (see [4, 5]). Questionable compositions of $BrCl\cdot n\ H_2O$ include n = 4 [2] or 5 and 10, see "Brom" 1931, pp. 339/40.

At atmospheric pressure, $BrCl\cdot n\ H_2O$ decomposes at about 287 K. The dissociation pressure at 273 K is about 125 Torr [4]. The measured equilibrium distribution of BrCl in aqueous solution and in the solid hydrate yielded the enthalpy of fusion of solid $BrCl\cdot n\ H_2O$ to BrCl(aq) to be $\Delta_{fus}H = -43.7\pm3.6$ kJ/mol (274 to 288 K) [1].

Two Raman bands for the BrCl stretching vibration in the BrCl hydrate were observed at 77 K. The band at 424 cm^{-1} was assigned to the $^{79}Br^{35}Cl$ and $^{81}Br^{35}Cl$ hydrate species and that at 414 cm^{-1} to the $^{79}Br^{37}Cl$ and $^{81}Br^{37}Cl$ hydrate species. The calculated isotope shift and relative band intensity agree with experimental data. Comparing the band shifts with that of gaseous BrCl indicates little interaction in the hydrate. A Raman band at 88 cm^{-1} was tentatively assigned to translational lattice vibrations [3].

Compared to gaseous BrCl, a slightly smaller force constant of 2.56 mdyn/Å and a slightly larger mean amplitude was calculated for the BrCl stretching vibration in $BrCl\cdot n\ H_2O$ [6].

References:

[1] Glew, D. N.; Hames, D. A. (Can. J. Chem. **47** [1969] 4651/4).
[2] Anwar-Ullah, S. (J. Chem. Soc. **1932** 1176/9).
[3] Anthonsen, J. W. (Acta Chem. Scand. A **29** [1975] 175/8).
[4] v. Stackelberg, M.; Müller, H. R. (Z. Elektrochem. **58** [1954] 25/39).
[5] Pauling, L.; Marsh, R. F. (Proc. Natl. Acad. Sci. U.S.A. **38** [1952] 112/8).
[6] Sanyal, N. K.; Verma, D. N.; Dixit, L. (Indian J. Pure Appl. Phys. **15** [1977] 242/5).

16.8 Monochlorobromine(1+), $BrCl^+$

CAS Registry Number: *[57142-99-1]*

The mass spectrum of (presumably) the Cl_2-BrCl-Br_2 equilibrium included the $BrCl^+$ ion with an appearance potential of 11.1 ± 0.2 eV (which gave an upper limit for the vertical ionization potential of the BrCl molecule, cf. p. 184) [1].

The standard heat of formation at 298.15 and 0 K and 0.1 MPa is recommended to be $\Delta_fH^\circ_{298.15} = \Delta_fH^\circ_0 = 1088$ kJ/mol [2]. A semiempirical MO calculation (MNDO) gave $\Delta_fH^\circ_0 = 1113.9$ kJ/mol [3]. An experimental value for the heat of formation from the Br^+ ion and the Cl atom, $\Delta_fH^{at}_0 = 854$ kJ/mol, is quoted and a value of $\Delta_fH^{at}_0 = 883$ kJ/mol was calculated semiempirically [4].

The He I photoelectron spectrum of BrCl covers the region of 6π, 5π, and 12σ electron removal, which corresponds to the formation of the $BrCl^+$ ion in its spin-orbit split electronic ground state X $^2\Pi_{3/2,1/2}(...(12\sigma)^2\ (5\pi)^4\ (6\pi)^3$; $\Delta E\ (^2\Pi_{3/2} - {}^2\Pi_{1/2}) = 2070\pm30\ cm^{-1}$) and in the excited states A $^2\Pi_{3/2,1/2}(...(12\sigma)^2\ (5\pi)^3\ (6\pi)^4)$ and B $^2\Sigma^+(...(12\sigma)^1\ (5\pi)^4\ (6\pi)^4)$. The vibrational structure in the first photoelectron band yielded the ground-state vibrational constants $\omega_e = 486\pm20\ cm^{-1}$ and $\omega_ex_e = 2.5\pm5.0\ cm^{-1}$ for the X $^2\Pi_{3/2}$ ionic state and $\omega_e = 510\pm20\ cm^{-1}$ and $\omega_ex_e = 2.5\pm5.0\ cm^{-1}$ for the X $^2\Pi_{1/2}$ ionic state [5]. Theoretically predicted values for the ground-state spin-orbit splitting (approximate method of Leach [6]) are $\Delta E = 2365\ cm^{-1}$ [7] and 2381 or $2182\ cm^{-1}$ (different bond lengths assumed for $BrCl^+$) [5]. A relativistic Hartree-Fock-Slater calculation of the ionization potentials of BrCl gave the spin-orbit splittings $\Delta E = 1887\ cm^{-1}$ for the ground state X $^2\Pi_{3/2,1/2}$ and $\Delta E = 823\ cm^{-1}$ for the first excited $^2\Pi$ state of $BrCl^+$ [8].

References:

[1] Irsa, A. P.; Friedman, L. (J. Inorg. Nucl. Chem. **6** [1958] 77/90).

[2] Wagman, D. D.; Evans, W. H.; Parker, V. B.; Schumm, R. H.; Halow, I.; Bailey, S. M.; Churney, K. L.; Nuttall, R. L. (J. Phys. Chem. Ref. Data **11** Suppl. No. 2 [1982] 2-1/2-392, 2-51).

[3] Dewar, M. J. S.; Healy, E. (J. Computat. Chem. **4** [1983] 542/51).

[4] Andreev, S. (God. Vissh. Khim. Tekhnol. Inst. Sofia **20** No. 2 [1972/74] 65/76; C.A. **83** [1975] No. 183865).

[5] Dunlavey, S. J.; Dyke, J. M.; Morris, A. (J. Electron Spectrosc. Relat. Phenom. **12** [1977] 259/63).

[6] Leach, S. (Acta Phys. Polon. **34** [1968] 705/14).

[7] Grimm, F. A. (J. Electron Spectrosc. Relat. Phenom. **2** [1973] 475/81).

[8] Dyke, J. M.; Josland, G. D.; Snijders, J. G.; Boerrigter, P. M. (Chem. Phys. **91** [1984] 419/24).

16.9 Monochlorobromate(1−), $BrCl^-$

CAS Registry Number: *[12589-83-2]*

Experimental results concerning a **free** $BrCl^-$ ion are not available. A valence bond pseudopotential calculation of the potential curves for the ground state X $^2\Sigma^+(...(12\sigma)^2\ (5\pi)^4\ (6\pi)^4\ (13\sigma)^1)$ and the three lowest excited states $^2\Pi(...(12\sigma)^2\ (5\pi)^4\ (6\pi)^3\ (13\sigma)^2)$, $^2\Pi(...(12\sigma)^2\ (5\pi)^3\ (6\pi)^4\ (13\sigma)^2)$, and $^2\Sigma^+(...(12\sigma)^1\ (5\pi)^4\ (6\pi)^4\ (13\sigma)^2)$ indicated the ground state to be bound and the excited states to be repulsive; the ground-state properties, equilibrium internuclear distance $r_e = 2.87$ Å, dissociation energy $D_e = 0.93$ eV, vibrational constants $\omega_e = 191\ cm^{-1}$ and $\omega_e x_e = 1.0\ cm^{-1}$, the coefficients for the ground-state potential curve fitted to a polynomial of r up to r^6, and the transition energies $\Delta E(^2\Pi, ^2\Pi, ^2\Sigma^+ \leftarrow X\ ^2\Sigma^+) = 1.73$, 2.54, and 3.13 eV at the ground-state equilibrium distance have been derived [1].

A **stabilization** and **orientation** of $BrCl^-$ ions as paramagnetic centers in alkali halide crystals has been achieved by X or γ irradiation of Br^--doped XCl crystals (X = Li, Na, K, Rb) at liquid nitrogen temperatures and suitable thermal treatment or optical bleaching. Various types of $BrCl^-$ centers of the V_K-, V_{KAA}-, V_F-, $H_A(Li^+)$-, $H_D^{\{110\}}(Ba^{2+})$-, $H_D^{\{100\}}(Sr^{2+})$- and $H_{DD}^{\{100\}}(Sr^{2+})$-type have been identified by electron spin resonance (ESR) and optical absorption measurements.

The V_K-type centers, with interstitial Cl^- and Br^- ions and a trapped hole oriented exactly along a [110] direction of the cubic host crystal, are formed in XCl crystals containing small amounts of the corresponding bromide (and with Pb^{2+} or Tl^+ ions acting as electron traps); their concentration is greatly enhanced by thermal or optical bleaching of the simultaneously formed $Cl_2^-(V_K)$ centers [2 to 8].

The V_{KAA}-type centers associated with two Na^+ ions in nearest-neighbor position to both Br and Cl and also oriented along a [110] direction are observed in Br^-- and Na^+-doped KCl crystals (no details on formation given) [7].

Warming up of the $BrCl^-$(V_K-type) centers produces the V_F-type $BrCl^-(v_+)$ centers which are associated with a cation vacancy in a nearest-neighbor position to both Br and Cl, are oriented in a (100) plane, and make a small angle (2.5° to 15.5°) with the [110] direction [6, 7] (conversion into the V_F-type $Cl_2^-(v_+)$ center is achieved by photoexcitation into the short-wavelength absorption band [9]).

Irradiation of KCl crystals doped with LiCl and KBr at 77 K and warming them up to 196 K results in the formation of the $BrCl^-(H_A(Li^+)$-type) centers. These occupy anion sites next to substitutional Li^+ ions, are oriented in a (110) plane, form an angle of 25° with the [100] direction, and have a "bent" molecular bond towards the Li^+ ion [10].

When KCl crystals (with Br^- impurities) doped with $BaCl_2$ or $SrCl_2$ are irradiated at 77 K and then warmed up to 223 K, two $H_D^{\{110\}}(Ba^{2+})$-type and two $H_D^{\{100\}}(Sr^{2+})$-type $BrCl^-$ centers and a $H_{DD}^{\{100\}}(Sr^{2+})$-type $BrCl^-$ center are formed; in the first case, $BrCl^-$ occupies an anion site next to a Ba^{2+} ion and an associated cation vacancy, lies in a (110) plane, and forms an angle of 13.5° (type 1) or 10.0° (type 2) with the [100] direction; in the second case, $BrCl^-$ occupies an anion site next to a Sr^{2+} ion and an associated cation vacancy, lies in a (100) plane, and forms an angle of 12.5° (type 1) or 4.5° (type 2) with the [100] direction; type 1 and type 2 typify a more substitutional and a more interstitial position of Br, respectively; the third center is a $BrCl^-$ ion on an anion site exactly oriented in a [100] direction and flanked by two Sr^{2+} ions at cation vacancies in a (100) plane [11].

Table 26
$BrCl^-$ Centers in $KCl{:}Br^-$ and $KCl{:}Li^+{:}Br^-$ Crystals. ESR and Optical Data.

	V_K-type [5, 6]	V_F-type [5, 6]	V_{KAA}-type [6]	$H_A(Li^+)$-type [7]
g_z	1.9839 ± 0.0001	1.9702 ± 0.0002	1.9845 ± 0.0002	1.9977 ± 0.0010
g_x	2.1239 ± 0.0005	2.177 ± 0.001	2.124 ± 0.002	2.061 ± 0.005
g_y	2.1350 ± 0.0005	2.188 ± 0.002	2.134 ± 0.003	2.066 ± 0.005
g_{iso}	2.0839	2.1117	2.0808	2.0416
$A_z(^{81}Br)$ in G	484.6 ± 0.1	484.9 ± 0.1	490.1 ± 0.5	511.5 ± 1
$A_\perp(^{81}Br)$ in G	115.5 ± 2	162 ± 5	121 ± 5	130 ± 15 (A_x) 148 ± 15 (A_y)
$A_{iso}(^{81}Br)$ in G	238.5	270	244	257
$A_z(^{35}Cl)$ in G	89.6 ± 0.1	82.1 ± 0.1	91.5 ± 0.5	102.7 ± 0.5
$A_\perp(^{35}Cl)$ in G	7.5 ± 1	8 ± 2	10 ± 3	10 ± 10
$A_{iso}(^{35}Cl)$ in G	34.9	32.7	37	41
λ_{max} in nm[a)]	382	368	–	–
λ_{max} in nm[b)]	~760	910	–	–
T_{disor} in K	93	45 and 55[c)]	–	–
T_{decay} in K	275	~320	–	235

[a)] Strong σ-polarized absorption band; assigned to the transition $^2\Pi(\ldots(12\sigma)^2\,(5\pi)^4\,(6\pi)^3\,(13\sigma)^2) \leftarrow X\,^2\Sigma^+(\ldots(12\sigma)^2\,(5\pi)^4\,(6\pi)^4\,(13\sigma)^1)$ [5, 6]. – [b)] Weak σ-polarized absorption band; assigned to the transition $^2\Sigma^+(\ldots(12\sigma)^1\,(5\pi)^4\,(6\pi)^4\,(13\sigma)^2) \leftarrow X\,^2\Sigma^+(\ldots(12\sigma)^2\,(5\pi)^4\,(6\pi)^4\,(13\sigma)^1)$ [4, 5]. – [c)] 115° and 73.2° jumps, respectively [6].

Due to the four isotopic $BrCl^-$ species and their abundancies and due to the nuclear spins $I(^{81}Br) = I(^{79}Br) = I(^{35}Cl) = I(^{37}Cl) = 3/2$, the ESR spectrum consists of $4 \cdot (2\,I_{Br} + 1) \cdot (2\,I_{Cl} + 1) = 4 \times 16$ lines with intensity ratios of about 3:3:1:1. For the V_K-, V_{KAA}-, V_F-, and $H_A(Li^+)$-type centers in KCl crystals the ESR parameters, i.e., the anisotropic g-tensor elements, g_z, g_x, g_y, hyperfine coupling constants, A_z, $A_x = A_y = A_\perp$, and their isotropic values, g_{iso}, A_{iso}, at 77 K are given in Table 26 together with the absorption maxima λ_{max} in the UV, visible, and near IR, and the disorientation and decay temperatures, T_{disor} and T_{decay}, as reported by Delbecq, Schoemaker, and co-workers [6 to 8] (see also [12]). ESR data for the H_D- and H_{DD}-type $BrCl^-$ centers in Ba^{2+}- and Sr^{2+}-doped KCl crystals from [11] are given in Table 27.

ESR data from unpublished studies on the $BrCl^-$ V_K-type centers in LiCl, NaCl, and RbCl, on the $BrCl^-$ V_F-type centers in NaCl and RbCl, and on the $BrCl^-$ V_{KAA}-type center in KCl (see Table 26) are quoted in [7].

Luminescence bands at 335 and 258 nm observed upon X-ray irradiation of mixed KBr-KCl crystals (10 to 90 mol% KBr) at 4.2 K have been tentatively assigned to $BrCl^-$ V_K-type centers [13].

Theoretical studies of the ESR parameters and optical transition energies of the $BrCl^-$ V_K-type centers in alkali halides using a simple MO LCAO wave function [6] and a valence-bond wave function [14] are available.

Table 27
$BrCl^-$ Centers in $KCl{:}Ba^{2+}{:}Br^-$ and $KCl{:}Sr^{2+}{:}Br^-$ Crystals. ESR Data from [11].

	$H_D^{\{110\}}(Ba^{2+})$ *) type 1	$H_D^{\{100\}}(Sr^{2+})$ type 1	type 2	$H_{DD}^{\{100\}}(Sr^{2+})$
g_z	1.971±0.001	1.969±0.001	1.963±0.001	1.989±0.001
$g_\perp$	2.20±0.05	2.18±0.05	2.20±0.05	—
$A_z(^{81}Br)$ in G	475±2	473±2	464±2	463±2
$A_\perp(^{81}Br)$ in G	70±30	140±30	150±30	—
$A_z(^{35}Cl)$ in G	79±2	78±2	77±2	87±2
$A_\perp(^{35}Cl)$ in G	10±10	10±10	10±10	—
T_{decay} in K	273	273	248	243

*) For type 2 of this center only a decay temperature of 243 K has been reported.

References:

[1] Tasker, P. W.; Balint-Kurti, G. G.; Dixon, R. N. (Mol. Phys. **32** [1976] 1651/60).
[2] Wilkins, J. W.; Gabriel, J. R. (Phys. Rev. [2] **132** [1963] 1950/7).
[3] Schoemaker, D. (Phys. Rev. [2] **149** [1966] 693/704).
[4] Schoemaker, D.; Delbecq, C. J.; Yuster, P. H. (Bull. Am. Phys. Soc. **9** [1964] 629/30).
[5] Delbecq, C. J.; Yuster, P. H.; Schoemaker, D. (Phys. Rev. [3] B **3** [1971] 473/87).
[6] Delbecq, C. J.; Yuster, P. H.; Schoemaker, D. (Phys. Rev. [3] B **7** [1973] 3933/44).
[7] Van Puymbroeck, W.; Lagendijk, A.; Schoemaker, D. (Phys. Status Solidi A **59** [1980] 585/95).
[8] Hausmann, A.; Pomplun, H. (Solid State Commun. **11** [1972] 867/9).
[9] Delbecq, C. L.; Schoemaker, D.; Yuster, P. H. (Phys. Rev. [3] B **9** [1974] 1913/20).
[10] Schoemaker, D.; Shirkey, C. T. (Phys. Rev. [3] B **6** [1972] 1562/72).

[11] Van Puymbroeck, W.; Schrijvers, N.; Bouwen, A.; Schoemaker, D. (Phys. Status Solidi B **112** [1982] 725/33).
[12] Morton, J. R.; Preston, K. F. (Landolt-Börnstein New Ser. Group II **9** Pt. a [1977] 5/289, 223).
[13] Wakita, S.; Hirai, M. (J. Phys. Soc. Japan **24** [1968] 1177).
[14] Jette, A. N.; Adrian, F. J. (Phys. Rev. [3] B **14** [1976] 3672/81).

16.10 Dibromine Dichloride, Br_2Cl_2

CAS Registry Number: *[12360-51-9]*

A T-shaped molecule Cl-Br(Br)-Cl was identified by its IR bands at 325.6, 323.8, 321.6, 319.4 cm^{-1} in an Ar matrix and at 319.6, 317.7, 315.8, 313 cm^{-1} in a Kr matrix after mixtures of Cl_2 and Br_2 in Ar or Kr were passed through a microwave discharge and condensed at 20 K. The isotopic shifts in the vibrations of the Cl-Br-Cl subunit were calculated with reasonable values for the force constants.

Reference:

Nelson, L. Y.; Pimentel, G. C. (Inorg. Chem. **7** [1968] 1695/9).

16.11 Van der Waals Complexes $Br_2\cdot(Cl_2)_n$, n = 1 to ≥10; $BrCl\cdot(Cl_2)_m$

CAS Registry Number: –

Van der Waals complexes $Br_2\cdot Cl_2$ up to $Br_2\cdot(Cl_2)_{n\geq 10}$ were observed in cross-beam studies of Br_2 and Cl_2 or $(Cl_2)_n$. There was no evidence for $BrCl\cdot(Cl_2)_m$ formation; see "Bromine" Suppl. Vol. A, 1985, p. 375.

16.12 $Br_2Cl_3^-$

CAS Registry Number: –

$Br_2Cl_3^-$ was detected by mass spectrometry of Xe-Cl_2 mixtures in a drift tube contaminated with Br_2. By varying the partial pressure of Cl_2 and simultaneously monitoring $Br_2Cl_3^-$, $BrCl_2^-$, Br_2Cl^-, Br^-, Cl^-, and Cl_3^-, the following reaction sequence was identified:

(I) $Cl^- + Br_2 + M \rightarrow Br_2Cl^- + M$ (III) $Cl^- + Cl_2 + M \rightarrow Cl_3^- + M$
(II) $Br_2Cl^- + Cl_2 \rightarrow Br_2Cl_3^- + M$ (IV) $Cl_3^- + Br_2 + M \rightarrow Br_2Cl_3^- + M$

M is either Cl_2 or Xe. Reactions (III) and (IV) are of minor importance.

Reference:

Huber, B. A.; Miller, T. M. (J. Appl. Phys. **48** [1977] 1708/11).

16.13 Bromochlorochlorine(1+) and Dichlorobromine(1+), $BrClCl^+$ and $ClBrCl^+$

CAS Registry Numbers: $BrClCl^+$ *[55426-80-7]*, $ClBrCl^+$ *[55426-81-8]*

Two structural isomers exist for ions of the general formula $BrCl_2^+$. Any assignment made relies on vibrational spectra and is tentative.

$BrCl_2^+$ ions were prepared by matrix isolation techniques [1] and by reacting Cl_2 with $BrSO_3F$ followed by solvolysis in SbF_5 [2, 3] or $Sn(SO_3F)_4$ [3]. CCl_3Br gas diluted with Ar was irradiated with 1.0 keV protons, and the products were deposited on a CsI window held at 15 K. IR bands at 492, 488, and 484 cm^{-1} were tentatively assigned to the isotopic combinations of the $ClBrCl^+$ isomer with two equivalent chlorines. Bands at 437 and 420 cm^{-1} were thought to arise from $BrClCl^+$ [1]. While nearly equimolar mixtures of $BrSO_4F$ and Cl_2 do not react, a more than tenfold excess of SbF_5 induces the reaction

$$BrSO_3F + Cl_2 + 2\,SbF_5 \rightarrow BrCl_2^+ + SbF_6^- + SbF_4SO_3F$$

After removing the excess Cl_2, the reaction mixture separates into two liquid phases. The $BrCl_2^+$ cation dissolved in SbF_5 is found in the more dense, colored phase [2, 3]. A similar route to $BrCl_2^+$ involves the reaction

$$2\,BrSO_3F + 2\,Cl_2 + Sn(SO_3F)_4 \rightarrow [BrCl_2^+]_2\,[Sn(SO_3F)_6^{2-}]$$

$BrCl_2^+$ dissolved in SbF_5 is deep red and diamagnetic. An intense Raman band at 430 cm^{-1}, a shoulder at 421 cm^{-1}, and a band at 167 cm^{-1} were assigned to the symmetric stretching, antisymmetric stretching, and bending modes, respectively, of the presumably bent cation [2, 3]. (This interpretation considers only the $ClBrCl^+$ isomer and differs from the IR band assignments made by [1].) The electronic spectrum of $BrCl_2^+$ in SbF_5 showed two bands, one at 305 nm assigned to the $BrCl_2^+$ cation and the other at 248 nm assigned to the SbF_6^- anion [2, 3]. Energy levels for valence MO's of an assumed L-shaped cation were calculated semiempirically [4].

At room temperature, $BrCl_2^+$ is stable in SbF_5 solution, but decomposes when the solvent is removed [2, 3].

References:

[1] Andrews, L.; Grzybowski, J. M.; Allen, R. O. (J. Phys. Chem. **79** [1975] 904/12, 910).

[2] Wilson, W. W.; Thompson, R. C.; Aubke, F. (Inorg. Chem. **19** [1980] 1489/93).

[3] Wilson, W. W.; Landa, B.; Aubke, F. (Inorg. Nucl. Chem. Letters **11** [1975] 529/34).

[4] Smolyar, A. E.; Charkin, O. P.; Klimenko, N. M. (Zh. Strukt. Khim. **15** [1974] 993/1003; J. Struct. Chem. USSR **15** [1974] 885/93).

16.14 Dichlorobromate(1−), $ClBrCl^-$

CAS Registry Number: *[14522-78-2]*

The $ClBrCl^-$ anion is more stable than the isomeric $BrClCl^-$ anion; see p.229 and some general arguments for trihalides given in reviews, for example [1, 2]. Therefore, one can assume that work on the $BrCl_2^-$ anion actually involves the $ClBrCl^-$ isomer, unless the species $BrClCl^-$ is explicitely mentioned.

Formation

$BrCl_2^-$ ions are easily accessible by reacting bromide ions with chlorine or chloride ions with bromine chloride.

Following the procedure of synthesizing caesium and rubidium salts, published very early, the respective bromides are dissolved in water, and the solution is saturated with chlorine gas. Salts precipitate upon cooling [3]. A variation, especially suited for tetraalkylammonium salts, is to dissolve the bromide in a minimum amount of glacial acetic acid [4, 5]. Chloroform is a suitable solvent when cations with large organic substituents are involved; see, for example [6, 7]. This method for the preparation of $BrCl_2^-$ salts is a simplified version of an earlier one where chlorides are dissolved in water [8, 9] or acetic acid [4], then bromine is added, and finally the solution is saturated with Cl_2. Dichlorobromates can also be prepared by a gas-solid reaction, for instance between Cl_2 and CsBr [10]. $BrCl_2^-$ ions in the gas phase were produced in a drift-tube mass-spectrometer apparatus from Cl_2, diluted with noble gases, and trace amounts of Br_2 [11, 12].

The 1:1 stoichiometry of the reaction between Br^- and Cl_2 in liquid HCl was verified by conductometric titration [13]. The stability of $BrCl_2^-$ ions in aqueous solutions is consistent with an estimated free energy of −46 kJ/mol for $Br^-(aq)+Cl_2(g) \rightarrow \{BrClCl^-(aq)\} \rightarrow ClBrCl^-(aq)$ [14].

In the preceding reaction, chlorine can be replaced by PCl_5 as demonstrated by the preparation of ammonium and phosphonium salts of $BrCl_2^-$ by heating PCl_5 and respective bromides [15].

The addition reaction $Cl^- + BrCl$ was used for high-yield syntheses of ammonium dichlorobromates from ammonium chlorides and bromine chloride in CH_3CN [16] or in a CH_2Cl_2-H_2O mixture [17]. The solubility of BrCl in H_2O increases with addition of Cl^- because of $BrCl_2^-$ formation [18]. Codeposition of KCl vapor and BrCl diluted in Ar led to several products (because of BrCl dissociation) among which matrix-isolated $ClBrCl^-$ was detected by IR spectroscopy (see below) [19].

Dissolution of Br_2 (and Cl_2) in aqueous solutions containing chloride (and bromide) ions leads to Br_2Cl^-, BrCl, Br_3^-, Cl_3^-, and $BrCl_2^-$ via several coupled equilibria (see "Bromine" Suppl. Vol. A, 1985, p. 455). An average equilibrium constant $K=[BrCl_2^-]/([BrCl][Cl^-])=5.3\pm0.3$ L/mol at 298 K and ionic strengths of 1.14 and 4 was derived from the measured vapor pressures when Br_2 and Cl_2 were dissolved simultaneously in Cl^--containing aqueous solutions [20]. Spectrophotometric measurements at ionic strengths up to about 8 and at temperatures between 283 and 318 K gave slightly smaller equilibrium constants that depended on the ionic strength, e.g., 1.41 and 1.07 L/mol at 298 K and I = 1.3 and 7.7, respectively. Equilibrium constants, extrapolated to zero ionic strength, and derived thermodynamic data are: K = 1.53 L/mol, $\Delta_rG=1.05$ kJ/mol at 298 K and $\Delta_rH=3.16\pm0.11$ kJ/mol, $\Delta_rS=14.2\pm0.4\ J\cdot K^{-1}\cdot mol^{-1}$ [21]. Earlier estimates of the free energy of reaction led to higher equilibrium constants at 298 K [14]: $Cl^-(aq)+BrCl(g) \rightarrow BrCl_2^-(aq)$, $\Delta_rG=-18$ kJ/mol, $K=1.5\times10^3\ atm^{-1}$ and $Cl^-(aq)+BrCl(aq) \rightarrow BrCl_2^-(aq)$, $\Delta_rG=-13.8$ kJ/mol, K = 260 L/mol.

Redox potentials measured on dilute aqueous solutions of Br_2 in the presence of Cl^- and Br^- were evaluated in terms of the equilibrium $Br_2+2\,Cl^- \rightarrow BrCl_2^- + Br^-$, and gave an equilibrium constant of 7.2×10^{-3} L/mol at 298 K [22].

A third method used for preparing $BrCl_2^-$ salts involves the reaction $BrO_3^- + 6\,HCl \rightarrow BrCl_2^- + 2\,Cl_2 + 3\,H_2O$. The yield of the reaction using pyridinium and ammonium salts in CH_3OH-H_2O was between 20 and 30%. An attempt to prepare $CsBrCl_2$ by this method was however unsuccessful [23].

Some $BrCl_2^-$ was detected by UV-VIS [16] and IR [24] spectrometry when BrCl was dissolved in polar solvents, such as CH_3CN [16] or pyridine [24], on account of the equilibrium $2\,BrCl \rightleftharpoons Br^+ + BrCl_2^-$. There was also some spectrophotometric evidence that $BrCl_2^-$ may result from Br_2Cl^- disproportionation in aqueous solutions via $2\,Br_2Cl^- \rightleftharpoons Br_3^- + BrCl_2^-$ [22].

Products from the radiolysis of CCl_3Br diluted in Ar by 1 keV protons contained $BrCl_2^-$ [25].

Thermodynamic Data of Formation

Based on measured equilibrium constants for $Cl^- + BrCl \rightleftharpoons BrCl_2^-$, the following data of formation were derived for $BrCl_2^-$ in aqueous solution at 298 K [21]:

$$\Delta_fG = -137.8 \text{ kJ/mol},\ \Delta_fH = -178.8 \text{ kJ/mol},\ S = 206.4 \text{ J}\cdot\text{K}^{-1}\cdot\text{mol}^{-1}$$

Molecular Properties and Spectra

The $BrCl_2^-$ anion is linear and symmetric in crystalline pyridinium dichlorobromate, as shown by X-ray diffraction. Br-Cl bond lengths of 2.368(4) Å are 0.2 Å longer than the sum of the covalent radii [23]. In analogy to better characterized trihalide anions, such as I_3^-, the structure of $BrCl_2^-$ depends to some extent on the environment. The UV-VIS reflectance [26], and IR and Raman spectra [27] of crystalline salts were interpreted in terms of a linear and symmetric ($D_{\infty h}$) anion in case of large and bulky cations, such as tetraalkylammonium ions. Spectra of crystalline salts with alkali cations, such as Cs^+ or Rb^+, were interpreted in terms of a slightly asymmetric ($C_{\infty v}$) configuration of $BrCl_2^-$ [26, 27]. The breakdown of selection rules, seen in vibrational spectra taken from solutions, gave some hints that symmetry lowering may also play a role in solutions [7, 27]. However, precise data on the structure of asymmetric $BrCl_2^-$ anions are not available. Semiempirical MO calculations favor a linear, symmetric $BrCl_2^-$ anion [28, 29] with a bond length of 2.37 Å [30].

Several semiempirical MO calculations were published for linear symmetric [26, 28, 30 to 35] and for asymmetric $BrCl_2^-$ [26]. $X\alpha$ calculations agree on the order of valence orbitals $2\pi_u$, $3\sigma_g$, $1\pi_g$, $1\pi_u$, $2\sigma_u$, $2\sigma_g$, $1\sigma_u$, $1\sigma_g$ and a σ_u^* LUMO [28, 31]. Bonding in $BrCl_2^-$ can be described by a delocalized three-center four-electron scheme developed for trihalides; see for example [28]. The calculated charge distributions [28, 30, 33, 36] were compared to a value of $-0.56\ e^-$ for each chlorine atom derived from measured NQR frequencies [36].

An ionization potential of 3.6 eV was calculated [31].

Chlorine-35 NQR frequencies of 24.00 and 24.11 MHz were measured for $(CH_3)_4NBrCl_2$ at 77 K [36]. An approximate nuclear quadrupole constant eqQ = 48 MHz could be reproduced by an SCF MO $X\alpha$ calculation (51 MHz). A calculated value for bromine-79 is 898 MHz [28].

Salts of $BrCl_2^-$ are yellow. UV-VIS spectra of solutions show two $BrCl_2^-$ absorption bands, a high-intensity band centered at ~240 nm and a low-intensity band at ~345 nm. Band maxima λ_{max} and extinction coefficients ε_{max} are as follows:

specimen	λ_{max} in nm	ε_{max} in $L\cdot mol^{-1}\cdot cm^{-1}$	Ref.
Cs^+ and $(C_2H_5)_4N^+$ salts in CH_2Cl_2	242	37300	[26]
	346	310	
$(CH_3)_4N^+$ salt in CH_3CN	237	48000	[16]
	342	380	

Lambert-Beer's law was obeyed in the concentration range 10^{-4} to 8×10^{-3} mol/L [16]. Average band maxima at 242 and 341 nm were reported for solutions of pyridinium dichlorobromate in CH_3OH, CH_3CN, H_2O, and 1,2-dichloroethane [23] and a maximum between 254 and 258 nm for $BrCl_2^-$ solutions in alcohols and water [33, 37]. Extinction coefficients measured earlier [37] are, however, much too low. Spectral data ascribed to $BrCl_2^-$ from an analysis of the overall extinction of aqueous Br_2 solutions containing Br^- and Cl^- [22] do not conform with the data above.

Diffuse UV-VIS reflectance spectra were recorded for powders of three salts and were resolved into the four bands D, C, B, and A corresponding to transitions into ${}^1\Sigma_u^+$, ${}^1\Pi_u$, ${}^3\Sigma_u^+$, and ${}^3\Pi_u$ states of $BrCl_2^-$:

specimen	λ_{max} in nm	remark	Ref.
$(C_2H_5)_4NBrCl_2$	239, 279, 342, 403	symmetric $BrCl_2^-$	[26]
$C_5H_5NHBrCl_2$	247, 276, 346, 420	symmetric $BrCl_2^-$	[38]
$CsBrCl_2$	228, 274, 350, 434	asymmetric $BrCl_2^-$	[26]

Compared to solution spectra, both long-wavelength bands gained considerably in intensity with respect to the short-wavelength bands. The assignment of optical transitions was made on the basis of simple MO schemes for both $BrCl_2^-$ configurations and relied on data for a series of trihalides. Moreover, in view of the strong absorption in solutions, bands D and C were assigned to the spin- and symmetry-allowed transitions $\sigma_u^* \leftarrow \pi_g$ and $\sigma_u^* \leftarrow \sigma_g$ for $BrCl_2^-$ of $D_{\infty h}$ symmetry and to $\sigma^* \leftarrow \pi_2$ and $\sigma^* \leftarrow \sigma_2$ for $BrCl_2^-$ of $C_{\infty v}$ symmetry [26]. No firm assignment could be reached for the triplet transitions (A and B bands) [26]. Photochemical cross sections in the vicinity of 350 nm were unusually high for spin- and/or symmetry-forbidden transitions [12]. Transition energies derived from MO calculations were between 5.5 and 5.20 eV for the singlet transition [28, 30, 33] and between 3.29 and 3.45 eV for the triplet transition [28, 30] in agreement with experimental (solution) spectra.

To a first approximation, symmetric stretching, bending, and antisymmetric stretching vibrations of $BrCl_2^-$ are at $\nu_1 \approx 275\ cm^{-1}$ (strong, polarized Raman band), $\nu_2 \approx 140\ cm^{-1}$ (weak IR band), and $\nu_3 \approx 225\ cm^{-1}$ (strong IR band), respectively.

IR spectra of $BrCl_2^-$ salts dissolved in dichloromethane [27], pyridine [24], and benzene plus acetonitrile [40] showed a strong band each at 225, 226, and 227 cm^{-1}, respectively. The position was independent of the cation. However, the IR-inactive, symmetric stretching vibration was also observed as a weak band at 280 cm^{-1} [40]. Raman spectra from $BrCl_2^-$ solutions in dichloromethane, nitromethane [27], pyridine [24], benzene plus acetonitrile [40], nitrobenzene, and water [39] showed a strong, polarized [24, 40] band between 273 and 280 cm^{-1}.

IR and Raman spectra [23, 27, 38 to 40] of solid salts and suspensions thereof were subject to several secondary effects which resulted in band shifts, band splittings, and the partial removal of some selection rules. The IR and Raman band positions below for solid $(C_2H_5)_4NBrCl_2$ containing a symmetric $ClBrCl^-$ anion, and for $CsBrCl_2$ containing an asymmetric $ClBrCl^-$ anion, exemplify this situation [27] (all wave numbers in cm^{-1}; FS = factor group splitting due to coupling in unit cell, IS = isotopic splitting):

ν_i	$(C_2H_5)_4NBrCl_2$		$CsBrCl_2$	
	IR	Raman	IR	Raman
ν_1	280 sh[a)]	285.6, 272.8 FS, IS[b)]	311 vw	303, 297, 293 IS[c)]
ν_2	144	–	147	156, 147 $\nu_2(B)$[d)], FS 171 $\nu_2(R)$[d)]
ν_3	238, 212 FS	237 vw	235, 204 FS	234, 208 FS

[a)] Active due to FS. – [b)] Isotopic splitting of 272.8 cm^{-1} band in a 9:6:1 intensity ratio at 272.8 cm^{-1} ($^{35}ClBr^{35}Cl^-$), 269.6 cm^{-1} ($^{35}ClBr^{37}Cl^-$), and 266.4 cm^{-1} ($^{37}ClBr^{37}Cl^-$) characteristic for a symmetric anion. – [c)] Isotopic splitting in a 9:3:3:1 intensity ratio characteristic for an asymmetric anion (weakest band not observed). – [d)] $\nu_2(B)$ in-plane and $\nu_2(R)$ out-of-plane mode for $C_{\infty v}$ $ClBrCl^-$.

Some cation-induced effects were noted before [40]. A first assignment of an IR band at 305 cm^{-1} for $CsBrCl_2$ in Nujol to the antisymmetric stretching vibration [39] was in error [40]. In view of the data given above, it is likely that the assignment of an IR band at 311 cm^{-1} to ν_3 of matrix-isolated $ClBrCl^-$ (Ar matrix at 15 K, K^+ cation) [19] (also compiled in [41]) is erroneous too.

General valence force field force constants for symmetric $BrCl_2^-$ are $f_r = 1.02$ [40], 1.08 [42] mdyn/Å and $f_{rr} = 0.47$ [40], 0.53 [42] mdyn/Å. Some results [39] are outdated. The sum $f_r + f_{rr}$ could be satisfactorily reproduced by a simple electrostatic model [43]. Force constants in a modified valence force field for asymmetric $BrCl_2^-$ are $f_1 = 1.38$ mdyn/Å (short bond) and $f_2 = 0.91$ mdyn/Å (long bond) provided the interaction constant for symmetric $BrCl_2^-$ is used. The low value of f_r illustrates the weakness of Br-Cl bonds, and the relatively high interaction constant the possibility that the ion could change from a symmetric to an asymmetric configuration [42]. The mean amplitude of vibration of the Br−Cl bond is 0.074 Å at 298 K [44, 45] (a lower value [46] relies on an outdated ν_3 frequency). Compliance constants [47] and approximate eigenvalues for Wilson's FG-matrix (outdated ν_3 frequency) [48] were published.

Electrochemical Behavior

The first oxidation and reduction potentials of $BrCl_2^-$ were measured by cyclic voltammetry at 298 K. For the $(C_4H_9)_4N^+$ salts dissolved in tetrahydrofuran and benzonitrile, values of >1.2 and 0.20 V (decomposition) vs. SCE (in THF) and 1.58 and −0.24 V vs. SCE (in BN) were obtained [49].

Chemical Behavior

When $BrCl_2^-$ salts are dissolved in water, the anion dissociates to a considerable extent; see the equilibrium constants given above. Solid $BrCl_2^-$ salts are thermally unstable, as recognized quite early [3, 8 to 10]. The vapor pressure over $RbBrCl_2$, for example, reaches 20 Torr at 290 K and 760 Torr at 366 K [3]. Two equilibria, $RbBrCl_2 \rightleftharpoons RbCl + BrCl$ and $2\,BrCl \rightleftharpoons Br_2 + Cl_2$, are involved [10]. A slow, probably autocatalytic decomposition of $BrCl_2^-$ dissolved in CH_3CN was inferred from spectrophotometry [16]:

$$2\,BrCl_2^- + 2\,e^- \rightarrow Br_2Cl^- + 3\,Cl^-$$
$$3\,Br_2Cl^- + 9\,Cl^- + 2\,e^- \rightarrow 2\,Br_3^- + 12\,Cl^-$$
$$Br_3^- + 6\,Cl^- + 2\,e^- \rightarrow 3\,Br^- + 6\,Cl^-$$

$BrCl_2^-$ is photochemically not stable. A solid phosphonium salt was noted to decompose under irradiation below ~500 nm [7]. $BrCl_2^-$ solutions are also sensitive to irradiation with visible or UV light [16, 23]. The photodestruction cross sections of gaseous $BrCl_2^-$ ions measured in a drift tube were small at wavelengths above 476 nm, but increased to 3.3×10^{-18} cm^2 at ~354 nm. By analogy to Cl_3^-, it was assumed that cross sections are related to a photodissociation rather than photodetachment process [12].

Isomerization of $ClBrCl^-$ to $BrClCl^-$ in aqueous solution is very unlikely under thermal conditions [14].

Reactions of $BrCl_2^-$ with Br^- or Br_3^- in H_2O [22] or with Br^+ in polar aprotic solvents [16, 24] are the reverse of some pathways of $BrCl_2^-$ formation mentioned above.

The reaction $CsBrCl_2 + 2\,ClONO_2 \rightarrow Cs[Br(ONO_2)_2] + 2\,Cl_2$ can be used for synthesis of $Br(ONO_2)_2^-$ salts [50]. An attempt to use a tetramethylammonium instead of caesium salt was, however, unsuccessful [51].

Tetraalkylammonium salts of $BrCl_2^-$ dissolved in aprotic organic solvents are convenient chlorobromination agents for a great variety of alkenes [52 to 56] (including styrenes [52, 55], 1-phenylpropenes [56], conjugated dienes [54]) and alkynes [57]. The stereo- and regiospecifity of these reactions is different when BrCl is used for chlorobromination and points toward an intermediate, three-centered π complex (Br–Cl bound at Br to the multiple bond) that is attacked by Cl^-; see, for example [57]. Bromochlorination of styrenes in protic solvents gives substantial amounts of side products which incorporate the solvent. The course of the reaction probably involves the attack of a bromonium intermediate by a chloride ion or a solvent molecule [52]. There was an attempt to determine the role which $BrCl_2^-$ plays in the kinetics of the reaction between diethyl fumarate or 2-chlorallylalcohol and Br_2 in Cl^--containing aqueous solutions [58].

The pyridine ring is brominated when melting pyridinium dichlorobromate [23]. Ammonium [17] and polymer-supported pyridinium salts [59] of $BrCl_2^-$ were used for brominating aromatics.

References:

[1] Popov, A. I. (in: Gutmann, V.; Halogen Chemistry, Vol. 1, Academic, London 1967, pp. 225/64).

[2] Wiebenga, E. H.; Havinga, E. E.; Boswigk, K. H. (Advan. Inorg. Chem. Radiochem. **3** [1961] 133/69).

[3] Ephraim, F. (Ber. Deut. Chem. Ges. **50** [1917] 1069/88).

[4] Chattaway, F. D.; Hoyle, G. (J. Chem. Soc. **123** [1923] 654/62).

[5] Popov, A. I.; Buckles, R. E. (Inorg. Synth. **5** [1957] 167/75).

[6] Kanai, K.; Hashimoto, T.; Kitano, H.; Fukui, K. (Nippon Kagaku Zasshi **86** [1965] 534/9, A 33).

[7] Zimmer, H.; Jayawant, M.; Amer, A.; Ault, B. S. (Z. Naturforsch. **38 b** [1983] 103/7).

[8] Wells, H. L.; Penfield, S. L. (Z. Anorg. Allgem. Chem. **1** [1892] 85/103; Am. J. Sci. [3] **43** [1892] 17/32).

[9] Wells, H. L.; Wheeler, H. L.; Penfield, S. L. (Z. Anorg. Allgem. Chem. **1** [1892] 442/55; Am. J. Sci. [3] **43** [1892] 475/87).

[10] Cremer, H. W.; Duncan, D. R. (J. Chem. Soc. **1933** 181/9, 187).

[11] Huber, B. A.; Miller, T. M. (J. Appl. Phys. **48** [1977] 1708/11).

[12] Lee, L. C.; Smith, G. P.; Moseley, J. T.; Cosby, P. C.; Guest, J. A. (J. Chem. Phys. **70** [1979] 3237/46).

[13] Salthouse, J. A.; Waddington, T. C. (J. Chem. Soc. A **1966** 1188/90).

[14] Scott, R. L. (J. Am. Chem. Soc. **75** [1953] 1550/2).

[15] Rozinov, V. G. (Zh. Obshch. Khim. **50** [1980] 1414/5; C.A. **93** [1980] No. 106244).

[16] Schmeisser, M.; Tytko, K.-H. (Z. Anorg. Allgem. Chem. **403** [1974] 231/42).

[17] Kajigaeshi, S.; Kakinami, T.; Moriwaki, M.; Tanaka, T.; Fujisaki, S.; Okamoto, T. (Chem. Express **3** [1988] 219/22).

[18] Stasinevich, D. S. (Zh. Fiz. Khim. **50** [1976] 815; Russ. J. Phys. Chem. **50** [1976] 487).

[19] Ault, B. S.; Andrews, L. (J. Chem. Phys. **64** [1976] 4853/9).

[20] Artamonov, Yu. F.; Gergert, V. R. (Zh. Neorg. Khim. **22** [1977] 18/22; Russ. J. Inorg. Chem. **22** [1977] 8/11).

[21] Kremer, V. A.; Onoprienko, T. A.; Zalkind, G. R.; Kashtanova, A. S. (Deposited Doc. VINITI 2363-76 [1976] 1/15).

[22] Bell, R. P.; Pring, M. (J. Chem. Soc. A **1966** 1607/9).

[23] Snyder, R. L. (Diss. Fordham Univ. 1968, pp. 1/160; Diss. Abstr. B **29** [1969] 2801).

[24] Ginn, S. G. W.; Haque, I.; Wood, J. L. (Spectrochim. Acta A **24** [1968] 1531/42).

[25] Andrews, L.; Grzybowski, J. M.; Allen, R. O. (J. Phys. Chem. **79** [1975] 904/12).
[26] Gabes, W.; Stufkens, D. J. (Spectrochim. Acta A **30** [1974] 1835/41).
[27] Gabes, W.; Gerding, H. (J. Mol. Struct. **14** [1972] 267/79).
[28] Bowmaker, G. A.; Boyd, P. D. W.; Sorrenson, R. J. (J. Chem. Soc. Faraday Trans. II **80** [1984] 1125/43).
[29] Evans, J. C.; Lo, G. Y.-S. (J. Chem. Phys. **45** [1966] 1069/71).
[30] Gabes, W.; Nigman-Meester, M. A. M. (Inorg. Chem. **12** [1973] 589/92).

[31] Gutsev, G. L. (Zh. Strukt. Khim. **30** No. 5 [1989] 41/7; J. Struct. Chem. [USSR] **30** [1989] 733/7).
[32] Smolyar, A. E.; Charkin, O. P.; Klimenko, N. M. (Zh. Strukt. Khim. **15** [1974] 993/1003; J. Struct. Chem. [USSR] **15** [1974] 885/93).
[33] Ohkubo, K.; Yoshinaga, K. (Bull. Japan Petrol. Inst. **19** [1977] 73/80), Ohkubo, K.; Aoji, T.; Yoshinaga, K. (Bull. Chem. Soc. Japan **50** [1977] 1883/4).
[34] Zakzhevskii, V. G.; Charkin, O. P.; Smolyar, A. E.; Zyubina, T. S.; Klimenko, N. M. (Zh. Strukt. Khim. **17** [1976] 763/74; J. Struct. Chem. [USSR] **17** [1976] 659/69).
[35] Gabes, W. (Chem. Phys. Letters **27** [1974] 183/6).
[36] Riedel, E. F.; Willett, R. D. (J. Am. Chem. Soc. **97** [1975] 701/4).
[37] Gilbert, F. L.; Goldstein, R. R.; Lowry, T. M. (J. Chem. Soc. **1931** 1092/103).
[38] Gabes, W.; Stufkens, D. J.; Gerding, H. (J. Mol. Struct. **17** [1973] 329/40).
[39] Person, W. B.; Anderson, G. R.; Fordemwalt, J. N.; Stammreich, H.; Forneris, R. (J. Chem. Phys. **35** [1961] 908/14).
[40] Evans, J. C.; Lo, G. Y.-S. (J. Chem. Phys. **44** [1966] 4356/7).

[41] Jacox, M. E. (J. Phys. Chem. Ref. Data **13** [1984] 945/1068, 983).
[42] Gabes, W.; Elst, R. (J. Mol. Struct. **21** [1974] 1/5).
[43] Gázquez, J. L.; Ray, N. K.; Parr, R. G. (Theor. Chim. Acta **49** [1978] 1/11, 7).
[44] Baran, E. J. (Monatsh. Chem. **104** [1973] 1653/9).
[45] Sanyal, N. K.; Verma, D. N.; Dixit, L. (Indian J. Pure Appl. Phys. **12** [1974] 393/4).
[46] Nagarajan, G. (Indian J. Pure Appl. Phys. **4** [1966] 351/4).
[47] Gupta, S. L.; Verma, U. P.; Pandey, A. N. (Acta Ciencia Indica **3** [1977] 340/2).
[48] Keeports, D. D. (Acta Chim. Hung. **124** [1987] 789/808, 792).
[49] Sakura, S.; Imai, H.; Anzai, H.; Moriya, T. (Bull. Chem. Soc. Japan **61** [1988] 3181/6).
[50] Wilson, W. W.; Christe, K. O. (Inorg. Chem. **26** [1987] 1573/80).

[51] Holthausen, R. (Diss. T. H. Aachen 1968 from [50]).
[52] Negoro, T.; Ikeda, Y. (Bull. Chem. Soc. Japan **59** [1986] 3519/22).
[53] Negoro, T.; Ikeda, Y. (Bull. Chem. Soc. Japan **59** [1986] 2547/51).
[54] Negoro, T.; Ikeda, Y. (Bull. Chem. Soc. Japan **58** [1985] 3655/6).
[55] Negoro, T.; Ikeda, Y. (Bull. Chem. Soc. Japan **57** [1984] 2111/5).
[56] Negoro, T.; Ikeda, Y. (Bull. Chem. Soc. Japan **57** [1984] 2116/20).
[57] Negoro, T.; Ikeda, Y. (Bull. Chem. Soc. Japan **59** [1986] 3515/8).
[58] Bell, R. P.; Pring, M. (J. Chem. Soc. B **1966** 1119/26).
[59] Zajc, B.; Zupan, M. (Tetrahedron **45** [1989] 7869/78).

16.15 Bromochlorochlorate(1 −), $BrClCl^-$

CAS Registry Number: *[96607-00-0]*

KBr vapor and Cl_2 in excess Ar were codeposited at 15 K. A single product IR band at 273 cm^{-1} was assigned to the antisymmetric stretching vibration of the $BrClCl^-$ anion [1] (listed in a review [2]). When end-on addition instead of insertion prevails under matrix isolation conditions, a strong UV band at 273 nm from the product of codepositing CsBr vapor with Cl_2 in excess Ar [3] may be ascribed to matrix-isolated $BrClCl^-$. Estimated free energies of ~13 kJ/mol for $Br^-(aq) + Cl_2(g) \rightarrow ClBrCl^-$ and of ~59 kJ/mol for the isomerization $BrClCl^- \rightarrow ClBrCl^-$ in aqueous solution [4] indicate the instability of the $BrClCl^-$ isomer.

References:

[1] Ault, B. S.; Andrews, L. (J. Chem. Phys. **64** [1976] 4853/9).
[2] Jacox, M. E. (J. Phys. Chem. Ref. Data **13** [1984] 945/1068, 983).
[3] Andrews, L.; Prochaska, E. S.; Loewenschuss, A. (Inorg. Chem. **19** [1980] 463/5).
[4] Scott, R. L. (J. Am. Chem. Soc. **75** [1953] 1550/2).

16.16 Dichlorobromate(2−), $ClBrCl^{2-}$

CAS Registry Number: *[63837-08-1]*

Doubly negative $ClBrCl^{2-}$ ions were detected as paramagnetic centers in KCl crystals which had been doped with Br^- and Na^+ and irradiated by X-rays at 77 K for several hours. Depending on the doping concentration and temperature, $ClBrCl^{2-}$ forms at three different positions in the crystal: 1) At temperatures up to 150 K, $ClBrCl^{2-}$ is oriented along symmetrically equivalent $\langle 110 \rangle$ axes. The molecule is slightly bent in the $\{100\}$ plane, and the BrCl bond directions deviate by 7° from the [110] axis. Bending is induced by a Na^+ impurity adjoining the central Br on one side. Spin-Hamiltonian parameters for this so-called $H_A(Na^+)$-type center are: $g_z = 1.9889 \pm 0.0003$, $g_x = 2.082 \pm 0.004$, $g_y = 2.106 \pm 0.007$, $A_z(Br) = +501.3 \pm 0.2$, $A_\perp(Br) = \pm 35 \pm 20$, $A_z(Cl) = +55.5 \pm 0.5$, $A_\perp(Cl) = +16 \pm 5$ (hyperfine interaction constants A in Gauss). 2) At high doping concentrations (1 mol%), a $H_{AA}(Na^+)$-type center forms which is stable below 185 K. This center contains a perfectly linear $ClBrCl^{2-}$ ion flanked by two Na^+ impurity ions. The orientation in the crystal is analogous to the $H_A(Na^+)$-type center. Spin-Hamiltonian parameters are: $g_z = 1.9893 \pm 0.0003$, $g_x = 2.076 \pm 0.003$, $g_y = 2.104 \pm 0.006$, $A_z(Br) = +502.1 \pm 0.2$, $A_\perp(Br) = \pm 41 \pm 12$, $A_z(Cl) = +56.2 \pm 0.5$, $A_\perp(Cl) = +16 \pm 1$. (Vectors z, x, y are along [110], [001], and $[1\bar{1}0]$ for both centers.) 3) The diinterstitial center which is stable up to 210 K is a linear $ClBrCl^{2-}$ unit stabilized by one Na^+ and one K^+ ion next to both Cl atoms, but located on opposite sides. The molecular axis is in the $\{110\}$ plane forming an angle of 22° with the [100] axis. Spin-Hamiltonian parameters are: $g_z = 1.9886 \pm 0.0003$, $g_x = 2.120 \pm 0.006$, $g_y = 2.076 \pm 0.006$, $A_z(Br) = +502.5 \pm 1$, $A_\perp(Br) = \pm 50 \pm 25$, $A_z(Cl) = +62.5 \pm 1$, $+49.0 \pm 1$, $A_\perp(Cl) = +15 \pm 15$ (both) (z along $\langle 100 \rangle + 22°$ in $\{110\}$).

$H_{AA}(Na^+)$-type $ClBrCl^{2-}$ has a strong optical absorption band around 380 nm and there was evidence for a weak band at 580 nm.

Reference:

Schoemaker, D.; Lagendijk, A. (Phys. Rev. [3] B **15** [1977] 5927/37).

16.17 Bromine Trichloride, $BrCl_3$

CAS Registry Number: *[12360-50-8]*

Br_2 and excess Cl_2 were diluted by a rare gas, passed through a microwave discharge, and the mixture was condensed at 20 K. The IR bands at 332.2, 330.2 (in Ar), 364.3, 362.1, 326.1, 324.0 (in Kr), and at 370.5 cm^{-1} (in Xe) were very tentatively assigned to a T-shaped

[8] Evans, J. C.; Lo, G. Y.-S. (J. Phys. Chem. **70** [1966] 20/5).
[9] Deiters, R. M. (Diss. Univ. Cincinnati 1967; Diss. Abstr. B **28** [1967] 2319).
[10] Salthouse, J. A.; Waddington, T. C. (J. Chem. Soc. **1964** 4664/6).

[11] Kohle, R.; Kuchen, W.; Peters, W. (Z. Anorg. Allgem. Chem. **551** [1987] 179/90).
[12] Salthouse, J. A.; Waddington, T. C. (J. Chem. Soc. A **1966** 28/30).
[13] Nibler, J. W.; Pimentel, G. C. (J. Chem. Phys. **47** [1967] 710/7).
[14] McDaniel, D. H.; Stitt, F. (private communication to [13, 15]).
[15] Pimentel, G. C.; McClellan, A. L. (Ann. Rev. Phys. Chem. **22** [1971] 347/85, 361/2, 383).
[16] Cotton, J. D.; Waddington, T. C. (J. Chem. Soc. A **1966** 785/9).
[17] Benoit, R. L.; Beauchamp, A. L.; Domain, R. (Inorg. Nucl. Chem. Letters **7** [1971] 557/62).
[18] Benoit, R. L.; Rinfret, M.; Domain, R. (Inorg. Chem. **11** [1972] 2603/5).
[19] Fujiwara, F. Y.; Martin, J. S. (J. Chem. Phys. **56** [1972] 4091/7).
[20] Fujiwara, F. Y.; Martin, J. S. (J. Am. Chem. Soc. **96** [1974] 7625/31).

[21] Pawlak, Z.; Mukherjee, L. M.; Hampton, A. A.; Bates, R. G. (J. Chem. Thermodyn. **15** [1983] 189/93).
[22] Yamdagni, R.; Kebarle, P. (J. Am. Chem. Soc. **93** [1971] 7139/43).
[23] Larson, J. W.; McMahon, T. B. (J. Am. Chem. Soc. **109** [1987] 6230/6).
[24] Larson, J. W.; McMahon, T. B. (Inorg. Chem. **23** [1984] 2029/33).
[25] Jiang, G. J.; Anderson, G. R. (J. Chem. Phys. **60** [1974] 3258/63).
[26] Jacox, M. E. (J. Phys. Chem. Ref. Data **13** [1984] 945/1068, 962).

17.3 $HBr \cdot Cl_2$

CAS Registry Number: *[51065-12-4]*

The formation of the complex $HBr \cdot Cl_2$ is described in Section "Hydrogen-Bonded Complexes" of HBr in "Bromine" Suppl. Vol. B1, 1990, p. 375.

17.4 $HBr \cdot HCl$

CAS Registry Numbers: *[15792-73-1, 62140-55-0]*

Data for the complex $HBr \cdot HCl$ are given in Section "Hydrogen-Bonded Complexes" of HBr in "Bromine" Suppl. Vol. B1, 1990, p. 373. For a formulation as $H_2Cl^+ \cdot Br^-$ see "Brom" 1931, p. 275.

18 Compounds of Bromine with Chlorine and Oxygen

18.1 $BrClO_2$ and $ClBrO_2$

These molecules were postulated to be intermediates in the reaction of chlorite and bromide in acidic solutions, which is initiated by (1) $HClO_2 + Br^- + H^+ \rightarrow HClO + HBrO$ and followed by an autocatalytic reaction (with HBrO as catalyzer) leading to a rapid production of ClO_2. The latter process was explained by the formation of $BrClO_2$ according to (2) $HClO_2 + HBrO \rightarrow BrClO_2 + H_2O$ and the successive reaction (3) $HClO_2 + BrClO_2 \rightarrow 2\,ClO_2 + Br^- + H^+$. The $ClBrO_2$ molecule, assumed to be formed by isomerization, (4) $BrClO_2 \rightarrow ClBrO_2$, can analogously react according to (5) $HClO_2 + ClBrO_2 \rightarrow BrO_2 + ClO_2 + Cl^- + H^+$. The sum of reactions (2), (4), (5), and (6) $BrO_2 + HClO_2 \rightarrow ClO_2 + HBrO_2$, and (7) $HBrO_2 + HClO_2 \rightarrow HBrO + ClO_3^- + H^+$ accounts for the stoichiometric decomposition of the chlorite caused by the HBrO catalyzer, (8) $4\,HClO_2 \rightarrow 2\,ClO_2 + ClO_3^- + Cl^- + 2\,H^+ + H_2O$.

Other possible reactions of $ClBrO_2$ were predicted by analogy with other X_2O_2 intermediates, (9) $ClBrO_2 + H_2O \rightarrow BrO_3^- + Cl^- + 2\,H^+$ and (10) $ClBrO_2 + Br^- + H^+ \rightarrow BrCl + HBrO_2$.

In analogy to the Cl_2O_2 molecule, the structures of the two isomers were assumed to be Br-O-Cl-O or Br-Cl-O_2 and Cl-O-Br-O or Cl-Br-O_2.

Reference:

Schmitz, G.; Rooze, H. (Can. J. Chem. **65** [1987] 497/501).

18.2 Bromine Perchlorate, $BrOClO_3$

CAS Registry Number: *[32707-10-1]*

$BrOClO_3$ was prepared for the first time by Schack et al. [1] by reacting $MClO_4$ ($M = NO_2$, Cs) with $BrSO_3F$ and Br_2 with $ClOClO_3$. In the first case, a prepassivated stainless steel cylinder was loaded with preweighed amounts of either NO_2ClO_4 or $CsClO_4$ followed by the condensation of a less than equimolar amount of $BrSO_3F$; the reaction was allowed to proceed at 253 K for five days or longer and, after fractional distillation (separation of unreacted $BrSO_3F$ and the by-products $FClO_2$ and $FClO_3$), $BrOClO_3$ was trapped at 209 K. In the second case, the cylinder was loaded at 77 K with Br_2 and $ClOClO_3$, and the mixture was allowed to react for five days at 228 K. After recooling to 195 and 209 K, the volatile products were pumped off and trapped at 195, 161, and 77 K. $BrOClO_3$, $ClOBrO_2$, and Cl_2 were separated by fractional condensation. The second method involving the oxidation of bromine proceeded quantitatively and yielded a purer product than the fluorosulfate reaction. The formulation as $BrOClO_3$ is based on its quantitative synthesis, elemental analysis, and IR spectrum (see below). Further support was obtained from its reactions with HBr and AgCl (see p. 238).

$BrOClO_3$ is a red liquid, which is unstable at ambient temperature, attains a vapor pressure of 5 Torr at 250 K, decomposes slowly at ~253 K, and freezes below 196 K. It is shock- and light-sensitive and should be handled using safety precautions.

The IR spectrum of gaseous $BrOClO_3$ at room temperature and 20 Torr and of $BrOClO_3$ in an Ar matrix (1:400) at 4 K was recorded, on the basis of which (and by analogy with the IR spectra of $HClO_4$ and $FOClO_3$ and the IR and Raman spectra of $ClOClO_3$) a structure of symmetry C_s (i.e., one symmetry plane through Br-O-Cl=O) was derived for $BrOClO_3$ [1, 2], see **Fig. 8**. The geometrical parameters (used for a normal coordinate analysis, see below) are assumed to be those of $HClO_4$, r(Cl=O) = 1.408 ± 0.002 Å, ∡(O=Cl=O) = $112.8° \pm 0.5°$, r(Cl-O) = 1.635 ± 0.007 Å, ∡(O-Cl=O) = $105.8° \pm 0.7°$ [3, 4], and, by comparison

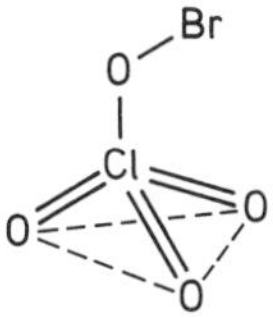

Fig. 8. Structure of $BrOClO_3$ [1].

with other molecules, r(O-Br) = 1.85 Å, ∢(Cl-O-Br) = 110°. These give the rotational constants A = 5626, B = 1155, and C = 1150 MHz [2].

For a six-atom molecule of C_s symmetry, a total of twelve fundamental vibrations are expected, eight of them belonging to the symmetry species A′ and four to A″. The IR spectrum of gaseous $BrOClO_3$ at 400 to 2400 cm^{-1} showed (besides a weak combination band at 2300 cm^{-1}) six bands of "very strong" to "medium" intensity identified with eight of the twelve fundamentals (see below): 1275 (vs), 1039 (s), 683 (m), 648 (s), 570 (ms), and 509 (m) cm^{-1}. The matrix spectrum is very complex in the 1280 to 1240 cm^{-1} region showing the two strong isotopic pairs (^{35}Cl-^{37}Cl) of the antisymmetric O=Cl stretching vibrations (ν_1, ν_9, which are degenerate in the gas-phase spectrum) and some weaker features caused presumably by matrix site splittings and/or by Fermi resonances between fundamentals and overtones; splittings were also observed in the 1040 and 650 cm^{-1} region, as well as the separation of two fundamentals (ν_5 and ν_{10}) around 570 cm^{-1} and an additional fundamental at 387 cm^{-1}. The following fundamental vibrations based on the gas-phase and matrix IR spectra [1, 2] were derived for $BrOClO_3$ [2]:

vibration	approximate description	wave number in cm^{-1}
ν_1 (A′)	antisymmetric ClO_3 stretching	1279
ν_2 (A′)	symmetric ClO_3 stretching	1039
ν_3 (A′)	O-Br stretching	683
ν_4 (A′)	Cl-O stretching	648
ν_5 (A′)	ClO_2 scissoring	572
ν_6 (A′)	ClO_3 umbrella	509
ν_7 (A′)	ClO_3 rocking	278[a)]
ν_8 (A′)	BrOCl deformation	159[a)]
ν_9 (A″)	antisymmetric ClO_3 stretching	1262
ν_{10} (A″)	antisymmetric ClO_3 deformation	566
ν_{11} (A″)	ClO_2 torsion	387
ν_{12} (A″)	OBr torsion	b)

a) Wave numbers calculated by using force constants for a simple valence force field. — b) Neither observed nor calculated.

Using the geometry for the $BrOClO_3$ molecule assumed above, the following stretching and bending force constants and stretching-stretching and stretching-bending interaction constants of a simple valence force field were calculated [2]:

f(Cl=O) = 8.8 mdyn/Å	f(O-Cl) = f(O-Br) = 2.65 mdyn/Å
f(O=Cl=O) = 1.9 mdyn·$Å^{-1}$·rad^{-2}	f(O=Cl-O) = 1.1 mdyn·$Å^{-1}$· rad^{-2}
f(Cl-O, O-Br) = 0.35 mdyn/Å	f(Cl-O, O=Cl-O) ≈ 0.2 mdyn·$Å^{-1}$· rad^{-1}

The heat capacity C_p^o and the thermodynamic functions S°, $-(G^o-H_0^o)/T$, and $H^o-H_0^o$ have been calculated for $BrOClO_3$ as an ideal gas (rigid-rotor harmonic oscillator approximation) and tabulated for T = 298.15 K and T = 0 to 2000 K at 100 K intervals; the fundamental frequencies and estimated rotational constants given above were used to establish the partition function; excerpted values are as follows [2]:

T in K	C_p^o	S°	$-(G^o-H_0^o)/T$	$H^o-H_0^o$ in kJ/mol
	in $J \cdot mol^{-1} \cdot K^{-1}$			
0	0	0	0	0
100	50.384	266.487	225.831	4.067
200	70.806	307.662	257.040	10.125
298.15	87.437	339.209	279.060	17.933
400	99.659	366.765	297.972	27.497
600	113.813	410.145	328.457	49.016
800	120.947	443.973	353.268	72.567
1000	124.851	471.420	374.242	97.182
1500	129.173	523.021	415.764	160.883
2000	130.813	560.438	447.470	225.936

The reaction of $BrOClO_3$ with O_3 yielded $O_2BrOClO_3$ (see below) [5]. The reaction with HBr yielded Br_2 and $HClO_4$, which were identified by vapor-pressure and IR measurements. The reaction with AgCl was examined only qualitatively; it gave $AgClO_4$ and ClO_2 identified by their IR spectra (and the disappearance of the $BrOClO_3$ spectrum) as well as Br_2, Cl_2, and other noncondensable gases at 77 K [1]. $CsBr(ClO_4)_2$ was obtained in the reaction of $BrOClO_3$ with $CsClO_4$ (see p. 239) [6]. Reactions with fluoroalkyl halides were reported in [7], reactions with perhaloolefines in [8].

References:

[1] Schack, C. J.; Christe, K. O.; Pilipovich, D.; Wilson, R. D. (Inorg. Chem. **10** [1971] 1078/80).
[2] Christe, K. O.; Schack, C. J.; Curtis, E. C. (Inorg. Chem. **10** [1971] 1589/93).
[3] Clark, A. H.; Beagley, B.; Cruickshank, D. W. J. (Chem. Commun. **1968** 14/5).
[4] Clark, A. H.; Beagley, B.; Cruickshank, D. W. J.; Hewitt, T. G. (J. Chem. Soc. A **1970** 1613/6).
[5] Schack, C. J.; Christe, K. O. (Inorg. Chem. **13** [1974] 2378/81).
[6] Christe, K. O.; Schack, C. J. (Inorg. Chem. **13** [1974] 1452/5).
[7] Schack, C. J.; Pilipovich, D.; Christe, K. O. (Inorg. Chem. **14** [1975] 145/51).
[8] Schack, C. J.; Pilipovich, D.; Hon, J. F. (Inorg. Chem. **12** [1973] 897/900).

18.3 Bromine Oxide Perchlorate, $O_2BrOClO_3$

CAS Registry Number: *[52225-67-9]*

$O_2BrOClO_3$ results from the oxidative oxygenation of the bromine atom in $BrOClO_3$ by O_3 via

$$BrOClO_3 + O_3 \rightarrow O_2BrOClO_3 + 1/2\ O_2$$

The O_3 was condensed at 77 K onto freshly prepared $BrOClO_3$ in CF_3Cl solution and the reactor warmed to 228 K. After 72 hours, $O_2BrOClO_3$ was separated from the solvent and unreacted starting material and O_2 by fractional condensation. The formation of $O_2BrOClO_3$

was observed only in the presence of the solvent which acts as a moderator; the reaction of neat O_3 and $BrOClO_3$ at temperatures between 195 and 228 K caused complete degradation to the elements.

$O_2BrOClO_3$ is a bright orange solid that does not melt below 238 K. Since it begins to decompose at higher temperatures and owing to its nonvolatility, no other properties of the compound could be determined.

A slow displacement reaction with FNO_2 was observed at 228 K

$$O_2BrOClO_3 + FNO_2 \rightarrow NO_2ClO_4 + [FBrO_2]$$

giving NO_2ClO_4 in quantitative yield, but $FBrO_2$ decomposed to the elements. The reaction served as an additional proof for the composition of $O_2BrOClO_3$.

Reference:

Schack, C. J.; Christe, K. O. (Inorg. Chem. **13** [1974] 2378/81).

18.4 Bis(perchlorato)bromate(1 −), $[Br(OClO_3)_2]^-$

CAS Registry Number: −

The synthesis and characterization of $[Br(OClO_3)_2]^-$, the first example of a perchloratobromate ion, was reported in 1974. Following the reaction of CsBr with excess $ClOClO_3$, the anion was prepared and isolated as its caesium salt, $Cs^+[Br(OClO_3)_2]^-$, a faint, yellow, crystalline solid which is stable at room temperature, hygroscopic, and readily hydrolyzes in water. It was obtained in 96% yield, when a stainless steel cylinder, loaded with powdered CsBr and $ClOClO_3$ at 77 K and warmed to 228 K, was stored at that temperature for two years and, after warming to room temperature, the volatile products (Cl_2, Cl_2O_6, and unreacted $ClOClO_3$) were pumped off. When the reaction of CsBr with $ClOClO_3$, carried out under similar conditions, was examined after six days, no $CsBr(OClO_3)_2$, but solid $CsClO_4$ and volatile Cl_2, $BrOClO_3$, and $ClOClO_3$ were obtained. After a reaction time of two months, the solid product consisted of 32 mol% $CsBr(OClO_3)_2$ and 68 mol% $CsClO_4$. These experiments show that CsBr interacts with an excess of $ClOClO_3$ relatively fast at 228 K according to

$$CsBr + 2\ ClOClO_3 \rightarrow CsClO_4 + BrOClO_3 + Cl_2$$

followed by the much slower second step

$$CsClO_4 + BrOClO_3 \rightarrow CsBr(OClO_3)_2$$

Accelerating the second step by raising the reaction temperature was not feasible owing to the thermal instability of the halogen monoperchlorates. No attempt was made to search for a proper solvent to increase the reaction rate since the halogen perchlorates are incompatible with most solvents and the reaction is so slow.

Raman and IR spectra of $CsBr(OClO_3)_2$ were recorded at room temperature. They confirmed a structural model of the $[Br(OClO_3)_2]^-$ anion containing two covalent, monodentate perchlorato groups in trans position and an approximately linear O-Br-O arrangement. This structure belongs to point group C_2 with the twofold symmetry axis perpendicular to the Cl-O-Br-O-Cl plane and passing through the Br central atom; see **Fig. 9**, p. 240.

For the eleven-atom molecule of point group C_2 (symmetry species A and B), a total of 27 fundamental vibrations are expected including eighteen (9 A, 9 B) involving the $-ClO_3$ modes and nine (4 A, 5 B) involving the Cl-O-Br-O-Cl skeletal modes. Besides four combi-

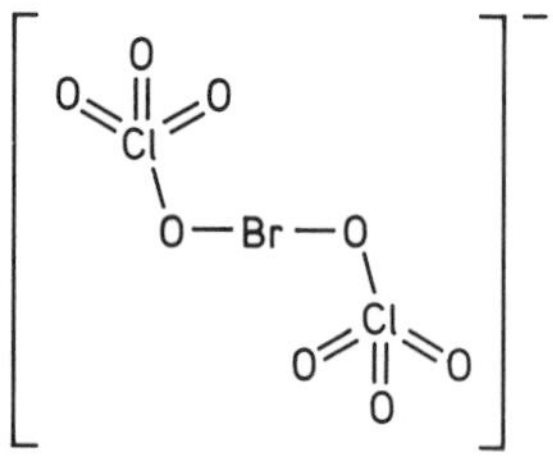

Fig. 9. Structure of the $[Br(OClO_3)_2]^-$ anion.

nation bands at 2930, 2360, 2040, and 1663 cm^{-1} in the IR spectrum, the following IR and Raman bands were assigned to fourteen of the $-ClO_3$ fundamentals and to four skeletal stretching modes (wave numbers in cm^{-1}):

IR	Raman	fundamental vibration in point group C_2
1300	1289	antisymmetric ClO_3 stretching, in phase (2 A)[a]
1115	1105	antisymmetric ClO_3 stretching, out of phase (2 B)[a]
1076	1078	symmetric ClO_3 stretching, in phase (A)
947	947, 933, 904 [b]	symmetric ClO_3 stretching, out of phase (B)
720	719	antisymmetric Br-O-Cl stretching, out of phase (B)
633, 622	633, 625 [c]	symmetric Br-O-Cl stretching, out of phase (B)
581	584	ClO_2 scissoring (A, B)
572	578	antisymmetric ClO_3 deformation (A, B)
558	558	ClO_3 umbrella (A, B)
	466	antisymmetric Br-O-Cl stretching, in phase (A)
	450	symmetric Br-O-Cl stretching, in phase (A)
	407	ClO_3 rocking (A)
	396	ClO_3 rocking (B)

[a] The difference of ~ 10 cm^{-1} between the IR and Raman band centers shows that these vibrations are not degenerate. — [b] Crystal field splitting. — [c] Intensity ratio of 3:1 indicates isotopic (^{35}Cl-^{37}Cl) splitting.

Reference:

Christe, K. O.; Schack, C. J. (Inorg. Chem. **13** [1974] 1452/5).

19 Compounds of Bromine with Chlorine and Nitrogen (and Hydrogen)

The formation of haloamines upon water chlorination and especially the formation of bromoamines and bromochloroamines in chlorinated seawater (due to the high bromide concentration of the latter) is a problem encountered, for instance, in cooling water systems of sea-side power plants. To study the environmental effects of water chlorination (in view of the persistence and toxicity of the haloamines in aquatic organisms), data are needed on the stability, reaction, decomposition rates, and concentrations of these compounds in water of varying pH, salinity, and ammonia and bromide content.

19.1 Dibromochloroamine, NBr_2Cl

CAS Registry Number: *[85896-86-2]*

NBr_2Cl is formed in ethereal solutions of NH_2Cl and HBrO in addition to NHBrCl (see below), if the HBrO/NH_2Cl molar ratio is increased from about 0.5 up to about 4. The UV spectrum shows a peak at 248 nm with $\varepsilon_{max}=6800\ L\cdot mol^{-1}\cdot cm^{-1}$, a shoulder centered at about 315 nm ($\varepsilon\approx750\ L\cdot mol^{-1}\cdot cm^{-1}$) and significant absorption out to 400 nm ($\varepsilon\approx230\ L\cdot mol^{-1}\cdot cm^{-1}$). The main peak for an aqueous solution of NBr_2Cl is expected at about 240 nm with $\varepsilon_{max}=6000\ L\cdot mol^{-1}\cdot cm^{-1}$ by comparison with NBr_3 and NCl_3.

Reference:

Haag, W. R.; Jolley, R. L. (Water Chlorination Environ. Impact Health Eff. **1983** Vol. 1, Chapter 4, pp. 77/83).

19.2 Bromodichloroamine, $NBrCl_2$

CAS Registry Number: –

This compound has not yet been investigated since it is not likely to be of great importance in water chlorination. An absorption peak at about 230 nm with $\varepsilon_{max}\approx 7000\ L\cdot mol^{-1}\cdot cm^{-1}$ is expected for $NBrCl_2$ in aqueous solution by comparison with NBr_3, NBr_2Cl, and NCl_3.

Reference:

Haag, W. R.; Jolley, R. L. (Water Chlorination Environ. Impact Health Eff. **1983** Vol. 1, Chapter 4, pp. 77/83).

19.3 Bromochloroamine, NHBrCl

CAS Registry Number: *[77352-23-9]*

The bromochloroamines cannot be obtained by simple mixing of aqueous solutions of ammonia and halogens at appropriate ratios and pH (as in the case of the chloro- and bromoamines), but there is spectroscopic and kinetic evidence for the formation of NHBrCl during the oxidation of Br^- by NH_2Cl; isolation in aqueous solutions to characterize some physical properties has not been accomplished, however [1 to 3].

The formation of NHBrCl in ethereal solution was demonstrated by Haag and Jolley [4] as follows: To an aqueous HBrO solution, diluted with a phosphate buffer to give a final pH in the range 7 to 8.5, an ethereal NH_2Cl solution ($\sim10^{-3}$ M) was added and the mixture shaken for 20 to 30 seconds. NHBrCl is formed according to $NH_2Cl+HBrO\rightarrow NHBrCl+H_2O$, if the HBrO/$NH_2Cl$ molar ratio is kept below 0.4; in this case, the

NHBrCl-NH_2Cl mixture is relatively stable in ether. The UV spectrum recorded between 200 and 400 nm showed two absorption maxima, a strong one at 220 nm with $\varepsilon_{max}=4430\ L\cdot mol^{-1}\cdot cm^{-1}$ and a weak one at 333 nm with $\varepsilon_{max}=373\ L\cdot mol^{-1}\cdot cm^{-1}$; the absorbances at 220 and 333 nm increased proportionally to the amount of bromine added, but the ratio of the peak heights was constant indicating the formation of the new Br-containing compound [4].

For ether extractions of reaction mixtures of NH_2Cl+Br^-, $HClO+NH_2Br$, and NH_2Cl+NH_2Br (with excess ammonia nitrogen) the formation of NHBrCl was also demonstrated spectroscopically ($\lambda_{max}=218$ nm observed, $\varepsilon_{max}\approx 2100\ L\cdot mol^{-1}\cdot cm^{-1}$ for aqueous solution estimated by comparison with NH_2Cl and NH_2Br) by [1 to 3].

The formation of NHBrCl in ether solution according to $NH_2Cl+Br_2\rightleftharpoons[NH_2BrCl^+Br^-]\rightarrow NHBrCl+HBr$ was attempted, but addition of bromine in ethereal solution to an ether solution of NH_2Cl resulted in slow and incomplete bromination [4].

The kinetics of the oxidation of Br^- by NH_2Cl to give NHBrCl was investigated in aqueous solutions of varying pH and salinity (chlorine-demand-free water, artificial and natural seawater). The reaction was found to be first order with respect to NH_2Cl, Br^-, and H^+ ion concentration over the pH range 6.8 to 8.2. The overall empirical rate constant $2\,k_1K_p=k=(2.8\pm0.3)\times10^6\ L^2\cdot mol^{-2}\cdot s^{-1}$ at 298 K is consistent with the following reaction mechanism [1 to 3]:

$$NH_2Cl+H^+\rightleftharpoons NH_3Cl^+ \qquad K_p$$
$$NH_3Cl^+ + Br^-\rightarrow NH_3Br^+ + Cl^- \qquad k_1,\ \text{fast}$$
$$NH_3Br^+ + NH_2Cl\rightarrow NHBrCl+NH_4^+ \qquad k_2,\ \text{slow}$$

The oxidation of Br^- by NH_2Cl is complicated by halogen exchange between $NHBr_2$ and NHBrCl and protonic dissociation of these species [5]:

$$NHBrCl+Br^-\rightleftharpoons NHBr_2+Cl \qquad K\approx 2\times10^4$$

and

$$NHBrCl\rightleftharpoons NBrCl^- + H^+ \qquad K\approx 1.3\times10^{-5}\ mol/L$$

Upon oxidizing N,N-diethyl-p-phenylenediamine (DPD) by a mixture of NHBrCl and NH_2Cl in aqueous solution, NHBrCl converts back to NH_2Cl in a fast reaction (followed by the slow reaction between DPD and NH_2Cl), as demonstrated by kinetic, titrimetric, and spectrophotometric measurements; this shows that the Br atom of NHBrCl is very labile and reactive [6].

References:

[1] Trofe, T. W.; Inman, G. W., Jr.; Johnson, D. J. (Environ. Sci. Technol. **14** [1980] 544/9).

[2] Trofe, T. W.; Johnson, J. D.; Inmann, G. W., Jr. (NUREG-CR-1116 [1980] 1/90; C.A. **94** [1981] No. 163334).

[3] Johnson, J. D.; Inman, G. W., Jr.; Trofe, T. W. (NUREG-CR-1522 [1982] 1/121, 80/98; C.A. **98** [1983] No. 204093).

[4] Haag, W. R.; Jolley, R. L. (Water Chlorination Environ. Impact Health Eff. **1983** Vol. 1, Chapter 4, pp. 77/83).

[5] Bousher, A.; Brimblecombe, P.; Midgley, D. (Water Res. **23** [1989] 1049/58).

[6] Valentine, R. L. (Environ. Sci. Technol. **20** [1986] 166/70).

20 Compounds of Bromine with Chlorine and Fluorine

20.1 Chlorofluorobromate(1 −), $ClBrF^-$

CAS Registry Number: *[25730-97-6]*

The nature of the chemical bonding in the series of linear, excess-electron compounds, $BrF_2^- - ClBrF^- - BrCl_2^-$, was studied by using localized molecular orbitals (LMO's), which were obtained from a semiempirical (NDDO) wave function [1]. The stability of a number of trihalide ions XYF^- with respect to their reference systems $XY + F^-$, among them the stability of $ClBrF^-$ with respect to $BrCl + F^-$, was the subject of a Hückel calculation [2].

References:

[1] Zakzhevskii, V. G.; Charkin, O. P.; Smolyar, A. E.; Zyubina, T. S.; Klimenko, N. M. (Zh. Strukt. Khim. **17** [1976] 763/74; J. Struct. Chem. [USSR] **17** [1976] 659/69).

[2] Wiebenga, E. H. (Stereochim. Inorg. Accad. Nazl. Lincei 9th Corso Estivo Chim., Rome 1965 [1967], pp. 319/31, 330/31).

20.2 Bromine Chloride Difluoride, $BrClF_2$

CAS Registry Number: *[24319-47-9]*

Hückel calculations used to interprete some polyhalide structures predicted the hypothetical $BrClF_2$ molecule to decompose according to 3 $BrClF_2 \rightarrow 2\ BrF_3 + BrCl + Cl_2$.

Reference:

Wiebenga, E. H. (Stereochim. Inorg. Accad. Nazl. Lincei 9th Corso Estivo Chim., Rome 1965 [1967], pp. 319/31, 327).

Physical Constants and Conversion Factors

Avogadro constant	N_A (or L) $= 6.02214 \times 10^{23}$ mol^{-1}
Faraday constant	$F = 9.64853 \times 10^{4}$ C/mol
molar gas constant	$R = 8.31451$ J·mol^{-1}·K^{-1}
molar volume (ideal gas) (273.15 K, 101325 Pa)	$V_m = 2.24141 \times 10^{1}$ L/mol
Planck constant	$h = 6.62608 \times 10^{-34}$ J·s
elementary charge	$e = 1.60218 \times 10^{-19}$ C
electron mass	$m_e = 9.10939 \times 10^{-31}$ kg
proton mass	$m_p = 1.67262 \times 10^{-27}$ kg

1 kg $= 2.205$ pounds
1 m $= 3.937 \times 10^{1}$ inches $= 3.281$ feet
1 m^3 $= 2.642 \times 10^{2}$ gallons (U.S.)
1 m^3 $= 2.200 \times 10^{2}$ gallons (Imperial)

Force	N	dyn	kp
1 N	1	10^{5}	1.019716×10^{-1}
1 dyn	10^{-5}	1	1.019716×10^{-6}
1 kp	9.80665	9.80665×10^{5}	1

Pressure	Pa	bar	kp/m^2	at	atm	Torr	lb/in^2
1 Pa = 1 N/m^2	1	10^{-5}	1.019716×10^{-1}	1.019716×10^{-5}	9.86923×10^{-6}	7.50062×10^{-3}	1.450378×10^{-4}
1 bar = 10^6 dyn/cm^2	10^{5}	1	1.019716×10^{4}	1.019716	9.86923×10^{-1}	7.50062×10^{2}	1.450378×10^{1}
1 kp/m^2 = 1 mm H_2O	9.80665	9.80665×10^{-5}	1	10^{-4}	9.67841×10^{-5}	7.35559×10^{-2}	1.422335×10^{-3}
1 at (technical)	9.80665×10^{4}	9.80665×10^{-1}	10^{4}	1	9.67841×10^{-1}	7.35559×10^{2}	1.422335×10^{1}
1 atm = 760 Torr	1.01325×10^{5}	1.01325	1.033227×10^{4}	1.033227	1	7.60×10^{2}	1.469595×10^{1}
1 Torr = 1 mmHg	1.333224×10^{2}	1.333224×10^{-3}	1.359510×10^{1}	1.359510×10^{-3}	1.315789×10^{-3}	1	1.933678×10^{-2}
1 lb/in^2 = 1 psi	6.89476×10^{3}	6.89476×10^{-2}	7.03069×10^{2}	7.03069×10^{-2}	6.80460×10^{-2}	5.17149×10^{1}	1

Work, Energy, Heat	J	kW·h	kcal	Btu	eV
1 J = 1 W·s = 1 N·m = 10^7 erg	1	2.778×10^{-7}	2.39006×10^{-4}	9.4781×10^{-4}	6.242×10^{18}
1 kW·h	3.6×10^6	1	8.604×10^2	3.41214×10^3	2.247×10^{25}
1 kcal	4.1840×10^3	1.1622×10^{-3}	1	3.96566	2.6117×10^{22}
1 Btu (British thermal unit)	1.05506×10^3	2.93071×10^{-4}	2.5164×10^{-1}	1	6.5858×10^{21}
1 eV	1.602×10^{-19}	4.450×10^{-26}	3.8289×10^{-23}	1.51840×10^{-22}	1

1 cm^{-1} = 1.239842×10^{-4} eV   1 Hz = 4.135669×10^{-15} eV

1 hartree = 27.2114 eV   1 eV ≙ 23.0578 kcal/mol

Power	kW	hp	$kp \cdot m \cdot s^{-1}$	kcal/s
1 kW = 10^3 J/s	1	1.35962	1.01972×10^2	2.39006×10^{-1}
1 hp (horsepower, metric)	7.3550×10^{-1}	1	7.5×10^1	1.7579×10^{-1}
1 $kp \cdot m \cdot s^{-1}$	9.80665×10^{-3}	1.333×10^{-2}	1	2.34384×10^{-3}
1 kcal/s	4.1840	5.6886	4.26650×10^2	1

References:

Mills, I. (Ed.), International Union of Pure and Applied Chemistry, Quantities, Units and Symbols in Physical Chemistry, Blackwell Scientific Publications, Oxford 1988.

The International System of Units (SI), National Bureau of Standards Spec. Publ. 330 [1972].

Landolt-Börnstein, 6th Ed., Vol. II, Pt. 1, 1971, pp. 1/14.

ISO Standards Handbook 2, Units of Measurement, 2nd Ed., Geneva 1982.

Cohen, E. R., Taylor, B. N., Codata Bulletin No. 63, Pergamon, Oxford 1986.